CHINA'S LOGIC

The Balance Development

CHINA'S LOGIC

The Balance Development

Lixing Zou

China Development Bank Research Institute, China

World Scientific

NEW JERSEY · LONDON · SINGAPORE · BEIJING · SHANGHAI · HONG KONG · TAIPEI · CHENNAI · TOKYO

Published by

World Scientific Publishing Co. Pte. Ltd.

5 Toh Tuck Link, Singapore 596224

USA office: 27 Warren Street, Suite 401-402, Hackensack, NJ 07601

UK office: 57 Shelton Street, Covent Garden, London WC2H 9HE

Library of Congress Cataloging-in-Publication Data

Names: Zou, Lixing, author.

Title: China's logic : the balance development / by Lixing Zou
  (China Development Bank Research Institute, China).

Description: New Jersey : World Scientific, [2017] | Includes index.

Identifiers: LCCN 2017015028 | ISBN 9789813222625 (hc : alk. paper)

Subjects: LCSH: Economic development--China. | Equilibrium.

Classification: LCC HC427.95 .Z68 2017 | DDC 338.951--dc23

LC record available at https://lccn.loc.gov/2017015028

British Library Cataloguing-in-Publication Data

A catalogue record for this book is available from the British Library.

Desk Editors: Chandrima Maitra/Lum Pui Yee

Typeset by Stallion Press
Email: enquiries@stallionpress.com

Printed in Singapore

Foreword

Human society is now in an age of profound adjustment, change and restructuring, and sustainable development is a major subject of strategic significance. Driven by technological and industrial revolution, human society's ways of production and life have changed profoundly, which not only reshapes the form and structure of human society but causes a profound and lasting impact on the living environment, pushing human civilization up to a new level. At the same time, all sorts of contradictions in human society have become more prominent and acute: contradiction between the past and the present, contradiction between being developed and being less developed, contradiction between human society and the environment, among others. While social change in the 21st century will be revolutionary, humanity will encounter unprecedented conflicts and faster changes, and countries and regions worldwide will meet with new opportunities and challenges.

Since the founding of new China, especially after the start of reform and opening up, China has engineered remarkable changes and achievements that have attracted worldwide attention. Today, China is regarded as one of the most influential powers in the world. Nevertheless, it is still a developing country whose level of comprehensive development is not high, with relatively low gross domestic product (GDP) per capita as well as rather weak capacity for innovation. After more than 30 years of rapid development at a double-digit annual average growth rate, China now is entering a "new normal" under the combined action of development laws, market changes in supply and demand, and internal and external factors. The country is experiencing a transition from high to moderate growth rates, from extensive to refined development, and from a developmental economy to a mature economy. In the new normal, China sticks to increasing the quality of economic development and economic efficiency as priorities, while attaching adequate importance to growth transformation and structural readjustment difficulties involved in reform; pursues innovation-driven development; tightens risk prevention; controls and strengthens social security; promotes sustainable, healthy economic development as well as social harmony and stability. In the 21st century, China is still in the primary stage of the socialist market economy, and as a big developing country it is facing a range of underlying problems: How to solve the imbalance of economic and social development, how to achieve national modernization and sustainable development, and how to realize rejuvenation of the Chinese nation and thereby contributing more to the entire human race. All these problems need in-depth research and deliberation.

I am very delighted that the author has succeeded in publishing this book after his long devotion to research on the balanced way of thinking in a way that combine theory with practice; provide an in-depth analysis on crucial factors influencing the sustainable development of China's economy and society, and exploring ideological methods, market mechanisms, innovation activities and international collaboration that promote sustainable development as well as measures for overcoming "weaknesses". In short, the author's work (i) gives an in-depth analysis of the laws of balance with emphasis placed on ideological methods, and employs the laws of balance as a device by which we understand and shape the world to explore paths of sustainable economic and social development, representing an active attempt in the new period to make Marxist dialectical materialism Chinese, modernize the traditional Doctrine of the Mean, and theorize the practice of the socialist market economy, (ii) considers China's modernization in global and historical contexts for an in-depth analysis of ideological, market, system, international and national factors that influence sustainable economic and social development, and puts forward some new ideas, views and measures, (iii) accentuates innovation by actively exploring innovation in thinking, market governance and mechanisms, displaying a spirit of independent thinking and of the emancipation of the mind, and (iv) pays particular attention to rural grassroots issues and proposes some important suggestions on planning for rural development. These studies and explorations are constructive for deepening reform and promoting the steady and healthy development of the Chinese economy. The author, who has long been engaged in research, is active in thinking and exploration, and neither goes with the flow nor regards himself as infallible, striving to do his part wholeheartedly. This book, which comprises his years of research findings as well as active explorations made from strategic, forward-looking and operable perspectives, displays his responsibility and passion as a scholar and the spirit of dedication to modern civilization. This is commendable. This book is thought provoking and refreshing, and though some propositions contained in it are debatable, it is of certain value for reference.

In conclusion, I would like to recommend this book to readers from all walks of life.

Jiang Zhenghua
Vice Chairman of the Standing Committee
of the 9th and 10th National People's Congresses
February 1, 2016

Preface: Sustainable Development from a Balanced Way of Thinking

Overview

- New normal, new perspective and new thinking.
- Follow and apply the laws of balance.
- Form and exploit the "Power of Thought, of Market, of Innovation, and of International Synergy" to promote sustainable economic and social development.

Balance is not just the motive power but also the purpose of development. This book looks into the laws of balance for the development of things while presenting a big picture of history, changes and challenges. It applies the balance law methodology to give a systematic analysis of major factors influencing sustainable development of the Chinese economy and society, and discusses the balanced way of thinking, market credit, innovation energy and international collaboration that boost sustainable development. While the theoretical system of socialism with Chinese characteristics needs the thinking of balanced development, solving the problem of imbalance requires a balanced approach as well as balanced interaction between man, nature and society, which is the big logic for economic development in the 21st century. The book strives to study the rules of balance, use a balance-oriented way of thinking to solve the imbalance in economic and social development, and offer some thoughts, paths, and policy measures for sustainable development, worth reading by economic workers, management personnel, university teachers and students, as well as people from various walks of life.

I

In my view, balanced development, democratic innovation and peaceful competition are the main themes of the 21st century; the global economy will be mainly characterized by the focus on flat organization, networking, synergies and regional growth; human society will move toward virtual reality, sharing economy and freelance work; the conflicts between tradition and modernity and between economic integration and political pluralism will be even sharper. Given such a prevailing trend, to renew the Chinese nation in the 21st century means to grasp six concepts, build four pillars and address three issues. The six major principles are legal system, innovation, thrift, green, opening up and sharing. Among these, legal system is the basic requirement. Innovation is the fundamental driver of the national rejuvenation, thrift is the important path, green is the true natural color, opening up is the international environment and sharing is the strategic goal. The four pillars are: a robust economic pillar built by deepening economic reform; a strong political party pillar built by pushing ahead with political reform; a powerful defense pillar by pushing forward with military reform; a favorable international environment created by cementing international ties. The three major issues are: sustainable economic development, national reunification and national governance modernization. I pay particular attention to applying balanced ways of thinking to study some issues about sustainable economic development.

In the course of global modernization, China started late, rose rapidly and got through countless difficulties in terms of economic development. As one of the ancient civilizations, China in the agrarian age was very successful and made great contributions to human civilization, but fell behind in the age of industrial civilization. During the late 19th century and the early 20th century, when developed countries had become industrial societies, China was still a semi-feudal and semi-colonial society, lagging far behind. Over the past century, the Chinese unceasingly explored paths of development from a feudal to an industrial society. After the founding of new China in 1949, facing a war-ruined country and grinding poverty, the Chinese mainly relied on its own efforts and overcame all sorts of difficulties in developing modern industry and agriculture, taking a firm step forward on the path to modernization. After the start of reform and opening up in particular, China focused on economic construction and accelerated modernization by pushing ahead with a strategy of "creating Special Economic Zones (SEZs), boosting the eastern regions, developing the western regions, invigorating the northeastern regions, and raising the central regions" and initiating the strategies of "Coordinated Development of Beijing, Tianjin and Hebei, Free Trade Zones (FTZs) Development, Yangtze River Economic Belt, and One Belt and One Road", and made achievements that garnered worldwide attention.

Today, China has become one of the most influential powers of the world. Meanwhile, we are soberly aware that China is still a country whose comprehensive development level is not high, and is still in the primary stage of modernization, and while the world's advanced countries are becoming knowledge-based economies, China on the whole is still in the development stage of industrial economy and facing prominent underlying contradictions ahead — for instance, inadequate resources for a big population, and very unbalanced development; huge gaps between urban and rural areas, between central and western regions, and between the rich and poor; inadequate capacity for industrial innovation owing to the market system and mechanism lagging behind; big production vs. few famous brands. How to overcome, in 30 years to come, environmental pollution, resource shortage, population aging among other issues? How to reform the underlying relations between the economic base and the superstructure along with the changes in productive forces and economic structure? How to cope with complex international situations, seize opportunities and speed up development? All these concern the survival and rejuvenation of the Chinese nation. China in the 21st century will still be a developing country and remain to be in the primary to intermediary stage of the socialist market economy. The Central Committee of the Communist Party of China (CPC) raised the Two Centenary goals: By 2020, the 100th anniversary of the founding of the CPC, China will have fully become a moderately prosperous society; by 2049, the centenary of the founding of new China, the country will have become a moderately developed nation. The above Two Centenary goals are merely stage-specific goals. The real development goal of China is to become a modern and strong country that will make greater contributions to the whole world. Through development in the 20th and 21st centuries, China aspires to truly achieve modernization, which is required by not only the rejuvenation of the Chinese nation, but also the development of human civilization. How to achieve this goal? As issues in China are quite complex, in-depth studies are required of the Central Committee's policies and specific implementation plans, in addition to more brainstorming, comparison and consultation. As a person who has long been engaged in economic research, I always feel obliged to do my bit to solve problems better, make development healthier, and make people happier and the country stronger. To solve China's issues we cannot stick to conventions, nor should we take rash actions, be complacent about what has been achieved, have blind faith in foreign things and rely on foreign help. To solve China's issues requires an appropriate way of thinking as well as priority, rhythm, balance and persistence.

In a developing country like China, to my way of thinking, the thinking methodology, market credit, innovation capacity, the Three Rural Issues, and an international vision are main factors that influence sustainable economic development. Thinking methodology is the core, market credit is the foundation, innovation activities are the motive force, the Three Rural Issues are the key, and mutual trust and collaboration represents the spirit. With a correct way of thinking, economic development, which may seemingly be slow, can actually be fast, and many problems can be readily solved; a well-built market credit system allows various factors of production to work better and production resources to be allocated more effectively, thus fueling greater vitality; increasing innovation activities will bring about unlimited energy to economic development because all things are created and determined by innovation and whoever is capable of innovation can secure a strategic high ground for development; the Three Rural Issues are fundamental issues and represent a weakness for not only China but also other developing countries, and overcoming this weakness is sure to boost economic growth; win–win collaboration is the basic starting point and purpose of global economic integration. Therefore, I have tried to explore in an in-depth manner the intrinsic relations between "balanced line, credit balance, innovation-driven growth, overall urban–rural planning, international cooperation" and sustainable economic development, as well as the intrinsic paths to such relations. I think that balance is the path that leads to sustainable economic development.

II

Balance is the law of motion of objective things, the soul of various notions and behaviors, and also a magic weapon for sustainable economic development.

Development always follows a cyclical pattern starting from imbalance to balance and then to rebalance, and in this cycle, the contradictory motion of and the change in things reinforce and constrain each other owing to a balance force which is a self-generating, self-balancing, voluntary, in-itself, for-itself and natural force that drives the absolute, eternal and dynamically stable motion of things. Under the action of the balance and counterbalance, things are in a motion that follows the balance law. The balance law is the foundation of the universe, the guarantee of life, the cause of existence, the basis of connection, the condition of activity, the origin of things and the base of development.

Development follows the balance law which has seven main characteristics. First is fundamentality, without which nothing can exist. In the chain of layered relations between things, for instance, there is always is a deepest layer where factors play a fundamental role in the change and development of things. Second is about structure. Things exist in structures. For example, the universe has a self-structure of intrinsic forces: the earth has a structure formed through its gravity, and the earth's gravity and the sun's gravity interact with each other to make the earth part of the solar system, which is part of the Milky Way. Third is about system. Things exist in the form of a system. For example, the biosphere is a strict system, and when a particular element in this system changes, a chain reaction and systematic change will occur. Fourth is about correspondence. Things exist corresponding to one another. One example is complexity vs. simplicity, where a complex world is certain to have a simple kernel corresponding to it. Fifth is about symmetry. Things exist symmetrically. Examples include positive and negative charges in physics, positive and reverse reactions in chemistry, prosperity and decline in economics, and positive and negative feedback in the financial market. Sixth is about reflectivity. Things reflect and influence one another. For example, people influence and are influenced by the environment; people's attitudes influence the development of things, which in return influences people's attitudes. Seventh is about rhythm. Things exist in rhythmic motion. For example, human life is physiologically, behaviorally and mentally rhythmic; human society is economically, politically and culturally rhythmic. Through the elements of internal connection, external connection, structure and feedback, the balance law passes the balance force to things and hence promotes their change and development.

As the law of balance works, there has to be a way of thinking that stands between materialism and idealism, a mixed economy underpinned by both public and private ownership, and an interaction mechanism that connects the government and the market. There has to be a middle ground between the two extremes. Under the action of the balance law, economic and social activities and changes in natural resources and the environment are certain to reach a state of dynamic balance, and the strength of human beings and that of nature are certain to be in a state of balance. In other words, if we do not follow the balance law, voluntarily comply with the trend of balance, and actively adopt balanced thinking and behavioral measures, human economic and social development will inevitably be punished by the natural force of balance that exists objectively. Therefore, we need to follow and apply the balance law consciously. Following the balance law on the path of development in the 21st century requires proper interaction between man and nature, active interaction between man and society, organic interaction between the government and the market, cooperative interaction between domestic and international markets, an economic line of balanced development as well as paying particular attention to the short boards in development.

Stressing the importance of the balanced way of thinking does not mean to negate the importance of other methods. Philosophically, the theory of balance is intrinsically associated with ontology, epistemology and dialectics, and in the economic sphere, the balance method and other analytical methods are complementary. We need to integrate multiple methods, including "balanced thinking network design, systems engineering, the law of continuity, marginal analysis and partial equilibrium, to enrich our tools, step up theoretical and practical innovation and boost reform and development".

III

I am trying here to employ the balance law method to build a comprehensive balanced growth model in the context of history, changes and challenges, and through it, analyze the correlation of the factors of "balanced line, credit balance, innovation balance, overall urban–rural balance, and international balance", with sustainable economic development. This model, neither relying on a raft of data as support nor describing necessary and definite quantitative relations between phenomena, attempts to give full consideration to some more important factors that need to be weighed up. It follows the balance law, considers problems comprehensively — including the combination of thinking and action, the connection between macro and micro aspects, overall urban–rural planning, interaction between the government and the market, and national and international synergy, elaborates on general trends of a probability nature, and stresses "balanced line + credit balance + innovation balance, + overall urban–rural balance, + international balance", in creating an ideological and policy system regarding sustainable economic development. From the perspective of balance law, the balanced development method and line is the strategic core of sustainable development; the market is king, and strengthening market credit system building is the strategic foundation of sustainable development; innovation activities — most notably a technological revolution — and innovation infrastructure building serve as the strategic motive force of sustainable development; deepening country-level governance reform, boosting county-level economic development, and properly solving the Three Rural Issues represent the strategic potential of sustainable development; boosting mutual trust and balanced diplomacy is of strategic significance to sustainable development.

Economic growth
  = Balanced line + credit balance
    + innovation balance + overall urban–rural balance
    + international balance (thought fixity power) (market activity power) (innovation power) (developing potential power) (international synergy power)

The balanced line means considering economic issues in a balanced way of thinking and hence forming a way of thinking about mixed economy development. China's economic development needs a mixed economy line of comprehensive balance, which entails active interaction between the market and the government, development in an integrated manner of the private and state-owned sectors of the economy, innovation both in supply and demand, organic combination of the real economy and the virtual economy, and collaborative operation of monetary, financial and industrial policies. In the process of economic and social transformation, balance is the soul of everything — theory, concept, behavior, structure and economic aggregate. The balance law requires that any activity should not go to extremes. Both the good and the bad of anything have a "limit"; speed will make way for quality, productivity for continuity, and efficiency for sustainability. Sow thinking, and reap action. The balanced line methodology is the central element of the development logic, representing the source of fixity for sustainable economic and social development.

Credit balance means putting credit first and taking the market as king, and by strengthening market credit system building, giving full boost to the activity of market participants, letting various factors of production play better roles, allowing resources for production to be allocated more efficiently, and promoting the comprehensive balance of the economy. In the market economy, credit serves as the foundation of economic development, and market credit system building represents the endogenous force of economic growth; whether the market is mature is determined by market credit which can be developed. With technological, economic and social development, the pattern of market credit activity is taking on new characteristics: as big data becomes an asset and instrument of strategic importance, the ways of credit evaluation and techniques for credit risk management based on big data are receiving ever-growing attention; Blockchain settlement and electronic money have become strong trends; the intangible value of innovation capacity, brand quality, rule making, etc. becomes prominent; criteria, rules and standards become important conditions for translating credit into capital; social capital accelerates the improvement of the central functions of financial capital, fundraising tools are getting popular, and ownership relations are gradually replaced by lease relations. In today's society, market changes are complex and involve many uncertainties; build credit and open up markets — win markets and you win the world; adjustments to international strategies are largely made to compete for markets. A big country has not only a competitive advantage in international markets but also broad leeway in the domestic market. Strengthening market credit system building and interaction between domestic and international markets help to promote balanced development.

Innovation balance means achieving balanced development through innovation in balance. Innovation comes from balanced thinking that pursues progress consciously or unconsciously; to innovate is to introduce new elements and effect new changes, turning dissatisfaction into satisfaction; innovation is also a result of balanced thinking, discovering in change what remains unchanged and finding simple laws from complexity. Therefore, innovation, original innovation in particular, is the most valuable of rare resources, the real driver and source of development, and the intrinsic path of balance to rebalance of things. Human society is entering an era of deep knowledge underpinned by deep technology where innovation is the means to satisfy people's ever-growing material needs as well as cultural pursuits. In today's world, the discovery of gravitational waves, gene decoding, intelligent machines, AlphaGo's triumph over the human race are important signs of a new technological and industrial revolution. Meanwhile, we are also clearly aware that innovation cannot overstep nature and involves risk. In seeking economic development through innovation, we need to pay particular attention to the balance between human society and nature, and to properly deal with the relationship between innovation and inheritance. On the one hand, we need to ignite as much as possible scientific and technological professionals' inner passion and desire for innovation by strengthening innovation supporting system building; to further open up channels by which non-living matter is converted into living matter through scientific and technological innovation, promoting new economic development; to further improve China's strategies for sustainable economic and social development through theoretical innovation; to further promote the development of rational organization through organizational innovation, enhancing the activity and creativity of market participants; to further improve the effectiveness of market supply and demand through supply–demand structure adjustment.

On the other hand, we need to concern about alienated technology, carry forward our ancient nation's spirit of commitment and ingenuity, learn all excellent cultures of the international community, work and research diligently, and be conscientious and responsible in delivering products and services.

Overall urban–rural balance means putting the Three Rural Issues at a prominent position in the sustainable development strategy, and by vigorously developing the county-level economy, invigorating the value of land assets, improving the level of agricultural production of a moderate scale and deepening the reform of rural governance to strengthen rural development in a way that lengthens the short board of the Three Rural Issues with the aim to promote the balanced development and urban and rural areas. China's rural areas are the basis of Chinese society, and how good the rural areas are determines how good Chinese society is. The Three Rural Issues are the biggest weaknesses of China and its greatest potential. Properly dealing with the Three Rural Issues has a fundamental and supportive role in the sustainable development of China and of the world as well. By the end of 21st century, even if China's modernization has reached the level of a moderately developed country, farmers will still account for a significant proportion (20–30%) of the country's total population. In quite a long period to come, the rural economy will have multiple forms such as state-run farms, collective farms and household farming firms, a moderately small-scale peasant economy will remain the main economic form in rural China, and agriculture will remain the fundamental industry of the national economy. Therefore, it will be quite a heavy task to tap the rural ecological, land and cultural value, improve the well-being of rural areas as a whole, provide rural poor residents with job opportunities and improve social security conditions, strengthen governance restructuring and development in modern rural communities centered around school districts, close the huge divide between urban and rural areas, and rebuild social relations between groups and between classes. We should make concrete improvements to the basic living and development opportunities, and development capacity of rural residents. Addressing these problems and completing these tasks will promote sustainable development for sure.

International balance means keeping planning, construction, trade and development on a collaborative and mutual-trust basis in the process of global economic integration; the Road and Belt Initiative for interconnection purposes for promoting — initially through infrastructure construction — balanced international economic and trade development as well as the balanced development of the country's south and north regions, and especially, deepening regional cooperation and boosting regional economic development with river basin economies as ties. The current international picture of "one superpower and multiple big powers" is the result of balance of various global contests. Facing complex and volatile political situations on the international stage of diversity and competition, we need to "size up the situation and practice integrity, encourage mutual respect, equality and mutual benefit, know ourselves and our adversaries and do more and talk less, secure a firm foothold in the domestic market and further explore international markets, and exercise balanced diplomacy and stress practical results". Joint efforts should be made to advance global governance reform, gradually improve international rules, and create a more enabling external environment to step up national modernization, contributing more to the development of human civilization.

Thinking creates the power of determination, credit boosts strength, innovation brings about energy, the Three Rural Issues means potential and collaboration produces synergy. When combined, these five aspects can boost sustainable economic and social development. Under the balance law, the human society will ultimately return to the line of balance. Balance is something necessary for history, theory, economy and development; it is the soul of thinking, market, innovation, modernization and international relations, and big balance is the dominant logic of economic development in the 21st century.

IV

Balanced development is to achieve the balance of development. For what do we pursue development? Development is to achieve a new balance out of imbalance. Achieving balance should be the purpose of our development. How to achieve the balance then? Only by following and applying the balance law can balance be achieved. Underdevelopment, too slow development, too fast development or overdevelopment are not balanced development, and can hardly reach a balance. Therefore, balance is not just the motive power but also the purpose of development. In dealing with relations between people and society, the economic thought of balance stresses the following aspects: overall planning, comprehensive balance and following and applying the balance law governing things in terms of strategic deployment, so as to promote long-term steady, healthy and balanced development of the economy and society; organic interaction between the market and the government in a way that not only gives full play to the important role that the market as an "invisible hand" has in economic activities, but also makes proper use of the important role that the government as a visible hand has in public affairs management; changing the driving force of development from factor input to innovation — including innovation in theory, market and technology — actively exploring ways of fine development, and improving the behavior of economic entities in a way that saves every resource, increases benefit through quality and seeks characteristic development; strengthening credit system building and promoting the interaction between domestic and international markets. In regard to the relation between man and nature, the economic thought of balance puts emphasis on the following: making rational use of renewable resources according to the speed of their reproduction; making rational use of non-renewable resources according to the length of their substitution; properly discharging all sorts of waste according to the absorptive capacity of the environment; rationally organize activities of production, circulation and consumption according to the needs of healthy development of human beings.

There exists and there is a need for economic games, political games, life games, etc. in the development of human society. The human race is great because of its ability to debate complex logic, seek simple principles and explore the value of correlation, as well as to understand, make and improve various rules. All sorts of rules share one thing in common: balance.

To follow the balance law requires making and improving a balanced economic line according to economic development trends and environmental changes with the aim to solve problems, and employing high technology and all effective means to build monitoring, warning, rescue and response mechanisms and policies aimed to overcome weaknesses and ensure economic security.

To follow the rules of balance and based on the characteristics of a modern Party, that is, "learning-, service-, innovation-, structure-, secularity-, and regulation-based", we need to build an interactive dual system which may include limited features but must maintain high quality, so as to channel top-notch talents that value ideals and true modernism to the most important departments. This requires creating and improving a balanced governance structure according to the characteristics of a modern political party, which embodies the law that "the people make history and the elite governs the country", strengthening the building of party discipline constraint mechanisms as well as mechanisms for communication of political views, in a way that accentuates the coordination of democracy and centralization, flexibility, and beliefs and principles that must be held fast to; and strengthening measures that help to communicate government orders.

This will help us deepen the political system reform and promote the transformation of the CPC toward a modern party for long-term ruling, while strengthening the core of our country leadership.

To follow the balance law requires establishing and improving a supervisory mechanism for the balance of responsibilities and rights in accordance with laws, creating a modern social relationship under which "everyone is equal in character but skills may vary from person to person", strengthening the building of a democratic, innovative and fair social environment, combating formalism, bureaucracy and hedonism, advocating the spirits of dedication, innovation and public interest, and enhancing the protection of common people's basic rights and interests. This requires us to, while following the Party's mass line, let well-trained outstanding talents play a leading role in the modernization drive. This requires giving full play to the roles of entrepreneurs, innovative people and elite people in various sectors, maintaining the status of enterprises as the carrier of national interests, and strengthening the country's economic security.

To follow the balance law, according to the common interests of the people on both sides of the strait, not only we need to profoundly judge the situation, make a hope for the people, but also guide and organize the people, keep the bottom line, strengthen economic cooperation, internal and external interaction. As of today, Taiwan society has become a gradual reality of mature society, Taiwan politics is no longer a Kuomintang party-led politics. China's mainland here, need to adapt to the the political ecology of Taiwan, the Communist Party of China needs in-depth analysis of the KMT decomposition, the reasons for the decline, as an important teaching. When it comes to cross-strait relations, opening to each other pragmatically as needed for the common and lasting well-being of the people across the strait and on the basis of 9.2 cross-strait common understandings, actively solving problems and guiding public opinion, being tolerant of history, ideology and systems, and stepping up the process of peaceful reunification of the nation starting with cross-strait economic integration, working together to defend the interests of the Chinese nation.

To follow the balance law requires building a balanced diplomacy theory and defense strategy according to the changes in international situations, global diversification trend, which is beneficial to many countries and regions, but may be detrimental to the super power. So, in the international affairs, we need to advocate an international relations concept of ecological thinking, establishing a broad international united front; in the defense side, we need to combine military and civilian, independent action with collaboration, attack with defense, technology with talents, and fixed strategic goals with flexible tactic means, and building a wall of steel for maintaining the central interests of the country.

Following the balance law helps to achieve sustainable economic development, to achieve national reunification, to improve the country's modern governance level, to boost the revival of the Chinese nation, and to contribute more to the development of human civilization.

V

This study of mine is intended to provide some ideas and policy suggestions for medium and long-term development of China. In order not to waste the reader's precious time, I tried my utmost to treat this study as a meaningful, fresh and well-structured piece of art while expounding on sustainable development in a balanced way of thinking. The Preface gives an introduction to the purpose, context and some important views of this study. The Introduction section, which is the core and soul of the study, provides a systematic analysis of the significance and characteristics of the balance law and balance methodology. The sections in each chapters focus on analyzing major factors influencing sustainable economic development and discussing some ideas and policy measures by using the balance methodology. The main text consists of 3 parts and 15 chapters; every part begins with a reading guide that introduces the part's major views and conclusions, and every chapter focuses on one issue, i.e. 15 issues in all for 15 chapters — all these issues are part of the subject of sustainable development. The entire study is problem driven in forming the basic framework of correlations between the factors of "balanced line, credit balance, innovation balance, overall urban–rural balance, and international balance" and sustainable economic development.

I hope this study can be something that triggers greater ideas about sustainable development.

About the Author

 **Dr. Lixing Zou** was born in Jiangxi Province of P.R. China. He was one of the "educated youths" to move to the countryside in the 1970s. He graduated from China Jingdazhang Ceramic University in 1982 to work in the Light Industry Press of China as Editor, holding a respected position as Division Chief as the Research Office of the State Council of China. He studied at Columbia University and American University in the US in the 1990s before joining the World Bank. Since his return to China in February 2003, he has been working in China Development Bank (CDB) as the Vice President of the Research Institute of CDB, and Senior Expert of CDB. In 2008, he was also a member of the World Economic Forum Global Agenda Council. Currently, he is Vice Chairman of China Social and Economic System Analysis Research Society. He has been involved in research on economic policies for more than 30 years, mainly focusing on international finance, technical economy and rural issues; he has also written research reports, drafted documents for the central government, speeches for leaders and published articles in newspapers at home and abroad. His recent publications mainly include *Financial Empire: The US Development and Enlightens* published by Hunan University Press (2009 edition); *China's Rise: Development-oriented Finance and Sustainable Development* published by World Scientific Press (2014 edition); *China's Base: County-level Economy and Society* published by World Scientific Press (2015 edition).

Contents

Foreword   v

Preface: Sustainable Development from a Balanced Way
of Thinking   ix

About the Author   xxv

Introduction: General Balance   xxxiii

Part I   1

On the Path to Balance   3

Chapter 1   The Balance History of Human Development   5

   I. Primitive Integration: Integration for Survival   6

  II. Historic Separation: Development along Multiple Lines   9

 III. Global Integration: Globalization of Interpersonal Interaction   21

 IV. Globalization Formed in the Interactions between Man
     and Nature   26

  V. Characteristics of the Current Globalization:
     Integration and Diversification   29

 VI. Summary: What We Learn from History   32

Chapter 2   The Balance Practice in China's Modernization   37

   I. Push Ahead Dauntlessly in the Last Two Centuries   38

  II. The 30 Years of a Planned Economy   44

 III. The 30 Years of a Market Economy   46

IV. Analysis of the Path of China's Modernization   51

V. Further Improve China's Development Theory
and Policy System   61

VI. Summary: What is the Level of China's Modernization   68

Chapter 3   The Modern Theory of Economic Equilibrium   71

I. Harbingers of Modern Economic Thoughts   71

II. Embracing the Era of Mixed Economies   76

III. Mixed Economy is a Natural Choice of China   79

IV. Promoting Healthy Development of China's
Mixed Economy   80

Case Study: Why Poverty Alleviation is a Tough and
Time-Consuming Campaign   84

Chapter 4   China's Road to Balanced Economic Development   89

I. China's Balance Sheet   90

II. The Underlying Causes behind China's Economic Slowdown   95

III. Features and Nature of the New Normal   98

IV. Economic Roadmap under the New Normal   102

Case Study: How Chen Yun Applied the Balanced Way
of Thinking   104

Part II   109

The Balance Market   111

Chapter 5   The Law of Equilibrium in Market Credit   115

I. The Nature of Credit   115

II. The Characteristics of Credit   117

III. Basic Laws for Credit Activities   120

Case Study: How to Assess the Infrastructure
of China's Financial Market   124

Chapter 6   Change in Credit Conditions and Perfection
of Policies   **127**

   I. Monetary Conditions and Policies   128

  II. Financial Conditions and Policies   129

 III. Credit Conditions and Policies   133

 IV. Four Credit Development Campaigns   136

Chapter 7   Market Uncertainty and Financial Risk   **149**

   I. International Market Uncertainty   149

  II. China's Market Uncertainty   151

 III. Change in Market Behavior   155

 IV. Motivation of Market Behaviors   158

  V. Identification of Financial Risks   160

 VI. Regulatory Philosophy and Mechanisms with Chinese
     Characteristics   168

Case Study: What is the Chongqing Land Quota   173

Chapter 8   Interaction between the Government
and the Market   **177**

   I. Dynamic Process of Government Function Adjustment   177

  II. Transformation of the Functions of Chinese Government 180

 III. China Development Bank (CDB) is an Important Bridge
    between the Government and the Market   186

Case Study: What is a Local Government Financing Vehicle 189

Chapter 9   China's Land Reform   **195**

   I. Rural Land System Reforms in China   196

  II. Discussions on Deepening Land Reform   208

 III. Optimizing Land Management Structure and System 210

 IV. Optimizing Rural Land Management Structure
    and System   213

V. Establishing a "Land to the Tillers" Market Operation
Mechanism   216

Case Study: What Are the Characteristics of Chinese Classical
Land Systems and Inspirations   219

Chapter 10   Innovation Balance and Manufacture of China   233

  I. Innovation Balance   233

  II. Comparison between Made in China 2025
and Germany's Industry 4.0   236

III. Economic Mode of the New-type Home-based
Manufacturing Industry   247

Part III   257

The Balance Diplomacy for the Opening Up   259

Chapter 11   Mutual Trust and Cooperation in Globalization   265

  I. The Significance of Mutual Trust and Cooperation   266

  II. The Spirit of Mutual Trust and Cooperation   267

**Chapter 12   The History and Prospects of China's
Foreign Trade   271**

  I. Government-led Economic and Trade Relations   272

  II. Market-oriented Economic and Trade Relations   277

III. Challenges that Cannot Be Ignored   280

IV. Improving the Ideological System of Economic Cooperation
for the New Era   285

  V. Correct Understanding of the Strategic Vision of the Belt
and Road Initiative   289

Case Study: How to Establish a Cross-border E-commence
Mode for Industrial Goods   292

Chapter 13 Changes in International Situation and the Countermeasures **301**

  I. Important Players in the International Geopolitical Landscape 302

 II. Effect of TPP and TTIP on the International Situation 317

III. Actively Planning International Layout 321

Chapter 14 Strengthening Asian–African–Latin American Cooperation **331**

  I. Tentative Idea on Establishing a Northeast Asian Development Bank 331

 II. Planning for Power Development Cooperation in Southern Africa 341

III. Thoughts on Cooperative Development of Caribbean Tourist Infrastructure 361

Case Study: How to Establish the BRICS Bank 366

Chapter 15 The Underlying Reasons for the Trade Issues of the US and China **375**

  I. Background 375

 II. Underlying Reasons for the Politicization of Economic and Trade Issues 376

III. Measures to Improve China–US Economic and Trade Relations 380

IV. Conclusion 383

Case Study: How Chinese and American Cultures Affect Decision Making 385

Epilogue 395

Index 399

Introduction: General Balance

Balance is the dynamic stability of contradictory motion, and also the basic law and objective demand of the development of things. The world today is undergoing profound technological revolution and deep social and economic changes. To cope with the complicated domestic and international situation, we need to study the law of balance and properly choose our focal and supporting points. This introduction is the core and soul of the book, which studies the balance law of the development of things; explores the connotation of balance law and its effect on the connection and interaction between things; reveals the basic characteristics of the balance law such as "being fundamental, structural, systematic, corresponding, symmetrical, reflective and rhythmical"; compares the association and difference between the balance law and other laws; analyzes the important role of balance law in international pattern, economic development and technological change. The introduction of General Balance consists of four parts. The first part outlines the phenomenon of balance and the origin of balance thought; the second part probes into the connotation of balance law; the third part reveals the important characteristics of the balance law; the fourth part analyzes the practical significance of the balance thought.

Why are things always in a progressive cyclic motion from imbalance to balance and then to rebalance? Why does the economy constantly change from one balanced state to another? The contradictory motion of things is balanced everywhere in the world. After the world has experienced more than 300 years of modern industrial movement characterized by a high consumption of carbon sources and China has undergone more than 30 years of rapid yet extensive development, today the Internet, big data and new biotechnologies are profoundly changing humanity's style of life and way of production. In this context, how should we respond to an increasingly paradoxical world? We need to have a balanced state of mind, a balanced method, and a balanced way of thinking (BWT).

I. Phenomena of Balance and the Origin of the Thought of Balance

Following the logic of "balance, focus and rhythm", the development of things is always in constant, progressive and cyclic motion "from imbalance to balance and then to rebalance". In order to better analyze such motion, we first summarize the phenomena of balance and the origin of thought of balance.

Balance is an omnipresent natural phenomenon. For example, the earth moves around the sun in accordance with the balance law (BL) of the universe; human beings live or die subject to the BL of life; the economy, influenced by the BL of the market, enjoys prosperity or suffers depression; capital price rises or falls according to the BL of exchange; people sometimes work very hard and sometimes not so hard under the influence of the BL of living; new substances are produced through chemical changes under the BL of forward reaction and reverse reaction. As natural phenomena, the balance of cosmic movement, the balance of market activities and the balance of chemical reactions are all concrete manifestations of BL. All things, whether microscopic or macroscopic, are governed by BL, and the phenomena of balance are ubiquitous.

The thought of balance originates from the phenomena of balance, rich, colorful, and advancing with the times. Understanding the balance of motion, exploring BL, and using it to serve the human society are not only an important field of humanity's ideological building, but also a long-term arduous task for the development of human society. German philosopher Hegel worked all his life to reveal the "law of quantitative change and qualitative change", "the law of unity of opposites" and "the law of negation of negation", and establish a dialectical cognitive approach, based on summarizing and assimilating the essence of classical European philosophy.[1] Marx and Engels improved and transformed Hegel's dialectics and created the theory of materialist dialectics.[2] However, they failed to reveal the characteristics of balance in the processes of "quantitative change & qualitative change, unity of opposites, negation of negation"; neither did they thoroughly analyze the balancing action in the contradiction and change of things.

Chinese philosophers have always been exploring BL. A "regular balance" concept was put forward as early as in the Yin and Zhou Dynasties where, as legend has it, Zhong and Li were the two persons, respectively responsible for heavenly and earthly affairs. When talking to his pupil Zigong, Confucius mentioned the "golden mean", saying that "to go beyond the limit is just as wrong as to fall short". Mozi started to apply the theory of dialectics to understanding things, believing that "extreme prosperity forebodes the beginning of decline". According to his theory, all things in the world have two opposite sides, and if one side becomes too strong, the whole thing will inevitably turn to decline.[3] Sun Zi's military work *The Art of War* stresses that before military operations, the decision maker needs to balance various factors such as the monarchy, the talent of army officers, weather and geographical conditions, the quality of weapons and soldiers, the reward and punishment system, etc. Only based on a balanced comprehensive analysis of all the conditions of the two opposite sides will one be able to pre-estimate the outcome of a war. Lao Tzu said, "being and not-being grow out of each another; being difficult and being easy complete each another; being long and being short are relative to each another; being high and being low are associated with each another; sound and tone are in harmony with each other". He emphasized that things are interrelated, interdependent, mutually transforming and mutually balanced. Confucianism stresses seeking what is in common from numerous different things and realizing the unity of all people, the unity of knowledge and action, the unity of body and mind, the

---

[1] Li Yuqing. *Hegel's Classic Articles*. Shanghai: Shanghai University Press, 2001, pp. 66–156.

[2] *Karl Marx and Friedrich Engels*. Beijing: People's Publishing House, 1972, pp. 342, 362, 926.

[3] Xu Xiyan. *Mohism Research*. Beijing: The Commercial Press, 2008, p. 82.

harmony between man and nature, and the oneness of human morality and Heaven, so as to achieve the perfect state of both internal and external balance.[4] Chinese Buddhism stresses that "a flower is a world and each leaf has a life" and requires people to lead a quiet life in order to realize self-cultivation, think about the future and better understand the world. It also lays emphasis on "precept, concentration and wisdom", and takes "precept" as the means, "concentration" as the pivot, and "wisdom" as the goal. It pursues "soberness, honesty, and purity" and requires people to stay sober, honest and pure instead of being confounded, deceitful and corrupt, so as to realize a perfectly balanced world. In his work "Annotations of the Doctrine of the Mean", Chinese master of neo-Confucianism Zhu Xi explains that being impartial is being in the middle, which is the fundamental principle of the world, and being changeless is being in harmony, which is the right way the world exists. According to his explanation, impartiality and changelessness are the nature of the world and the source of all changes. He stresses that excessiveness impairs impartiality and insufficiency leads to nothing. So, one should avoid the two extremes and use the middle to serve the people.[5] In particular, traditional Chinese medicine is governed by the thought of balance. It stresses both the ecological balance of "earth, water, fire and wind" and the human balance of "body, air, dryness and wetness". So, it is no wonder that a traditional Chinese doctor may cure a patient with headache by treating his feet, or treat a patient's external trauma by giving him internal medication. It can be said that all schools of thought in ancient China have their own thought of balance whether they be Buddhist, Taoist, Confucianist, Mohist, Legalist, military, the Yin and Yang, the Vertical and Horizontal, agricultural, or medical.

Deeply rooted in people's world outlook, methodology, epistemology and philosophy of life, the thought of balance in ancient Chinese philosophy is spontaneous and natural and is the essence of traditional Chinese culture. It used to play an important role in promoting social harmony and

---

[4] Meng Peiyuan. *Ten Ideas of Zhu Xi Philosophy*. Beijing: China Renmin University Press, 2010, pp. 54, 58, 60, 149.

[5] Zhu Xi. *Annotations of the Four Books* (two volumes in all). Translation by Jin Liangnian. Shanghai: Shanghai Classics Publishing House, 2006, p. 23.

progress under the social and historical conditions in ancient times. However, traditional Chinese philosophy attaches too much importance to Heaven and destiny. For instance, *"the Doctrine of the Mean"* says that the fate of humans is determined by Heaven,[6] which has restricted humanity's subjective initiative and therefore, hinders people from thoroughly understanding the BL in the contradiction and motion of things and actively using it.

Mao Zedong, great leader of the Chinese people of the 20th century, epitomized the thought of balance of both Chinese and foreign philosophers and wrote some glorious works such as *On Contradiction*, which guided China's new democratic revolution and socialist construction from one great victory to another. Due to the limitation of development conditions, Mao overemphasized the opposition of the two world views, but overlooked the association between materialism and idealism, regarding the theory of evolution as "vulgar".[7] This led to some extreme policies and events such as the "cultural revolution" and the "destruction of the old and creation of the new", making the Chinese society have no law, no virtue, no beauty, no production growth, no knowledge and no sympathy, and causing grave losses to the Chinese civilization.

In the first 30-some years after the founding of the People's Republic of China, the country developed highly centralized planned economy and the people's commune movement went to the extreme, leading to serious social and economic consequences. In the second 30-odd years of reform and opening up, however, China's market economy also went to the extreme in some areas such as basic education and basic research which were "commercialized" or "industrialized", causing severely negative impacts on the Chinese modernization process. Therefore, it is urgent for

---

[6]The first chapter of the *Golden Mean* begins with: "The natural endowment of human beings is called 'nature'; acting in line with nature is called 'Tao'; educating oneself in Tao is referred to as 'teaching'. Tao is something people can never do away with; if they can, it must not be Tao. Therefore, a noble man is cautious and self-restrained in places where no one can see or hear him. The more hidden a place is, the more conspicuous it may be; the more subtle something is, the more significant it may be. So, a noble man is self-disciplined even when no one is with him".

[7]Mao Zedong. *Selected Works of Mao Zedong* (Vol. I). Beijing: People's Publishing House, 1966, pp. 288, 289, 294.

the country to sum up experience, profoundly study and correctly use BWT, keep to the "middle path" in its modernization drive, and become steady and mature.

With the development of science, technology and modern society, great changes have taken place both in China and in the world, and in this context the role and objective regularity of BL have become more obvious. A thorough study of BL helps to develop a special ability that will enable us to think coherently and in a balanced way when faced with complex problems, which, of course, is conducive to the steady development of China. As we know, a branch of science is a knowledge system that needs the mutual influence, joint development and concerted promotion of countless knowledge explorers. When pointing out that the ancient Chinese ideological system of balance needs further development, we are implying that our exploration so far in this field is incomplete and needs to go further. The philosophy of balance tells us that in an open and dynamic environment, no branch of science can reach the level of completeness or perfectness. Problems always exist and everything is relative.

BL is an objective law that prevails everywhere. By studying BL, we mainly want to use it as an ideological weapon to understand the world, build our country, and promote sustainable social and economic development. By talking about BL in the new era, we aim to further adapt Marxist dialectical materialism to China's real conditions, further modernize China's ancient doctrine of the mean, and further theorize our practice of socialist market economy.

II. The Connotation of BL

Preliminary research shows that: in a multi-dimensional space and infinite time all things have both internal relations and external interactions; and in such relations and interactions there is always a contradictory movement of things or a balance force that reinforces and restrains the change of things to make things mutually restricted and drive them to develop in a progressive cycle of "imbalance–balance–rebalance". Such a balance force is a self-generating, self-balancing, free, independent, natural, and also an absolute, eternal and relatively stable force that drives the movement of things. Under the action of the balance force, things are kept in balanced motion, and this law is called BL. It is the fundamental of the universe, the guarantee of life, the reason for existence, the basis of connection, the condition for activity and the origin of all things. All things achieve development in a cycle of "imbalance–balance–rebalance". In this process, balance is not only the dynamic stability but also the law of contradictory movement of things. When the law of equilibrium works, it may result in diverse states and present various features. We will analyze the basic state of the action process of BL as below.[8]

The first state is the balance of motion, which refers to the "imbalance–balance–rebalance" cyclic motion of things. This is the fundamental of the universe, reflecting the overall relationship of all things in the universe. As the fundamental of the universe, BL governs all things in the universe and keeps them moving in the cycle of "imbalance–balance–rebalance" all the time. For example, space stars twinkle, the brightness of some stars at an alarming rate of change in the past 100 years weakened by 19%, the reason lies in the formation of the universe balance of super-alien structure movement; cosmic balance is like a super alien, constantly collecting and releasing energy, regularly adjust it around the planet. The vitality of the universe lies in that all things move from imbalance to balance and then to rebalance. The basic characteristic of all forms of life is that everything has its own balanced movement and life cycle. Different galaxies have different spatial and temporal scopes of balanced movement, different life forms have their own life cycles, but their balance and cyclic movements are constrained by the overall balance at the same time. The force of overall balance comes from the process of "imbalance–balance–rebalance", which is not only a process of releasing and absorbing natural forces but also a process of forming new natural forces (energy).

The second state is the position of balance, i.e. the balance between a thing and its environment. This is the guarantee of life, reflecting the balance between one thing and the surrounding things in the universe. As the guarantee of life, BL reveals the natural positions of things in the

---

[8] Wang Yebin. *General Equilibrium Theory*. USA: The National Academies Press, 2010, pp. 1–82.

environment. Everything has its own natural place in the environment, which shows that it is in harmony with the environment. The movement of things is not for the sake of movement but for reaching a natural position. In order to survive and reach their natural positions, all things can either make internal adjustments to adapt themselves to the present environment, or choose a new environment that is suitable for their survival. The natural position determined by natural forces is the most reasonable position of things. Artificially changing the natural position will disrupt the natural balance in most cases. To establish a new balance, the thing must either adapt to the present environment through self-adjustments or choose a new environment.

The third state is self-balance, i.e. the balance of internal structure and energy of things. This is the reason for existence, reflecting the balance of the central points of all things in nature. As the reason for existence, BL reveals the instinct of seeking self-existence of things, which is also the motive power for their self-balance. All things have an instinct for survival, a motive power for realizing self-balance, a natural position, and a natural purpose. For example, the system theory tells us that there is a "self-organization phenomenon"; atomic physics proves that electrons automatically occupy the lowest-energy orbit; and the "orderphilicity by living creatures" according to the game theory reveals that all living creatures have the instinct to seek regularity and order in a bad unknown environment. The rotation and revolution of celestial bodies are the process through which they realize survival. The wind, cloud, rain, snow, tornado, debris flow, earthquake, tsunami and volcano eruption on the earth are different forms of energy release through which the earth achieves self-balance. The present living environment of human beings is just the result of the earth's self-adjustment and self-balancing as mentioned above; the development of human society is also subject to the balance of power.

The fourth state is symmetrical balance, that is, the balance between systems. This is the basis of connection, reflecting the balance between one thing and another in the universe. As the basis of connection, BL reveals the symmetry of the universe. Everything appears coincidentally and has its opposites, and the latter's existence is the reason for the former's existence. Things and their opposites are mutually dependent and symmetrical. Even God has his opposite. Neither side can destroy the opposite side to exist independently. So, the two sides have no choice but to treat each other correctly. All things show symmetry at different levels in terms of time, space, mass and energy; and symmetry of the universe is the result of time–space unity as well as mass–energy unity.

The fifth state is synergetic balance, that is, the balance between a particular thing and the system it belongs to. This is the condition for activity, reflecting the balance between one thing in the universe and the corresponding system. As the condition for activity, BL unveils the synergistic effect of all things in nature. Everything in the world has synergy with its system, which is shown by the mutual dependence of a component and the system. In most cases, a functional component can work well only when non-functional components are normal. Though people run on two feet instead of on two hands, it will be difficult for them to run when their hands are tied together. Similarly, birds fly using wings rather than tail, but if the tail feather was cut off it would be impossible for them to fly. The ability to write a good essay is far beyond writing skills; the ability to play a role in a movie is far beyond acting skills; and usefulness lies in uselessness. Many wonderful things in the world are produced as the result of a synergy of useless things.

The sixth state is the ultimate sameness of things, that is, the balance of all things in a certain condition. This is the origin of all things, reflecting the fact that all things in the universe share the same source and same fate. As the origin of all things, BL uncovers the nature of ultimate sameness of things. In the universe, the ultimate sameness of things refers to the fact that all things originate from the same start point and will come to the same end point after countless balance cycles. All things are just the same at the beginning and at the end, which is the so-called "greatness in simplicity". In nature, all complex structures and processes are composed of some simple basic elements that interact with each other. For instance, the proteins and nucleic acids of different living things in the earth's biosphere are composed of the same amino acid and nucleotide, and the genetic materials of all life forms have the same central rule and genetic code. All things are identical at the atomic level; what is different is their structure.

The balance of motion, the position of balance, self balance, symmetrical balance, synergetic balance and the ultimate sameness of things are all evidence to prove the existence and functions of BL.

To better understand the law, we may construct a mathematical model to analyze the state of equilibrium. For example, we can describe the law of equilibrium by means of meta-synthesis and vector representation[9]:

$$\vec{F}_e = \sum_{i=1}^{m} d_i \left(\vec{a}, \vec{b}, \vec{c}, \ldots \right) + \vec{e}.$$

In this mathematical model it is assumed that: $\vec{F}_e$ is an integrated balancing force which can be decomposed into a number of vectors $d_i(\vec{a}, \vec{b}, \vec{c})$ along a, b and c axes from the origin of the coordinate system; for m ($m > 1$) vector elements of a thing $\vec{V}_1, \vec{V}_2, \vec{V}_3, \ldots$, if there exist m vectors $d_1, d_2, d_3, \ldots, d_m$, which are not all zero, then there will a mixed product of vector elements; $\sum_{i=1}^{m} d_i \vec{V}_1$, and $\vec{e}$ is the error caused by the omission of certain vector elements.

According to these assumptions, a thing will be in a balanced state when the balancing force is equal to the mixed product of the relevant vector elements. As all elements are in constant motion and change and there is always an error $\vec{e}$ that is neglected, things are always in a state of motion from imbalance to balance.

In brief, the law of equilibrium in the physical world is independent of human feelings. Therefore, we can not only get a direct understanding of the law based on general knowledge but also construct a mathematical model to analyze it. Although the model is a much simplified version of how it actually works, we can still tell whether something is in a state of equilibrium from the simple mathematical relationship between dependent and independent variables. After illustrating the law of equilibrium as objective

---

[9] See Graduate Management Department of China Academy of Aerospace Systems Science and Engineering, *System Engineering Lectures* (Vol. I). Beijing: China Astronautic Publishing House, 2013, first edition, p. 69. The integrated approach is a method of studying the same problem of a huge complex system (including the social system) from different aspects and perspectives. It is an interdisciplinary research method or way of thinking.

reality from the perspectives of knowledge and mathematical model, we will examine the intrinsic features of this law in the next section.

All in all, BL exists in the objective world no matter we are aware of it or not. In this regard, we not only get intuitive knowledge from the general understanding, but also can build mathematical model for analysis, although the model than the actual state of things to be much simpler, and also can get a mathematical reflection of things balance from this simple strain and independent variables in the relationship between interdependence. From the point of view of epistemology and mathematical model, we show the objective existence of the law of things balance. What, then, are the inherent characteristics of the objective law of equilibrium? In the next section, we will further analyze the ontological characteristics of the law of equilibrium.

III. Characteristics of BL

By summing up the mathematical reflection and intuitive reflection, we find that BL has the basic characteristics of being fundamental, structural, systematic, corresponding, symmetrical, reflective and rhythmical.

1. *BL is Fundamental*

Nothing can exist without a root. Being fundamental reflects the action process of the balancing force of things in the original causal relationship. The existence of a thing is the result of interaction between multi-level heterogeneous elements. Such interaction constitutes the causal chain of things. In a hierarchical causal chain, the lower level has an ascending causality relative to the higher level, showing whole emergence.[10] In the biological system for example, the vitality of living things is the result of interaction between different material elements in their body. The higher level has a descending causality relative to the lower level, presenting a property of reductive release.[11] For example, since the 1950s people have

---

[10] John Henry Holland. *Emergence: From Chaos to Order*. Translation by Chen Yu *et al.* Shanghai: Shanghai Scientific and Technical Publishers, 2001, pp. 222–236, 258–260.
[11] Stuart Kauffman. *The Origins of Order: Self-Organization and Selection in Evolution*. Britain: Oxford University, 1993, pp. 33–36.

been furthering their studies of biological phenomena by means of analysis, decomposition and reduction from the overall morphology level, organ level, cell level, molecular level, all the way to the atomic level. Based on this, they have discovered the double helix structure of DNA and revealed many mysteries of life. In the hierarchical causal chain of heterogeneous elements of things, objectively there is a deepest level that plays a fundamental role in the balance and development of things. A change at the deepest level is most likely to change the nature and development direction of things. Because the associated hierarchies of things are infinite in number, the deepest level is also infinite. Under certain technical conditions in a certain historical stage, humanity's understanding of a specific thing is limited to a certain level and such understanding is always relative rather than absolute. However, continuity exists between different levels. Nature never leaps or jumps. So, a seemingly unexpected event is actually the result of continuous accumulation. Science, for example, has continuity, where any new theory can find its connections with the relevant previous theories. Therefore, BL's characteristics of being fundamental including whole emergence, reductive release, infinity of hierarchy, and continuous accumulation decide that BL plays a fundamental role in the development of things.

2. BL is Structural

Things exist based on their structure. The structure reflects the action process of balancing force on self-organization of things. Everything is made up of smaller parts and elements and has an internal strength to integrate these parts and elements into a system. Self-organization is the process in which a thing develops from a disorderly state into an orderly state due to the interaction between its internal components.[11] In the field of computer science, especially in the study of artificial intelligence, the concepts of "independent variable", "self-adaptation", "self-evolution" and "self-learning" are quite similar to the concept of "self-organization" as mentioned above. Opposite to self-organization is hetero-organization. Self-organization and hetero-organization depend on, complement and restrain each other. All organizations grow, operate and evolve in a certain environment. The environment plays a special structuralizing role in the formation, survival and evolution of systems. The structuralizing role makes the whole thing greater than the sum of all parts. This is because a change in the way the components of a thing are arranged may lead to a change in the whole thing. In a larger system, the components of things have an endogenous force that drives the things to form subsystems. For instance, the universe has an endogenous self-forming force, where the earth forms the earth system by means of its gravitational force, the earth system is made part of the solar system due to the interaction between the earth's gravitational force and the sun's gravitational force, and then the solar system constitutes part of the Milky Way in a similar manner.[12] The human society also has an endogenous self-forming force. Tradition and change, innovation and conservation, deregulation and regulation are an "endogenous part" of the system of a thing. In the market system, each enterprise is not only an individual entity that has its own corporate behaviors but also a member of the market that concerns and influences other enterprises. The self-forming process of things is a "balance–rebalance" process, in which imbalance is there all the time. The earth's orbit around the sun is not a perfect circle; in the development of human society there is always imbalance and such imbalance cannot be completely eliminated by laws or public policies no matter how good they are. Imbalance is absolute and balance is relative, both being structural.

3. *BL is Systematic*

Things exist in the form of a system. Systematization reflects the process in which the balancing force of things integrates system elements. The physical world, biological world and the human society all have their own rigorous integrity. When there is a change in any one of the elements of the system, a chain of changes or a systemic change may occur under the action of the balancing force. System elements are in constant interaction to keep the system in constant motion. Every system has its own life, with internal elements associating and interacting with each other to give life to

---

[12]The *ABC* Spanish Daily Newspaper reported on March 21, 2014 that NASA released a 360° panoramic digital photo of the Milky Way, which was made by mosaicking 200 photos taken by the Spitzer space telescope over the past 10 years. Seen from the panoramic image, the Milky Way is a flat spiral and the sun is located on one of the spiral arms.

the system and make it effective. The interaction between system elements through direct feedback can produce direct system effects. For example, a driver's perception of road conditions will produce a safety effect. Indirect system effects can also be generated when system elements respond to each other across time and social space via the system. For instance, global warming reduces the habitat of frogs and other amphibians and drives them to smaller areas where they are more exposed to various diseases, thus indirectly yet sharply reducing their population. The behavioral effects of system elements are always diverse and unpredictable. Since the attributes of a system are different from those of its components, it is impossible for people to deduce system properties from the properties of its components. System properties are new characteristics determined by the structure of the system. All systems have the characteristics of co-evolution with their environment: living things evolve together with nature; human society evolves together with the physical world. Systems are open. Under the action of balancing force, they constantly exchange materials and energy with the environment, maintain their life processes and move from a disorderly state to an orderly state. Material and energy exchange also takes place between the system's internal elements, which promotes the system's dynamic balance. The system itself can be regarded as either a dependent variable or an independent variable. For example, with the deepening of globalization, the international system's impact on some countries often exceeds the influence of these countries' domestic political and social structure on the external world.[13] The dependent variable and independent variable of a system can be converted to each other. For example, the correlative relationship in statistics can be converted into a functional relationship. The system may achieve local gains or suffer local losses, but its integrity remains. Systems often exhibit a nonlinear relationship, and complex behaviors do not necessarily require complex factors. Systems tend to interact with one another in an "action–response" mode. For this reason, the balance and motion of things are systematic.

---

[13] Robert Jervis. *System Effects: The Complexity in Political and Social Life*. Translation by Li Shaojun, Yang Shaohua and Guan Zhixiong. Shanghai: Shanghai People's Publishing House, 2008, pp. 118, 154–179; also refer to the web page of the Santa Fe Institute (SFI) of the US.

4. *BL is Corresponding*

Things in existence are corresponding to each other. Such correspondence reflects the action process of balancing force on the unity of opposites of things. The most typical correspondence is complexity and simplicity. Our complex world has a simple kernel, the former and the latter forming the two opposite ends that correspond to each other, so the world is a unity of complexity and simplicity. We can understand the corresponding characteristics of BL by looking at the relationship between complexity and simplicity.

Corresponding to simplicity is complexity, and the latter comes from the heterogeneity and interactivity of individual behaviors. Heterogeneity makes the result of interaction more complicated and changeable, and interactivity makes heterogeneity more diverse. Under the action of numerous, individual, local and variable heterogenic and interactive behaviors, the world is extremely complex.[11] The complexity of the objective world is often beyond our current understanding and research ability. A more in-depth study of such complexity has to touch the epistemology and the outlook on development of philosophy on a larger scope, at a deeper level and from a more complex relationship. But from the perspective of balance, no matter how complicated things may be, they show a high degree of synthesis on the basis of a high degree of differentiation. Such synthesis is manifested in at least four aspects. The first is integrity. For example, in the economic field, individuals and their environment constitute a whole; plans and the market interact with each other to form a whole. The second is compatibility. Though systems may be different from one another, each of them is the result of the accommodation and coordination of a larger number of subsystems and therefore, can be managed using similar theories and methods. The third is homogeneity, which centrally reflects the overall characteristics of things and their development trend. For instance, according to the law of large number the probability for an incident to happen follows the "central limit theorem".[14] The fourth is centrality (or oneness), which means that in complexity everything is dominated by the change in its core (the core change). A core

---

[14] Li Songxue. *A Frontier Approach to and Empirical Research of Quantitative Economics.* Beijing: China Social Sciences Press, 2013, pp. 230–233, 250.

change is the result of the action of internal causes, and is also the principal contradiction. A complex thing becomes simple when we have tackled its principal contradiction. So, complexity corresponds to simplicity.

Simplicity reflects the abstract internal relationship of things. Simplification embodies human beings' subjective initiative and is the right way we face and explore the complex world. For instance, the methods of inductive logic and generalization, etc., are effective approaches for us to profoundly understand the objective world. Moreover, simplification can also help us to build a specific world. For example, in computer science the numbers of "0" and "1" constitute a unique computer world. However, people's understanding of the objective world is always subject to some limitations. In the process of simplifying complex things, we are often faced with distorted or incomplete interpretations, but this cannot negate the objective existence of simplicity or the value of simplification. Things are both complex and simple; contradictions are both universal and special. Every movement or activity, be it physical movement or ideological activity, has its own internal special contradiction which constitutes its special nature that is different from that of any other things. What is essential is fundamental and is the principal contradiction. Once the principal contradiction is seized, all problems can be readily solved. From small details we can see a person's moral quality; small changes create the world; the more complex a problem is, the simpler its solution will be; the more profound a theory is, the simpler it will be. The means of scientific research is actually very simple: starting from observation, turning to theory, and then making induction. In a certain sense, it is a manifestation of human beings' following the rule of simplification of things. The process of induction and summarization is the process of simplification. In the long process of following and adapting to natural environmental changes, humanity has developed an instinct of being good at simplification (induction and summarization). It is with such an instinct that we are able to choose and utilize simpler methods to perceive and predict the dynamics of the world, reliably forecast environmental changes, and effectively foresee the possible consequences of our actions. Without simplification (induction and summarization), our interaction with the environment would be totally in a mess, we would have no reason to believe that time and space will be the same in the future as in the past, and we would not be able to know whether rice is nutritious or poisonous. Simplification (induction and summarization) is not necessarily 100% reliable, but tends to reflect the objective law of change of things. Things are extremely rich in variety, but their movement is bound to follow certain simple rules of nature. Simplification is essential not only in our observation of things and summarization of experience, but also in our thinking about the logical concepts of theory, ideology, and spirit. It should be noted, however, that different areas have different simplification methods. For example, the method of bisection is effective in the study of natural phenomena, but may be misleading in the study of social sciences. Yet, in social sciences there must be simple methods to replace the bisection method, except that they are to be fully discovered, understood and utilized.

Based on the two corresponding processes of complication and simplification, we explain the problem of correspondence of balance. The fact that things exist in correspondence to each other helps us to explain not only the concrete but also the abstract relationship between them. In the world as we understand, spirit and matter influence and rely on each other and we cannot say which of the two comes first or is superior to the other. For humanity, a world of matter without spirit is just as meaningless as a world of spirit without matter; pursuing truth and dominating society complement each other. It violates the law of correspondence of balance to only stress spirit without paying sufficient attention to matter, and *vise versa*.

5. *BL is Symmetrical*

Things exist symmetrically. Symmetry reflects the action process of balancing force on the internal structure of things. In comparison with correspondence, symmetry does more to reflect the physical form of the structure of things. Examples of symmetry include negative and positive charges in physics, negative feedback and positive feedback in the financial market, and negative reaction and positive reaction in chemistry. The physical forms of some things may not be completely symmetrical, but their internal properties and structures are symmetrical. For example, the time period of economic boom is relatively long but the time period of economic recession is relatively short during which a large number of companies go bankrupt, forming the symmetry of prosperity and depression as called in economics. Truth is the opposite of error; failure is the mother of success. A mistake itself is not terrible; what is really terrible is that one is not aware of his mistake and cannot learn from the mistake. From the above discussion we can see that the balance and movement of things have the feature of symmetry.

6. *BL is Reflective*

Things in existence reflect each other. Reflection is the process in which things reflect and influence each other. People affect the environment, and the environment affects people; people's views may affect the development of things, and the development of things may have an impact on people's views; if you treat friends as enemies, then they may turn into real enemies. Human beings have a natural cognitive aptitude which is manifested in the form of consciousness; and they also have a natural control ability which is embodied by behavior. There is a two-way connection and transformation between people's consciousness and behavior. With the rapid development of science and technology, people's dominating functions are constantly activated, and their ability to understand and control the forces of nature constantly improves. People's subconscious mind may form a kind of latent energy, scatter a kind of latent behavior, and cause a kind of social outcome. Such an outcome may be the same as or different from expected, but the subconscious mind has played an important role in it, and this is the reason why people's views may affect the development of things. Under the action of BL's reflection, things will move in the direction of the subconscious mind, which is the burden of consciousness.

7. *BL is Rhythmical*

Things are in rhythmical movement. Rhythm reflects the process in which things experience periodic changes in a certain manner under the action of the balancing force. Things move, change and develop in a regular, orderly and rhythmic manner in the dimensions of height, width, depth, time and space. Changes in rhythm are the origin of development of things. All things without exception develop in a certain rhythmic way, such as the ups and downs of mountains, the cycle of sunspot activities, the ripples of gravitational waves, the rotation and revolution of Earth, the birth and death of humans, the leisure of agrarian society, the busyness of industrial society, the rhythm of art, the pace of work, etc. Rhythm is the way in which the vitality of thing lasts. It is a source of explosive power, a turning point in the development of things, a mode of sustainable development of the system, and the optimum state of interaction between system components. Everything has a qualitative prescription and quantitative prescription, and undergoes qualitative changes and quantitative change.[15] In his work "Das Kapital", Marx pointed out that "pure quantitative changes will turn into a qualitative change when reaching a certain point" (see Footnote 2). The "certain point" mentioned herein is actually a kind of rhythm with which things interact with and convert into each other. At the "edge of order and chaos" there is a point where orderliness and disorderliness meet each other. This point is also a sort of rhythm which is the best location for the system's self-organization.[16] Between the quality and quantity of things there are qualitative and quantitative provisions, reflected in qualitative and quantitative changes in two states and Tempo (rhythm). Tempo is the speed at which things happen and also the reflection of how things evolve. The change of tempo is the way how the vitality of things is sustained, a source of outburst and a turning point. It is also a model of sustained development for a system and the optimal state of interaction among the constituent elements of a system. It also can be said that the rhythm is the origin of the development of things and the manifestation of the inherent law of the movement of things. The rhythm change is the sustaining way of the vitality of the thing. It is the source of the explosive force, the turning point of the change of the thing, the mode of the system's continuous development, the optimal state of interaction of constituent elements. For all living and non-living things, change is eternal and balance is dynamic; and the process from change to balance follows the law of rhythm.

For example, human growth is rhythmic in physiological behavior and thinking. Physiological rhythm is the functional coordination of human

---

[15] Marshall P. Mahler. *Economics*. Beijing: China Machine Press, 2009, p. 18.
[16] The Santa Fe Institute (SFI), USA. Also see Wotje Waldrop, pp. 410–413.

tissues, organs and systems to enable the human body to realize orderly material, energy and information exchange within the body so that human beings can effectively adapt to the environment. Behavioral rhythm is the coordination between human behavior and the external environment to enable the human body to achieve orderly material, energy and information exchange with the external world so that human beings can rationally select and utilize the environment. The rhythm of thinking is the degree of correct reflection of the essence and the law of change of objective things in the human brain to enable human beings to adjust and improve, in a timely manner, their way of thinking and development.

The human society is rhythmic in economy, politics and culture. In a proper economic rhythm, people adjust and control the allocation of various production factors so as to realize orderly production and consumption and the primary distribution of social wealth. In a certain political rhythm, people adjust and control various consumption policies and plans so as to fully utilize resources, coordinate various relations, balance the interests of all parties, alleviate social contradictions, maintain social stability and realize secondary distribution of social wealth. In an appropriate cultural rhythm, people adjust and guide the relationship between social practice and ideological theory to make the society more systematic in theory, more identical in concept, more continuous in thought, more connective in psychology, and more harmonious with nature so as to fulfill the planned social goals, unify social wills, coordinate social actions, concentrate social wisdom, unite social forces, promote social development and realize tertiary distribution of social wealth.

A comprehensive overview of these important characteristics of BL shows that: the fundamental nature of things is the origin of their changes; the structural nature of things displays the hierarchy of their changes; the systematic nature of things reflects the overall characteristics of their changes; the corresponding nature of things manifests the internal relations among their changes; the symmetrical nature of things reflects the structural form of their changes; the reflective nature of things explains the process interaction between them; and that the rhythmic nature of things describes the steps of their changes. These characteristics further demonstrate that the connotation of BL is extremely rich. In order to gain a deeper understanding of BL, we will analyze the consciousness of BL in the following section.

IV. BL Awareness and BWT

Like in the objective world, dynamic balancing activities also exist in the subjective world. We refer to this kind of dynamic psychological balancing activity as "balanced way of thinking (BWT)", which highlights the importance of relationships between different things and the avoidance of extreme circumstances. To master and use BWT, we will first review some of the balanced states in the mental world and analyze the nature of BWT, especially the relationship between BWT and other ways of thinking.

1. *Balance of the Mental World*

The mental or psychological world is a wonderful world. Descartes said: "I think; therefore I am". One knows his existence when being aware of his own mental activities.[17] Wang Yangming, a great thinker of China in the Ming Dynasty, sought the ultimate truth of things all his life and eventually discovered that "the ultimate truth" is in the mind. He advocated the unity of body and mind and the unity of knowledge and action. In his belief, whether one deeply understands something can be seen from his action; and in turn, whether his action is correct affects his understanding of the thing.[18] According to Wang's theory, thoroughly knowing something and correctly doing it is an organic unity.

The mental world and the material world are interrelated. For instance, the sum of the three interior angles of a triangle is 180°, which is a correct reflection of the mental world to the material world. The subjective world is reflected in people's minds where contradictory movement of ideas is formed to promote the development of ideas.[19] Ideas are bigger than events. The subjective world also has active functions, guiding people's actions and affecting the changes of things. A change in people's mood and idea may lead to a change in the environment and affect the object of interaction. If blackness is dominant, then redness and greenness will gradually turn into blackness. In real life, some shanty towns are inhabited

---

[17] *Readings in Western Philosophy* (Vol. I). Beijing: The Commercial Press, 1982, pp. 369–374.
[18] Wang Shouren. *Complete Works of Wang Yangming*. Shanghai: Shanghai Classics Publishing House, 1992, p. 50.
[19] Tom Rockmore. *Kant and Idealism* (Translator's Preface). Translation by Xu Xiangdong. Shanghai: Shanghai Translation Publishing House, 2011.

by human bodies but others by human hearts. Both the two kinds of shanty towns need to be improved simultaneously.

The state of mind is like the market. Mental activities are associated with the environment: a person's attitude tends to be positive/negative when he is positively/negatively evaluated. The spiritual power generated in BWT is manifested in various forms, such as the power of belief, the power of notion, and the power of cognition. Deep in a person's heart, there is an internal cognitive ability, i.e. inspiration or spiritual quotient (SQ). A person's SQ enables the person not only to understand his intrinsic value but also to dominate his connection with the outside world. In addition, it promotes the coordination of people's attitude and behavior. "Precious as freedom and love are, both can be thrown away for the sake of mission". The poem profoundly illustrates the mightiness of cognitive power. In terms of belief, people are willing to believe and seek unconventional procedures in order to realize their goals. In the political arena, people may take "the enemy's enemy as a friend" for political purposes, and may groundlessly expand their scope of collaboration with a third party so as to achieve a balance of emotion and power. However, in the presence of BL, overcorrection will sooner or later lead to over-correctness, and going to extremes is unwise. One's ideal and goal must be based on reality; otherwise he is very likely to be a utopian.[20]

2. *The Sense of BWT*

To master and use BWT, the key is to have an awareness of BWT. This awareness is an instinct and inner impulse, both of which can be aroused and cultivated. To develop the awareness of balance, we need to understand, from the perspective of BL, understand the concept, category, idea as well as the scope and culture of balance, all of which are associated with the sense of balance. As a reflection of the essence of things in human minds, the concept of balance is an important part of people's rational knowledge and is characterized by the combination of extendibility and integrity. As a basic relationship of the essential attributes of things

---

[20] Chen Yun. *Selected Works of Chen Yun* (Vol. III). Beijing: People's Publishing House, 1995, pp. 34–43.

reflected in human minds, the category of balance is the logical form of people's rational thinking and is characterized by the combination of cascades and networks. People's understanding of the balance of things cannot be achieved overnight, or once and for all. It is an infinite progressive development process of thinking. As the people's subjective sense of things, the idea of balance is a collection of balance-related knowledge formed in practice and has a dominant effect on human behavior. This idea can make significant contributions to understanding and solving problems, but there is no such a thing as idealism in a general sense in this context.[21] Ideas can be innovated, and the idea of balance is an innovative one. The culture of balance is deeply rooted in people's heart. It is a kind of freedom with restraints as premise and such freedom needs no reminding. It tells people to lead a meaningful life for the good of themselves and the whole society by giving up something and gaining something. It stresses that people should have a sense, spirit and ability of balance in daily life.[22] People can form and enhance their ability and instinct of balance by establishing a correct concept, category, idea and culture of balance.

3. *The Relationship between BWT and Other Ways of Thinking*

To master and use BWT, we should correctly handle the relationship between BWT and other ways of thinking. If we put the activities of balance of the mental world into the philosophical category, and compare them with the activities of traditional philosophy, modern philosophy and contemporary philosophy, we will find that BWT is both associated with and different from the thinking of ontology, epistemology and methodology and that there is a process from ontology to epistemology, then to methodology, and then to the theory of balance.

　　While trying to understand the world, people are not satisfied with the level of their understanding themselves. As they always want to discover the trend and the law of development of things, it is natural that they put forward new ideas and raise new questions that are different from those of

---

[21] Descartes. *Descartes's Speculative Philosophy*. Translation by Shang Xinjian. Beijing: Jiuzhou Press, 2006, pp. 325–363.

[22] Sigmund Freud. Freud's Philosophy of Life. Translation by Zhou Bin. Beijing: Jiuzhou Press, 2010, pp. 1–47.

the traditional philosophy. Although these new ideas and new questions may not be all correct or even may be totally wrong, they are still of certain value for people to ponder on, thus leading to the appearance of epistemology. Are new questions and new ideas put forward from the perspective of epistemology necessarily in line with the development direction of modern science and human civilization? This is a very difficult thing. Without a logical and scientific method, it would be impossible to really understand the trend and the law of development of things. So, there appears methodology which studies the methods of cognition and consciousness, such as the logic of thinking, the method of innovation and the way of expression. Methods determine effects, and different methods produce different effects. Famous American psychologist Howard Gardner beliefs that people have at least 168 possibilities of changing their way of thinking from the three dimensions of content, field and factor.[23] Among these possibilities, BWT is the most important.

BWT means to use BL in the thinking process, with the balance of things as the object of study. It is a way as well as a feature of thinking where balance is not only the motive but also the purpose. It is a scientific and dialectical way of thinking, or a way of thinking characterized by system analysis. It is a core-level way of thinking for studying both the natural world and the human society. As Engels pointed out, "to accurately describe the universe, the evolution of the universe and humanity, and human brain's reflection of such evolution, the only way is to use a dialectical method and constantly observe the interaction between formation and disappearance, and between forward changes and backward changes".[24] Based on the general dialectical method, BWT further stresses the importance of following the requirements of BL. In other words, BWT is to follow the law of equilibrium based on dialectics so it helps people to better grasp the movement of things. As a continuation and development of the general dialectical method, BWT helps the human brain to better describe and reflect the movement of things. In today's world which is more economically and socially complex, BWT has more

---

[23] Howard Gardner. *Change of Thinking*. Beijing: Chinese People's University Press, 2009, pp. 16–27.

[24] Zhan Shichuang (Editor-in-chief). *Newly Complied History of Chinese Philosophy*. Beijing: China Books, 2002, p. 27.

Table 1.   The Relationship between BWT and Other Ways of Thinking

	Object	Method	Characteristics	Relationship
Traditional philosophy	The problem of ontology	Intuitive method	Direct perception	Everything is in philosophy
Modern philosophy	The problem of cognition	Psychological method	"I think; therefore I am"	Nothing is in philosophy
Contemporary philosophy	The problem of methodology	Analytical method	Everything can be analyzed	Philosophy is in everything
The philosophy of balance	The problem of balance	BWT	Balance is not only the motive but also the purpose	Everything is in balance

special significance than other methods. The relationship between BWT and other ways of thinking is summarized in the Table 1.

Seen from Table 1, changes in human thinking reflect the process of gradual deepening of human understanding of the world. With the development of methodology and BWT, the rationality of humanity putting forward new views and new questions has greatly increased, humanity's understanding of new things has deepened, and human civilization has entered a new realm. Today, a balanced and integrated philosophical thinking is being formed, such as unity of the macro world, the identity of living and non-living things, the systematic point of view, etc., and a new ideological revolution is brewing.

In this revolution, a balanced philosophical way of thinking may provide us with some new insights. For example, we may acquire new understanding about the relationship between spirit and matter, such as which comes first and which comes second. As is known to all, there are two opposite philosophy systems each of which has its own understanding of the relationship between the spiritual world and the physical world, leading to the formation of dualism. The world outlook that believes nature is original and comes before human consciousness and spirit is classified as materialism. The world outlook that believes human consciousness, idea or spirit is original and comes before the physical world and that nature is the result of human thought or spirit is categorized as idealism. Hegel is considered to be representative of idealism. Marx inherited the core

Hegelian philosophy. Regarding nature and society as a process and observing them during the process, Marx founded the theory of socialism. The theory played a guiding role in the socialist revolution of many countries including the Soviet Union, but they suffered severe setbacks in later years. Why were tremendous achievements made and why were huge setbacks suffered? Is it really necessary to artificially determine whether matter comes first or spirit comes first? Is there a middle zone between the two propositions? According to BL, matter and spirit are interactive with each other and equally important. Of course, we must not hesitate to adhere to dialectical thinking, but at the same time we must not ignore the role of spirit, because matter and spirit are inseparable and interrelated. While criticizing Hegelian idealism, Marx stressed that matter is primary. This innovation enriched humanity's philosophy, yet it regarded spirit as an absolutely opposite matter, which is very likely to lead people's thinking and action from one extreme to the other. Mao Zedong should be considered a thoroughgoing materialist, but he erroneously launched the "Great Cultural Revolution" in his late years. Why? Because the "thoroughgoing" materialist went to the extreme to reach the opposite of materialism: idealistic utopian socialism. The economy is bound to fail if "planned to the extreme". Similarly, when reaching the extreme, the theory of the omnipotence of productive forces will become a prisoner of idealism or a victim of commodity fetishism. We have painful lessons and bitter experience in this respect.

4. *The "Degree" of Balance*

By comparing BWT with other ways of thinking, or by comparing BL with other laws, we find that BL has an important and unique feature — "degree", in addition to other important features. Balanced thinking is a branch of science that studies "degree". It emphasizes the importance of "degree", and reveals the interval and turning points of the development process of things so as to provide a scientific basis for active intervention. An appropriate degree of balance is a kind of wisdom with which people grasp the interrelations among things and make trade-offs; it is just the right extent rather than a simple compromise or golden mean; and it is a quantitative measurement which can be expressed as the value or interval detected. Although the *Book of Changes* established an image–number system which expressed Chinese ancestors' dialectical thinking and decision-making process of obtaining images from numbers, it is far from meeting the needs of modern society for balance measurement theory and technology.[24] In this sense, the mathematical model of balance set up in the second section of this book has a very important practical significance and can be used as a tool to quantify the degree of BWT.

Compared with the general dialectics, BL intensifies the connotation and function of "degree". When the movement of things exceeds a certain "degree", the original balance will be broken and a new balance will be approached. The transformation of things needs a certain number of conditions, so we can design a visible pre-warning interval according to these conditions and take proactive measures to promote the transformation of things in the desired direction. Compared with the law of unity of opposites, BL places more emphasis on the middle zone between or the intersection of the two opposites instead of the two extremes of them. Antithesis is a form but not all forms of struggle between conflicting parties. There is not only unity of opposites, but also non-opposite connections between things. The rules of balance does not only focus on the connection between the two ends of things but also the process of connecting, or the connection in the middle section. In the middle zone, there is usually bigger room for the development of things, where there are a variety of possibilities rather than merely two — "black"and "white". It is undesirable to conclude that something is either "good" or "bad" since there may be divergent views about what seems to be "good" and what seems to be "bad" may also have its worth. Because there are phenomena and essence that are neither good nor bad between the good and the bad. Things move amid antithesis and cooperation. Compared with the law of negation of negation, BL pays more attention to the comprehensive utilization of abandonment and retention. Neither complete abandonment nor complete retention is in conformity with the requirements of BL. Negation contains certain elements of affirmation and affirmation includes some elements of negation. Between the two there is a space for mutual inclusion. Compared with the law of interchange of quality and quantity, BL attaches more importance to the balance between quality and quantity of things, because an imbalance between the two will lead to a stagnation of

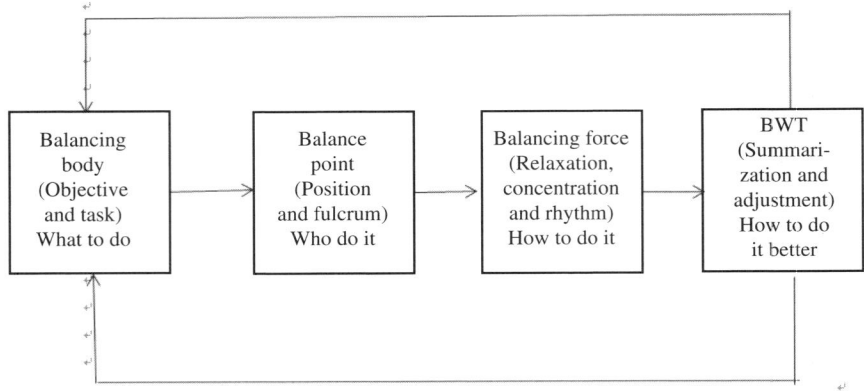

Figure 1.   Application of BL

things. In the process of developing a market economy, "quantity" is impossible to fully meet the demand for "quality" and people should not be encouraged to indulge in the endless pursuit of material enjoyment.[25]

Balance does not go against winning a competition because balance is a dynamic status and it is the basic pathway towards brilliant achievements in a certain period. The continuous changes to balance will result in continuous adjustments, improvements, and innovations, and ultimately lead to the greatest achievements possible in a certain period. The CCTV program "Challenge the Impossible" offers many typical cases. It is fair to say that every impossibility that turns into possibility is the result of thinking and doing in compliance with the rules of balance.

The life of theory consists in practice. As a methodology, BWT is of both general and special significance for human beings to understand the movement of things including daily study, work and life. We use Figure 1 to outline the significance of BWT for daily routines, as well as the four steps to follow and apply BL.

Firstly, by using the BWT approach, we can better understand the boundary and balancing body of things in the real environment, rationally identify the tasks and objectives, and clearly know what we are going to

---

[25] Shang Xinjian. *Works on Seeking the Truth*. Beijing: Peking University Press, 2012, pp. 55–110.

do. People's knowledge is dynamic and will be deepened or improved with the change of conditions. Seeing things from the perspective of balance helps people to deepen their dynamic understanding of objective things and correctly grasp the development trend of things.

Secondly, by using the BWT approach, we can find the balance point of the movement of things and the foothold based on which our work can be carried out; we can clearly know our position and use the fulcrum of the lever; and we will know "whom to rely on" and "who to do it". This is true of a person, a department, an organization, a country, a region, and the whole world, especially of economic work where some things but not everything should be done. To this end, we must find the right people and the right team.

Thirdly, by using the BWT approach, we can form a good balance force and solve the problem of "how to do it". A balance force cannot be formed at too fast or too slow a pace. It is usually formed at a relaxing, focused, and rhythmic pace instead. A good balance force can only be generated after we have made a comprehensive consideration, focused on a certain key point, and attentively carried out a certain activity in a relaxing environment and calm mood.

Lastly, by using the BWT approach, we can make summaries and adjustments at any time if necessary, and answer the question of "how to do it better". When we have developed the habit of BWT, we will be able to adapt to the inertia of the movement of things and the pace of their progress, break the old balance and create a new one, and promote the virtuous circle and sustainable development of things. Although BWT is something belonging in ideology and superstructure, it is indispensable to the sustainable development of economy. Indeed, China needs abundant thoughts of balanced development that reflect both inheritance and development, are inclusive, and are of guidance to modern social and economic activities.

V. Summary

The balance-oriented way of thinking is a theory that examines how things of conflicts influence one another and reach dynamic balance. Based on general dialectics, the balance-oriented way of thinking pays attention to both the results and the process of cognition and practice, so BWT is a continuation and development of general dialectics.

Balance is the fundamental law of the change of things, and the change of things is the change of dynamic balance. Balance is eternal, which contains the laws of nature and the true meaning of social ecology. Balance is a kind of beauty which is both classical and *avant-garde*, both flashy and simple. Balance is a kind of strength which not only reflects the history but also foretells the future. We pursue balance not only to accumulate heritage, make achievements, enter the realm of freedom, but also to seek truth.

Balance is seen everywhere in this world. No pains, no gains; no things can be done once and for all, and nothing is perfect; human life is both short and eternal; vicissitudes of life are both an experience and a part of nature. In the ideological world, the more one knows, the more he is unfamiliar with; balance must be achieved between supply and demand and between liberty and equality; we must have a reliable expectation for the future society. No thinker's conclusion can fully withstand the test of history; and no theory can be universally accepted. Yet, the achievements made by ideological workers and other people in trying to understand the world can give us enlightenment. The development of thought always needs the hard work of ideological explorers and the expansion of noble hearts.

In the present era full of contradictions, technological revolution, economic adjustment, and social change are not only owing to but also for global balance. Our world is a dynamically balanced world, where internal and external structures are interrelated and various factors act on each other. When one balance is destroyed, the interior and exterior of things need to establish a new relationship and various forces need to readjust their direction and intensity in order to achieve a new balance. The strength for connection and re-adjustment is the force of BL.

In the economic field where a balance should be realized between economy and reform, we particularly need to follow and apply BL.[26] In our subsequent chapters, we will mainly discuss how to apply the BWT approach in the economic work so as to achieve sustainable development of the economy. We hope to propose new ideas and new measures for the sake of balanced development of the country and the world.

---

[26] Lixing Zou. On the Law of Balance. *Research on Financial and Economic Issues*, 2015, (2), pp. 6–11.

Part I

On the Path to Balance

Under the law of equilibrium, the development of human society, international economic theories and practices, and China's modernization process show a tendency towards balance. This is because the world we live in is in dynamic equilibrium and all things are interrelated, act as checks on each other and are subject to the law of equilibrium.

The world history of globalization shows that human society follows a path to balanced development characterized by "the shift from oneness to diversity and from diversity to an organic whole".

Since the mid-18th century, there have been heated debates among economic theorists and practitioners over the relationship between public and private ownership and that between market competition and government intervention, and two schools of thoughts have emerged accordingly. Roughly speaking, development in the late 18th and 19th centuries was mainly driven by private ownership and market competition; the 20th century saw a bigger role of government intervention and public ownership; a tide of mixed economy is rising with the advent of the 21st century. A mixed economy is one where private and public ownership coexist, and the market and the government both play their due roles. This trend is the result of historical and social developments under the law of equilibrium.

China's modern history has lasted for about two centuries. The 1st century following the First Opium War was marked by national salvation. In the New Democratic Revolution, the CPC drew valuable lessons from previous democratic revolutions and integrated Marxism–Leninism with China's realities. As a result, Mao Zedong Thought was created, the "three big mountains" — imperialism, feudalism and bureaucrat-capitalism were overthrown, and the People's Republic of China (PRC) was founded. More than 60 years have passed since the founding of the PRC. Today's China, the world's second largest economy, has reached a new stage in terms of comprehensive national strength, people's living standards and international influence. However, the task of building socialism with Chinese characteristics is yet to be accomplished in a satisfactory way. China's economic and social development is unbalanced and unsustainable. In the first 30 years following the founding of the PRC, political movements occurred one after another for some time and there was a "Great Leap Forward" achieved based on planning in the form of mass campaign with politics at the core. Its purpose was to attain the ideal of communism, but haste makes waste. In the second 30 years, launching of the policy of reform and opening up, the role of the market was once again overemphasized, and there was a market-based campaign encouraging all members of society to pursue wealth. Its purpose was to build a moderately prosperous society in all respects, but the result was a crisis in faith and trust, and the absence of leading values such as confidence, responsibility and vision. Lessons in the two aspects are both bitter. In the next 30 years, holistic approach and overall balance should remain the focus and soul of economic work. Under the new normal, China needs to adopt a balanced approach to economic growth and seek innovation-driven sustainable development. Speed should give way to quality, productivity to continuity and efficiency to sustainability. The system of theories about socialism with Chinese characteristics includes not only the primary stage of socialism, the Party's basic line and the three-step development strategy but also the thought on balanced development. Chen Yun's economic thinking on overall balance is a perfect example of following and applying the law of equilibrium.

Historical inertia and patterns have a significant influence on how society works and on the global modernization process. At present, the human society is entering the stage of globalization on a deeper level featuring the interaction between man and nature and between man and society. Many new problems have come up and commands our close attention, such as economic globalization vs. cultural diversity, the drive towards social equity vs. wide wealth gap, global integration vs. governance diversity, etc. We should take the initiative to discuss the balance mechanism of society. Adapting to social changes and trends, keeping up with the time, incorporating different cultures while maintaining one's own characteristics, building the constructive capacity for communicating with other people and nations, interacting with the nature and people in an appropriate manner, being brave to explore new grounds, and pursuing shared development are the basic conditions and driving forces for sustainable development amid globalization for both mankind and China.

Chapter 1

The Balance History of Human Development

How did human society develop in a balanced manner in history? What is the impact on the present and the future? History is a mirror. The deeper you look into the past, the farther you will be able to see into the future. Much of today's complicated international situation can be attributed to historical factors. When humans lived on hunting and gathering, there was a "primitive globalization". In the agricultural age, communication between people existed mainly along several separate lines. Modern globalization did not start until Columbus discovered the new continent. And it then went through periods respectively featured by colonial trade, steam engine, electric motor, automation, finance and information technology. Entering the 21st century, globalization deepened, mankind faced exacerbating shared difficulties such as climate change and shortage of resources, and the conflict between economic globalization and cultural diversity intensified. This chapter analyzes how human society moved from one single source to pluralism and then to integration, exploring the law behind the historical balance and its implications for the present.

Global modernization rests on the basis of history. Looking through the human history, we can see roughly seven stages: (1) anthropoid stage (2 million years ago) when man walked on two legs; (2) ape-man stage (1.8–0.3 million years ago) when man started making simple tools; (3) homo sapiens stage (300,000–40,000 years ago) when man had some intelligence to improve their living standards; (4) ancient stage of hunting and gathering (40,000–12,000 years ago); (5) middle ancient state (12,000–5,000 years ago) when man produced food; (6) ancestor stage (5,000 years ago–the invention of steam engine) of written civilizations; (7) modern human age (the first Industrial Revolution–till now) with modern industrial techniques. Corresponding to human evolution, the human society followed a pattern of "unity, separation, unity", moving from one source to pluralism and then to integration. Ever since the creation of mankind by Mother Nature, our ancestors for the sake of survival were required to share the same mind, form groups to fight against the harsh conditions and to reproduce. During the Stone Age, man made stone tools and were more capable of coping with harsh conditions and became more productive. People moved to different places around the world, established different groups and cultures of diversity, and developed along separate lines. In the Bronze Age,[1] man created writing systems and adopted agriculture. The discovery of the new continent in 1492, in particular, accelerated the pace of integration. Today, we are going through in-depth globalization and some new features are worth paying close attention.

I. Primitive Integration: Integration for Survival

Humans, as well as all other species, are a result of nature and its laws. According to history and paleoanthropology, around 3.75 million years ago, in Africa, the Great Rift Valley in East Africa and today's Ethiopia in particular, the climate was agreeable and man started to walk on two legs. Walking on two legs set free the two hands of the anthropoids and enabled them to grab food and hold it in hand while walking fast. This gave them an advantage in the competition for survival and offered them a chance to evolve toward modern humans.[2] However, during the evolution of anthropoids toward modern humans, they gathered in a small area, collected food and fought against beasts as a team. Working together and the early form of integration played a crucial role in the birth of human society, for

---

[1] Man in different places of the world entered the Bronze Age at different times. In Mesopotamia and Europe, bronzeware was in use in 4,000–3,000 BC; in India and Egypt, it was 3,000–2,000 BC; in China, it was 3,000 BC.
[2] Philip Lee Ralph, Robert E. Lerner, Meacham Standish and Edward Burns. *World Civilizations: Their History and Their Culture.* Translated by Zhao Fengfeng *et al.* Beijing: The Commercial Press, 1998, pp. 12–13.

only by doing so were they able to survive and thrive.[3] Therefore, the direct predecessor of human, which was included in the category of human rather than apes, is called handy, literary meaning "able man". Primitive integration and joint efforts are the true origins of human society.

About 2 million years ago, primitive handy knew to some extent how to hunt collectively, how to use animal bones and trees, how to sharpen a piece of stone and how to use these simple tools to hunt, dig tubers, fell plants, smash food and cut off rotten meat. Thus, they had ample food supply. Superior to plants and animals, primitive man knew how to use tools to collect food and thus obtained more and varied foods. This was a precondition of human history and a milestone on the way toward civilization.

About 1.6 million and 300,000 years ago, handy matured in its later period and is known among historians as workman.[4] Later-stage handy was larger and more intelligent than its predecessor, with 40% larger cranial capacity.[5] They could use and control fire for heating and keeping dangerous animals off; more importantly, they could use language for communication, which is especially superior to animals, marking their progress towards homo sapiens.

Primitive homo sapiens lived around 300,000–125,000 years ago. They launched the first globalization process of mankind. When hunting, they chased collectible food from the North to the Middle East and then some continued west to reach today's Europe and became Neanderthals as we call them today, while some others turned east to reach today's Siberia and Asia, and became what is known to us today as Denisovans. In 1856, traces of ancient humans were discovered in the Neanderthal River Valley of Germany and named the discovered type of humans as the Neanderthals. In 1911, Marcellin Boule, a French paleontologist, published the first scientific description of Neanderthal species: they had shifted from a

---

[3] Keith Thomas. *Man and the Natural World Changing Attitudes in England 1500–1800*. Translated by Song Lili. Nanjing: Yilin Press, 2008, p. 18.
[4] Felipe Fernández-Armesto. *The World: A Brief History*. Beijing: Peking University Press, 2010, p. 7.
[5] Keith Thomas. *Man and the Natural World Changing Attitudes in England 1500–1800*. Translated by Song Lili. Nanjing: Yilin Press, 2008, p. 15.

caricature of caveman to a remarkably sophisticated species; they built tools, made jewelry, were probably stronger than us and maybe just as smart as us. They were homo sapiens of the later stage, i.e. around 150,000–40,000 years ago, which was the time of the latest Ice Age. The Earth cooled off greatly about 150,000 years ago and the ice top covered vast areas, making it suitable for walking and migrating. On the tundra-covered Eurasian continent, many mammals stored large amounts of fat to defend themselves against the harsh coldness during the long Ice Age, and humans at that time thus had sufficient prey and could spare more time to study the nature and make tools.[6] Humans in this period were highly skilled at tool-making, invented 60 large tools including knife, chisel, drill and spearhead. They could build shelters, were good at hunting, enjoyed higher living standards and became more reproductive. As a result, the population surged and they expanded with faster steps either intentionally or unintentionally in the face of climate change and pressure for survival. They moved while hunting and continued with global expansion.

Entering the Neolithic Age (40,000–12,000 years ago), men invented bow and arrow, and became more efficient in hunting and also more reproductive. They were quite "modern" actually. They enjoyed arts, had ambitions, had religion, public forums and political practices. Also, they were similar to us in terms of intelligence and physical power. Since it was still Ice Age then (around 10,600–8,000 years ago, temperature was volatile, and ice gradually melted, and the Ice Age gradually ended), they continued to explore new areas, and scattered around the world, building new homes here and there, basically forming the world order today.

Figure 1.1 shows the process of how human evolution spread. Generally, the earliest humans were in Africa and moved to other places later as evolution continued. By the Ice Age, especially in the Neolithic Age, migration accelerated, basically forming our world order today. This process can be regarded as a kind of globalization.

The Ice Age, especially the Neolithic Age, was a great time for human development. "In this age, at all places inhabited by man, the culture shared the same basic elements. All men lived on hunting and gathering, used similar techniques, ate similar foods and had similar levels of

---

[6] *Ibid.*, pp. 19–22.

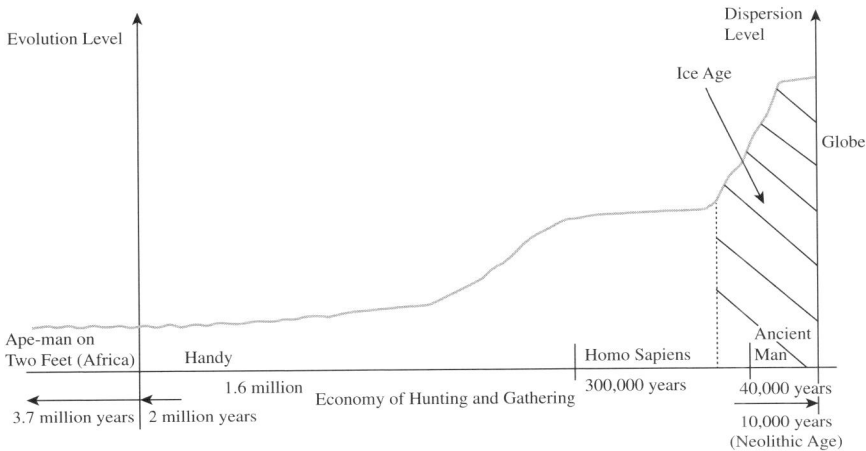

Figure 1.1.   Human Society: From One Source to Pluralism

material culture. And, based on the information available to us, they had similar religious rituals".[7] Therefore, in a sense, we can say this was a time of primitive globalization. Primitive globalization evolved with the interactions between man and nature.

II. Historic Separation: Development along Multiple Lines

As man spread out around the world, different species, races and societies were formed and the human society developed along separate lines over a prolonged process. On the ancient Earth, each tribe constituted a separate traditional society, each tradition equaled one unique way of life, and each way of life represented one civilization.

1. *Species Formed in Migration*

Before the Neolithic Age, man lived mainly on gathering or hunting. Such a way of life means that man at that time must continuously try to find vegetarian food, or, as hunters, must migrate along with its prey. For

---

[7] Felipe Fernández-Armesto. *The World: A Brief History.* Beijing: Peking University Press, 2010, p. 24.

survival and development, man migrated to different places around the world, tapped into every inch of livable land, adapted to local climate, settled down at various regions, and adopted with different species.

History shows that there were generally three branches of man: ancient African man, ancient Asian man and ancient European man (Heidelberg man). According to archaeological findings, the first and the third started to differ significantly as early as in the Middle Pleistocene (Paleolithic Age). Evolution may have been more complicated in the region where China is located. Man here was neither the African branch (homo sapiens) nor Heidelberg, rather they combined the features of East Asian, European and African man of the middle Pleistocene era.

About 500,000 years ago, ancient Indian man emerged on the sub-continent of India.[8] Between 1926 and 1930, archaeologists found over 40 pieces of bones of a Peking man in a cave of Zhoukoudian, 40 km or 25 miles southwest of Beijing, which dated back to no less than 500,000 years. This means that man did live in the vast Southeast Asia at that time. Archaeologists believe that activities of homo sapiens date back to about 65,000 years in mainland China.[9]

About 40,000–30,000 years ago, man in the eastern hemisphere migrated to the western hemisphere via the land bridge that then existed between Siberia and Alaska. And traces of anthropogenic cultivation were found to date back to 30,000 years in Australia.[10] Historians believe that some Asian tribes migrated to North America via the Bering Strait around 30,000 years ago, then to Mexico 21,300 years ago and to Chile 13,000 years ago.[11] Migration never ceased. About 10,000 years ago, Aryans living in western Russia migrated in three groups. One group moved west to today's Europe and was the ancestors of some of today's European people; one moved eastward to Persia and was the ancestors of some of today's Persians; the remaining group moved southeast over the Hindu

[8] *India: Brief History of a Civilization.* Chapter 1, p. 1.
[9] Philip Lee Ralph, Robert Standish Meacham and Joseph Edwards. *World Civilizations:Their History and Their Culture.* Beijing: The Commercial Press, 1998, p. 16.
[10] *Ibid.,* p. 18.
[11] Liu Wenlong. *General History of Big Countries: General History of Mexico.* Shanghai: Shanghai Academy of Social Sciences Press, 2008, p. 7.

Kush Mountains to northwestern India and was the ancestors of some of the Indians today.

Evolution differed by environment: Geographic differences resulted in the differentiation of humans. Anthropoarchaeologists found that some distinctive characteristics, e.g. the relative area of the first palatial motar, differed greatly across races and geographic locations.[12] Different environments also resulted in different modern human behaviors. The evolution of ancient human behavior shows temporal and spatial diversity. Based on available materials, about 100,000 years ago, the characteristics of modern human started to appear in southern China. The modern behavior of Chinese modern man, though more or less influenced by external humans and technical import, is mainly the result of continuous local evolution.[13] According to the findings of anthropoarchaeologists, stones sheets and stoneware developed in systematic continuation in northern China, while modern man in Africa and western Eurasia, and in east Eurasia followed different paths of evolution. This means the evolution of modern man occurred in different regions.[14]

2. *Diversity of Human Civilization*

About 10,000 years ago, agriculture appeared and man's way of life changed radically. Instead of wandering about, they settled down and formed villages. Life in settlements deepened the development of human society along multiple lines and accelerated the progress of human civilization. From the agricultural age to the discovery of the new continent by Columbus, though communication continued, e.g. October 12, 2016, BBC

---

[12] Institute of Vertebrate Paleontology and Paleoanthropolgy, CAS. *Acta Anthropologica Sinica*, Vol. 33, No. 3, 2014, p. 400; in October 2015, *Nature* published an article by Liu Wu, Wu Xiujie *et al.* of the Institute of Vertebrate Paleontology and Paleoanthropolgy, CAS, describing 47 pieces of fossil of human teeth with full characteristics of a modern human discovered in Daoxian County, Hunan Province and concluding the existence of modern humans in that area 80,000–120,000 years ago.

[13] Modern behavior is taken by anthropologists as a major distinction between modern and ancient man.

[14] *World Civilizations: Their History and Their Culture*, p. 517.

reported that Western contact with China began long before Marco Polo, the multiple lines of development remained relatively independent from one another. And this served as the basis for cultural diversity. To a large extent, civilizations and cultures of different regions were influenced by the geographic environment, different natural conditions led to varied ways of life, and various ways of life gave birth to different cultural spirits.[15] Human civilization and culture developed on the basis of certain natural and geographic conditions and are clearly marked by various geographic conditions.

(1) Different Writing Systems and Legislature: Between 4,000 BC and 1,000 BC, different writing systems were invented in different areas of the ancient world, laying the foundation for the diversity of human civilization.

One of them is the sphenogram of Mesopotamia: Between 3,500 BC and 2,500 BC, in the area between the Tigris River and the Euphrates River, or Mesopotamia, especially Sumer at the southernmost of Mesopotamia, human society, for the first time ever, was civilized and the first city was formed. The Sumerians started to carve images on rocks and invented cuneiform script. They made stylus with triangular tips with reeds and carved cuneiform scripts on clay tablets so as to keep record of dialogues, proverbs, myths and hymns. The script was born in the company of credit lending. Characters carved on clay sheets recorded lending that occurred in 2,500 BC and mentioned obligation contribution. This means exchange was already an important social activity at that time.[16] The Sumerians also invented the solar calendar, marking the 12 lunar cycles or 12 months as a year and used this system to calculate agricultural cycles. These major inventions made important contributions to the ancient Mesopotamian civilization.

Another writing system is hieroglyph of Egypt: In the Nile river basin, in the predynastic period (around 3,100 BC) of ancient Egypt, man had

---

[15] Thomas R. Trautmann. *India: Brief History of a Civilization*. Oxford: Oxford University Press, p. 7.

[16] William. N. Goetzmann and K. Geert Rouwenhorst (eds.). *The Origins of Value* (revised edition). Translated by Wang Yu and Wang Wenyu. Shenyang: Northern United Publishing & Media (Group) Company Limited, pp. 18–19.

settled down and started farming. Besides stone tools, they had also bronze tools and invented hieroglyphs. Hieroglyphs are made up of pictographs, syllables and letters and were written on flattened and sun-dried papyrus.[17] Papyrus can be rolled up and easily carried around and passed on to others, thereby greatly enhancing the administrative efficiency of Pharaohs. Therefore, this was the most important progress in the political history of ancient Egypt. Between 2,700 BC and 2,600 BC, the pharaohs were able to organize over 70,000 people to collect and transport 2,500 tons of rocks for the building of the pyramids (some were more than 200 feet in height). This was undoubtedly a huge management project and documentation played a significant role in this process. Influenced by the Egyptian Empire, the ancient Roman Empire that rose later witnessed the emergence of signed contract of tax partnership in written form. This is sort of a prototype of today's companies and was a great progress for human civilization.[18] Also, the hieroglyph of Egypt is the basis of modern Western writings. It is fair to say that every single letter used today in the West was derived from the writing system of the ancient Egypt.

The third is the hieroglyph of India: Around 4,000 BC, in the Indus river basin, people started to use wheels to make ceramics, raise various kinds of animals, and created highly developed Dravidian cities. By about 3,000 BC, the Indus river valley entered the civilized age and that civilization well rivaled ancient Egypt and Sumer. During this period, stone seals appeared as a kind of tool used to keep records of daily lives. These are known among historians today as the Indus script. It "includes 270 hieroglyphs and letters and has nothing to do with any other writing system".[19] The script of ancient India keeps record of hymns and prayers that were orally passed down. Rigveda (hymn of wisdom), for example, included over a thousand poems from 1,500 BC to 900 BC and is a major source of information for a long period of the Indian history. In the 5th century BC, *Mahabharata* and *Ramayana*, epics of ancient India, both mentioned China and recorded it as "Cina" with Latin letters. The ancient Indian culture was early in having contacts with the outside world and introduced

---

[17] *World Civilizations: Their History and Their Culture*, p. 95.
[18] *Ibid.*, p. 32.
[19] *Ibid.*, p. 195.

China to the West. Although the Indus Valley civilization declined and was cut-off later, it was more or less absorbed and carried on by later civilizations and continued its influence on the Indian subcontinent.

The fourth is the Chinese writing system: In the evolution of human civilization, despite regime changes and invasions, the Chinese nation has generally carried through history since the Neolithic Age. Different from the civilizations in the Far East, the Mediterranean basin and the Indus Valley, the Chinese civilization has a unique source and a long history. Folklores about Fuxi, Shennong and Huangdi are still widespread in China even today. About 6,000–7,000 years ago, Chinese ancestors started to use bronzeware. In 1975, a bronze knife was unearthed from the ruins of Majiayao culture Linjia, Dongxiang, Gansu Province, which dated back to around 3,000 BC, showing that China had entered the Bronze Age by then. The three dynasties of Xia, Shang and Zhou lasted 1,800 years and bronzeware was used as ritual wares and weapons throughout the period. Script was carved into bronze *ding* and other artifacts to keep records of major events, forming a special bronze culture of the Chinese nation. Rivaling the bronze culture was scripts carved on ceramics of Liangzhu Culture. They also show the life of ancient Chinese people. The oracle bones of the Shang Dynasty, in particular, laid down basically all rules in Chinese. The Chinese writing, besides pictographic elements, also contains ideographic and phonographic elements. And two different signs and concepts may be combined to form new meanings. For example, when the sun and the moon are combined, the character means light and brightness. In *Shuowen Jiezi*, Xu Shen of the Eastern Han Dynasty summarized the construction of Chinese characters into six categories: pictograms, simple indicatives, compound indicatives, phono-semantic compounds, derived characters and borrowed characters. Each character is like a small picture, showing both pronunciation and meaning, as well as the relationship between the two and an overall form. Therefore, the Chinese scripts are compound characters providing complicated information in combination with an emphasis on relationship between things. Such a unique writing system gave rise to a unique socioeconomic system, e.g. an ancient currency system totally different from the ones in the Mediterranean and West Asia. The oldest currencies of China were bronze coins. These coins did not bear different face values but were used by counting the number. Upper level units were indicated by stringing coins together. The value of each single coin was determined by the government. This is a major characteristic of the Chinese civilization.[20] As a carrier of social development, Chinese characters have carried on for thousands of years and is now one of the writing systems with maximum users in the world.

The fifth is the Mayan calendar and the Minoan Civilization: In Central America, the local Mayans created the Mayan calendar in the 6th century BC. The calendar started in 3,114 BC and ended on December 21, 2012, telling the past and the future of the human world. Through calculation, the Mayans learned the accurate length of a solar year as 365.2420 days with an error of merely 0.0002 day compared with today's calculation of 365.2422 days. This is to say that over 5,000 years, the error will be only one day. Such high capability left later generations in great awe.

In addition, there were also some other writing systems and traces of ancient civilizations. For example, on the Greek islands on the Aegean Sea, a great civilization was at its prime time around 2,000 BC. It was created by Minoans living on the Crete Island and is thus known as the Minoan Civilization. They invented a unique linearograph "A", created the first flushing toilet, and built huge palaces with swimming pools. Facing the Minoan Civilization, we have to acknowledge that many things originally deemed modern actually have their origins deep in history.

(2) Different Religions and Philosophy: From 1,500 BC to 1,000 BC, Judaism, Christianity, Islam and Buddhism basically became the classical religions in their respective regions. The first three are somewhat connected: they all originated from today's Middle East, all worship Abraham as prophet, all believe firmly in internal nourishment of the spirit, all are believed to be Western religions, and all practice monotheism. However, each of the three believes in one creating prophet, believes itself to be the

---

[20] William. N. Goetzmann and K. Geert Rouwenhorst (eds.). *The Origins of Value* (revised edition). Translated by Wang Yu and Wang Wenyu. Shenyang: Northern United Publishing & Media (Group) Company Limited, p. 65.

only one in possession of the complete and concrete explanation to the one and only divinity, and claims itself to be clearly different from the others. Buddhism, as an oriental religion, differs even more vastly from Western religions, though not without connection to any of them. To some extent, the emergence and development of oriental religions went parallel to the dissemination of Western religions.

Judaism believes Moses to be its creating prophet, Jehovah the only divinity, and *torah* its sacred scripture: According to Judaism, the Jewish ancestors suffered all kinds of oppression and Jehovah selected Moses, a Jewish national hero, granted him the Ten Commandments, and demanded him to lead the Israelis to escape from the chasing troops of the Pharaohs, leave Egypt, cross the Red Sea, travel through the Sinai Desert, enter the "promised land" of Jerusalem, a city "flowing with milk and honey", and establish the state of Israel and the temple of Judaism. Emphasizing on staying away from evils and doing good deeds, martyr-dom, and a natural lifestyle, Judaism has a profound influence on the Western civilization and ethics. Unfortunately, the state of Israel was ruined successively by the Babylonians (597 BC) and the Romans (70 AD) and Jews were expelled from Jerusalem. For nearly 2,000 years, Jews did not have a motherland and wandered around the world in exile. During World War II, 6 million Jews were massacred by the Nazis. On November 29, 1947, the United Nations adopted a resolution to establish a Jewish state and an Arab state to the west of Jordan. On May 14, 1949, Jews announced the establishment of the State of Israel in the Palestinian region, marking the second return of Jews around the world to their home. Leave Egypt, establish a state, exile, return, exile again and return once more — such experience had profound impact on the religious ideas of the Jews.

Christianity is derived from but different from Judaism: The central figure of Christianity, Jesus and most of his followers are all pious Jews. The early presbyters and prophets of Judaism, e.g. Moses, Ruth, David, etc., are also respected by Christians. The first part of the Christian Bible, i.e. the Old Testament, is consistent with Jewish scripture. However, Jesus and his followers created the second part of the Bible, i.e. the New Testament. According to the New Testament, God is the only divinity and He says, "God is the heavenly father and people are brothers". This is regarded as a blasphemy by Jesus to Jehovah and Jesus was therefore crucified. Nevertheless, the followers were still moved by the miracles and blessings of Jesus, and believed him as the savior and thus became Christians. In 317 AD, the Eastern Roman Empire of Constantinople was committed to Christianity and issued the renowned *Edictum Mediolanense*, designating Christianity as the state religion of the Roman Empire. Subsequently, Christianity spread around Europe and other regions, and was divided based on different interpretations into the Roman Catholic Church in the West and the Orthodox Eastern Church in the East. In 1,500 AD, on the basis of the Roman Catholic Church, the Christian reformists represented by the German monk Martin Luther divided the European church and created Protestantism. Protestants included Christians who were neither Catholic nor Orthodox. Protestants detested some practices of the Church in the late Middle Ages, such as bigot, cheating and the salvation of money (selling indulgences) and disagreed with some basic Christian doctrines. For Protestants, a core belief is faith-based salvation, which means one can be salvaged only for his/her faith; anyone who has the faith is a priest; the Pope and all clergymen are just ordinary people and may make mistakes. Though both follow the Bible, Protestantism differs greatly from and poses great challenge for Catholics, exerting a significant influence on the modern world.

Islam believes Mohammed as the creator and prophet, Allah the sole divinity and the *Quran* a must-read classics: Between 570 and 632 AD, having received revelations from Allah, Mohammed established Kaaba in Mecca as the sacred temple, and called on Muslims to follow Allah and unite all Arabians with the faith in Allah to end chaos and internal wars. The basic doctrines are: faith in Allah, faith in angels, faith in classics, faith in afterlife, faith in predestination, completing the five basic acts of testimony, prayer, alms-giving, fasting, pilgrimage, comply with the *Quran* and the rules made by Prophet Mohammed. Islam does not require sacred rituals or any priest. Each and every person who believes in Allah is the direct responsibility of Allah without any intermediary. After Mohammed died, his followers separated into Sunni and Shia sects. The former claims itself to be the orthodox, has more followers around the world and selects its leaders by following the opinion of the majority.

The Shia is the minority, with most followers distributed in Iran and Lebanon. It believes that only Ali and his descendants are the legitimate successors of Mohammed. Islam advocates universal mercy for all human beings, including the overall revelations of morality, society and the ultimate destiny, aims at building a world of perfect harmony between daily routines and religious discipline, and plays an important role in the social and political lives of the world.[21]

Buddhism originated in India and developed in China: In the 6th century BC, Sakyamuni attained complete enlightenment on sin and suffering under a *bodhi* tree in the state of Kapilavastu (in today's Nepal), and created Buddhism with the central concept of virtue. Buddhism spread to China later and took its root here. Comparatively, Chinese religions, including Buddhism and Taoism, are less developed than Christianity and Islam and show less need for ultimate concern and their followers have less reverence. However, the traditional Chinese culture is immeasurably inclusive. After coming to China, Buddhism became a part of the Chinese culture. Bodhisattva, for example, turned from a male figure into a merciful female. With mutual influence and promotion, Buddhism, Confucianism and Taoism constituted an ideology with Chinese characteristics.

Chinese Buddhism emphasizes spiritual nurturing, moral enlightenment, progress and education. It believes that everyone holds his/her destiny in his/her own hands, that craving is the source of all sufferings, that the only solution to sufferings is being selfless and caring more for others. It advocates the law of *karma*, believing that every cause has an effect, good deeds bring good results, while bad deeds bring bad ones. It calls on people to shoulder responsibilities for themselves, others, families, nations, countries, the world and everything. It reveals the truth that the *karma* of everyone combined creates the world and encourages people to serve others and work hard. It bears a significant meaning for a harmonious world. In the process of the localization of Buddhism in China, the eight Chinese sects, including Zen and the Pure Land Sect played important roles. The Sixth Patriarch of Zen, Hui Neng, holds that Buddha is in

---

[21] Mang Lishu, James D. Whitehead and Evelyn Eaton Whitehead. *Religious Sociology: Religions in China*. Beijing: Current Affairs Press, 2010, p. 50.

your mind. Such enlightenment applies to everyone. Everyone has Buddha in his/her mind. If one can be aware of his/her true mind and discover where his/her heart really lies, he/she will naturally obtain the enlightenment of Buddhism. The Pure Land Sect believes that by delving into the name of Buddha, one will be able to reach the pure land of ultimate happiness via the willpower of Buddha. These two sects reformed and replaced the asceticism and the sole focus on classic teachings of Indian Buddhism, took the middle way between indulgence and abstinence, and had a decisive influence on the spiritual world of Chinese people. The spiritual balance advocated by the Chinese Buddhism actually echoes the findings of modern biology.

While Buddhism was spreading, Taoism was created locally in China. In the 3rd century BC, or earlier than that, Zhuangzi proposed the idea of adapting to the nature and Laozi composed Tao-te Ching to interpret the Tao that features transcendence and pursues a path that integrates life and nature. In addition, Confucianism, as a set of moral rules and a religion-like value system, offers moral guidance and ultimate pursuit for people; also, as a kind of humanistic philosophy and political ideology, it has profound and lasting influence on the political systems and cultural life of China. In the era of nation states, the world view of Confucianism collided and mediated with nationalism and contributed to world peace and development. In the process of historical evolution, Confucianism, Buddhism and Taoism mingled and integrated, constituting unique spiritual wealth of Chinese people. The Chinese civilization created on the basis of the three religions is an ancient and outstanding civilization featuring harmonious relationship between people and nature, exerting great influence on human society. As Confucianism, Buddhism and Taoism spread to Europe and America, they contributed to the diversity of the Western society and culture.

From the above, we can see that mankind developed along multiple lines. Such diversity is manifested on various levels and many aspects. Similar phenomena may come into being separately under vastly different geographic conditions and at places far from each other. For example, stories of sudden enlightenment appeared in the religious reforms in both the East and the West. In the 15th century, Martin Luther of Germany obtained enlightenment in the bell tower of an abbey and acquired the truth of justification by faith. This enlightenment played a fundamental

Table 1.1.   Characteristics of the Three Major Religions

Characteristics	Religion		
	Christianity	Islam	Buddhism
Founder	Jesus	Mohammed	Sakyamuni Sinicized later
Divinity	God	Allah	Buddha
Scripture	Bible	*Quran*	Buddhist texts
Admittance	Baptism	No baptism	Refuge to Buddhism; Faith at heart
Doctrines	God is the Heavenly Father and people are brothers	Allah is the sole dominator, Mohammed is the missionary of Allah, taking care of both sides	Virtue, no greedy cravings, pursuing nirvana

role in the reform of the traditional Christianity. Hui Neng, the Patriarch of Chinese Zen, obtained sudden enlightenment that Buddha is in every individual's mind in the process of reforming Buddhism. The two cases are similar as they both put mind in the most important position. This means that man may have roughly the same thoughts despite differences in environment and time, and achieve similar greatness along different lines. Table 1.1 summarizes the differences between the three major religions of human society and gives a clearer picture of the development along separate lines.

In the development process of different religions, we can roughly see two lines: one is the path of Western religions and the other is that of oriental religions. Both Western and oriental religions nourish people's spiritual world and offer spiritual support. This is common in various religions. However, the differences between the two religions are obvious too. For example, Western religions believe that men are born sinful and need to be salvaged by God or Allah. Thus, they are likely to cultivate conflicting personalities. On the contrary, oriental religions believe that men are born to be virtuous and need to get rid of greedy cravings, have heartfelt faith in Buddha, and achieve nirvana, putting the emphasis on the nurturing of collective personality. Generally, humans are the same as other life forms as we all have the instinct of pursuing survival. You cannot say such an instinct is good or bad. It is a sign of life and shall not be judged with a binary view of good or bad. This instinct gives rise to both good and evil. When you do too much for survival, your private cravings expand and the evil reveals itself; when you remain humble while pursuing survival, you contribute to society and do good deeds. It is right to advocate the good and encourage contribution and devotion, but we should not thus negate reasonable personal interest. Affirming and protecting personal interest is a manifestation of human instinct and is necessary for human development. To contain the evil, we need both external constraints and internal refraining. Both Western and oriental religions help people build their internal power to refrain. During modern development of the West, industrialization shifted the emphasis of religions toward its role of a social lubricant to maintain a relatively healthy society. In contrast, China paid little attention to this aspect and in the process of social changes, people's spiritual world was somewhat fragile and empty, and life was impulsive and nearsighted. This is something of great concern and we need to take an appropriate attitude toward it.

III. Global Integration: Globalization of Interpersonal Interaction

Entering the Agricultural Age, though the world continued to develop along separate lines, communication and interaction did occur, driven by the nature of the people. Pushed by monsoon and ocean currents, our ancestors may have however established connections between the eastern and western hemisphere long ago. In the Bronze Age, they successively learned metallurgy techniques and the use of metals, and thus were better equipped for communication. In the 2nd century BC, the Silk Road connected Eurasia and North Africa, and the Maritime Silk Road linked East Asian coasts, South Asia and its islands, and East Africa. However, due to environmental and technical reasons, e.g. devastating epidemics caused by the infection of some diseases in new groups of people, population fell sharply along the Silk Road.[22] Therefore, communications conducted by our early ancestors did not have much influence and development along

---

[22] Jerry Bentley and Herbert Ziegler. *Traditions & Encounters*. Translated by Wei Fenglian, Zhang Ying and Bai Yuguang. Beijing: Peking University Press, 2007, pp. 307–312.

separate lines. As a result, societies were diverse, with various independent social groups and political entities, including nomads, tribes, kingdoms and empires. Each of them then had a self-sufficient economy, unique local culture and an independent identity. Neither did they change themselves easily, nor did they compare themselves with others.[23] After 1492, when Columbus discovered the new world, both the technological level and people's willingness to interact surged, and 90% of the world population became closely connected, giving rise to a high tide of global integration. If previous communications are the prelude of a grand drama, the discovery of the new world is the opening. More and more products and knowledge became globally accepted, global trade expanded, the global markets were more and more connected, political, economic, legal, social and cultural interactions grew.

The process of moving from pluralism to integration is shown in Figure 1.2. Generally, there had always been communication and interaction, but they were not quite influential in the early days. Entering the Industrial Age, communication and interaction intensified and modern

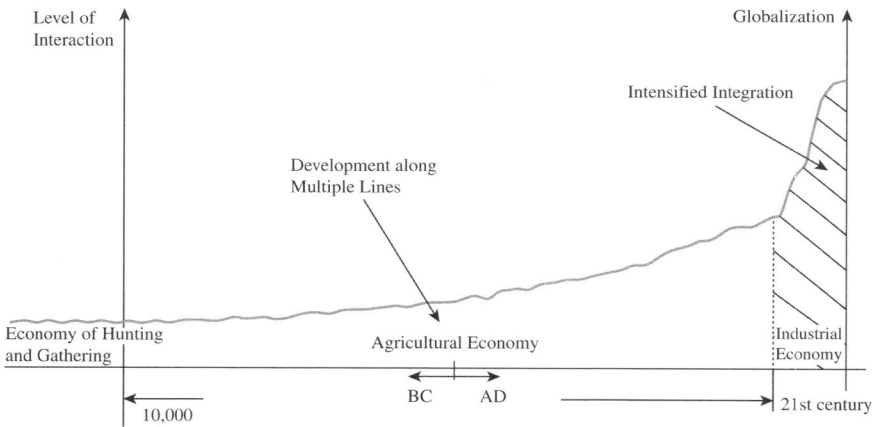

Figure 1.2.   Human Society: From Pluralism to Integration

---

[23] Piotr Sztompka. *The Sociology of Social Change*. Translated by Lin Juren *et al.* Beijing: Peking University Press, 2011, p. 82.

globalization started. Modern globalization was formed through interpersonal interactions.

The modern global integration can be divided roughly into six stages, each with its own driving force.

The first stage (1492–1831), driven by colonial trade, started from the discovery of the new world by Columbus and ended when steam engine was invented. It mainly involved colonization, agriculture and handcraft trade. Milestones are as follows: Marine powers rose after the Renaissance, supported by Protestantism, launched grand colonization efforts in North America, Africa, Australia and Asia, ushering in an era of global travel via intercontinental transportation routes; modern commercial companies such as the East India and West India, owned by the Netherlands and Britain,[24] created new modern models of trade and pushed the sea transportation capacity up from 730,000 tons to 14,600,000 tons between 1570 and 1850, up by 20 times[25]; corporate modernization and regular commercial vessel schedules gave rise to a modern trade system. However, trade was mainly for handcrafts and agricultural products during this stage and its total volume was limited.

The second stage (1831–1882), driven by industrial technologies, saw globalization propelled by steam engine-powered industries and transportation. Milestones are as follows: The invention of steam engine brought about new production models and radically improved the efficiency of production; railway and relevant equipment made transformation mechanical and improved the efficiency of transportation radically; joint ventures outsourced production tasks to different divisions and improved the management and funding capacity radically. Global trade now involved not only traditional handcrafts and agricultural products but also machines and railway equipment. Machine-made products were

[24] Stock company is a delicate organizational form. It is based on an abstract legal person, more stable and less influenced if one member leaves. Investors do not involve in management directly and may authorize someone else to do it. Such companies offer brand new possibilities for a prosperous market and a flourishing economy.

[25] Angus Maddison. *The World Economy: A Millennial Perspective*. Translated by Wu Xiaoying *et al*. Beijing: Peking University Press, 2003, p. 88.

traded in large volumes in particular, promoting radical development of global trade.

The third stage (1882–1931), featured by electronic technologies, witnessed globalization driven by the electrification of industries, transportation and home appliances. Milestones are as follows: Thomas Edison successfully launched the power plant in New York and the electrification of production, transportation and home appliances thus started; electronic products, electric equipment, heavy chemical industry products and steel products became an important part of global trade; telegraph, steel rails and steel ships strengthened the transportation capacity and shortened the time of a single trip; a specialized vocational management system (e.g. Taylorism) gave rise to huge factories and greatly expanded the scale of production and trade.

The fourth stage (1931–1980), featured by automation, was the globalization driven by the automation of transportation, and civil and military products. Milestones are as follows: Ford automobile production line promoted massive automobile production and consumption worldwide, and the Ford management system accelerated management modernization around the world; radio, highways, airports and flight routes connected the global market closely; automobiles, tractors, tanks, internal-combustion engines, aircraft, oil refineries, petroleum, natural gas and synthetic chemical products became the mainstay of global trade.

The fifth stage (1980–2000) saw globalization driven by finance, especially mergers and acquisitions (M&A). Milestones are as follows: Financial groups promoted cross-sector and cross-industry global M&A on an unprecedented scale, showing a clear trend of powers joining hands; in 2000, multinational M&A had a total value of USD 11 trillion, with 109 cases above USD 1 billion in value, setting a historical record. M&A intensified within the financial sector, with 12 cases involving more than USD 290 billion of assets between 1995 and 2000; this expanded the overall scale of banks, lowered their costs, improved the comprehensive earnings and extended their international business.

The sixth stage (2000–now), featured by IT, is a witness to globalization driven by IT-based economic development. Milestones are as follows: Led

Table 1.2.   Global Integration: Process and Characteristics

Stage	Driving force	Characteristics
1492–1831	Colonial trade	Discovery of new world, agriculture, handcraft trade
1831–1882	Steam engine-powered industries	Large-scale mechanization of production and transportation
1882–1931	Electrification	Taylorism of production and sales, electric equipment
1931–1980	Automation	Highways, heavy equipment and petrol trade
1980–2000	Global M&A	Transnational M&A, super banking
2000–now	Computer	Information network, global production and distribution
Future	Combination of biotech and manufacturing	Intelligent robot, biomedical products, bio-industrial products, bio-agricultural products

by IT, a lot of technological breakthroughs have occurred in clusters; the information highway built the ties of global markets; computers, software, electronic equipment, biotechnology and chips have became the mainstay in world trade; regional organizations such as the European Union (EU), the Association of Southeast Asian Nations (ASEAN) and North American Free Trade Agreement (NAFTA) are now playing major roles in promoting regional and global trade; financial institutions have launched multi-business operation and one-stop services, serving their customers from the cradle to the tomb and promoting world integration; global production and distribution networks have deepened international division of labor; international network has developed, involving all countries and regions in it.

Table 1.2 gives a summary of the process and driving forces of modern globalization. A major feature of this round of global integration is interpersonal interactions. Interactions occurred between people from the eastern and the western hemispheres, and those from developed and underdeveloped countries, involving man-made, processed and cultured products. Compared with the primitive globalization of mankind, this is a great progress of human society and a new stage in the development process of human society.

IV. Globalization Formed in the Interactions between Man and Nature

Throughout human history, man has gone through the Paleolithic Age, Neolithic Age, Bronze Age, Iron Age, Agricultural Age, Industrial Age, Electric Age, Automation Age and we are now in the Information Age. The Information Age is still booming. Informatization is combined with industrialization, and it may give rise to some new industries and offer more space for the development of informatization. For example, wireless technologies will be a major breakthrough for the future development of the industrial sector. Germany launched its Industry 4.0 Strategy, aiming a future where factory equipment and personnel will be connected via wireless network and managed in a visualized way through automation software. As productive technologies change, changes also occur in the ways of production, organization and life of the human society. Every change in the system of human society is the result of changes in material production technologies achieved in interpersonal interactions and interactions between man and nature.

By the latter half of the 21st century, man may achieve major breakthroughs in biosciences and new materials and important developments may also occur in bioengineering, bionic engineering and inorganic life forms. A number of new vaccines and diagnostic reagents will be in use for the prevention and diagnosis of major contagious diseases which are major threats to people's health and life. Vaccines and reagents for cancers, in particular, may give rise to an entire bio-pharmaceutical industry aimed at treating cancers with vaccines to replace the current medication-based approach of fighting cancer with toxins. Also, man-made organs and regenerated organs may also help solve some complicated cases, thus promoting people's health and ushering in an era of new biology. In this new era, biological logic will be stored in computer chips, robot modules, pharmaceutical production processes, software designs and corporate management approaches. Inanimate objects may be enlivened, and technologies and industries of medical biology, information biology and industrial biology may become decisive forces in economic and social development.[26] Table 1.3 describes possible major events in the coming 300 years.

---

[26] See General Office of the Ministry of Science and Technology. *The World Advanced Technology Development Report 2006*. Beijing: Science Press, 2007, pp. 57–70.

Table 1.3.   Possible Major Event in the Coming 300 Years

Time	Major progress
15 years later	3D holographic virtuality will enter people's life and people will be able to communicate face to face over long distances; the first aerodynamic wingless electromagnetic aircraft (first-generation flying saucer) will come out in the USA; solar glass that absorbs solar energy by applying a layer of coating on ordinary glass will enter our homes; man-made retina will be put into wide use.
30 years later	Quantum computer will gradually be put into application and quantum power generator will be created; hydro-fueled car will be put into massive production; human will set foot on the Mars; undersea tunnel will be built in the Bering Strait, and global highway will connect over 100 countries around the world.
60 years later	The first experimental positron reactor will be created, marking the start of the era of antimatter energy; people will "migrate" to the future by freezing themselves for enjoyment.
80 years later	Evaluated tube maglev train will be in commercial operation around the world, putting Beijing and New York in a one-hour transportation circle; nanometer robots will be the fourth type of blood cells in human bodies to cleanse the grease and clogs and keep blood vessels in good condition.
100 years later	There will be a "brain cap" that simulates all sensual information and doctors will feel what a patient feels when he/she puts on this cap; computer will be able to build a man-made brain.
120 years later	We will have the first space carrier and small spacecrafts, which will take off from the space carrier, totally changing the military structure of the world.
140 years later	Man-made intelligence will be as good as a human brain, making human–computer intelligent dialogue possible. The Turing test will become reality in some areas, and intelligent robots will replace workers on large scales.
180 years later	Man will develop the Moon on a large scale, use the rich reserve of He-3 there for power generation through nuclear fusion and use the power on the Moon by sending it up via huge Infrared (IR) transmitters.
220 years later	Major difficulties like radiation feedback will be solved technically, and people will enter the era of time and space travel, with real "pre-history parks" coming into being.
250 years later	Physiological research on human brain will breakthrough the critical point, memory will be transplanted, brain chips will be made and our education system will be changed fundamentally.
280 years later	There will be fewer and fewer people going without genetic change, and human society on the Earth will be dominated by post-humans.

Source: Entrepreneurs' Club, May 10, 2015.

The new biology era may be the start of a new cycle of human development. While trying to solve the problem of infertility by creating sperm with embryonic stem cell, we may give rise to a conflict between manmade life and human life. Intelligent robots that carry biological chips will be more and more similar to human beings in terms of balance, vision, voice and other aspects, and may even be able to do more than what humans can do and replace people or largely reduce the physical or intelligent workload of people. People create history and labor opens up the future.[27] Industrialization and informatization, while improving labor productivity, cause severe damage to the natural environment we live in as well as the resources. Every year, 300 million mu of land is degraded and becomes infertile, and soil erosion involves 3.56 million km^2 of land in China. Also, industrialization and informatization have negative effects on human history and culture. For example, pollution through nuclear waste will cause long sufferings, massive industrialization and urbanization cause damage to historic sites and cultural heritage and underground railway buildings may result in irreversible loss of historic sites. In addition, industrialization and informatization will also greatly reduce people's workload. This can be good or bad. This is because less labor will definitely hamper human development. That is to say, though biotechnology and other advanced technologies since the industrial revolution have, to an extremely large extent, freed people from physical work, reduced diseases and improved labor productivity, when physical labor reduces, electronic devices and information have caused more damage to mankind. Examples include more cases of cell phone disease, TV disease, Internet disease and device disease, more negative effect of high and new technologies, possible decline of physical functions and shortened human life. This will be a huge challenge to human development. Human survival and development is an eternal theme. We must value the role of the nature and the effect of the law of balance. We must think it over, reduce the use of various devices, return to nature, increase physical and intelligent labor, including manual work, farming and doing daily work with manpower, and start a new era of manual work. Manual work of this new era may be totally different from the traditional manual work. In the new era,

---

[27] Important speech by Xi Jinping at a meeting with representatives of the national model workers. Xinhuanet, Beijing, April 28, 2013.

handcrafts will adopt natural designs, integrate technology with the nature, combine interpersonal interaction with the interaction between man and nature, and take place over the Internet. This will be a hallmark of the civilization of the new biology era. Today, new home-based manufacturing has appeared over the Internet and it may be the herald of the new manual work era. The new manual work era will start a brand new cycle of human development and may witness totally unexpected divisions.

V. Characteristics of the Current Globalization: Integration and Diversification

Entering the 21st century, globalization has deepened and people around the world are involved in comprehensive interactions and accelerated developments. However, conflicts come along too. For example, shared problems such as global warming, resource shortage and population growth are now serious issues. Moreover, new conflicts that rose in the wake of the financial crisis of 2007 have also intensified and call for a close attention.

First, there is the conflict between market integration and political diversification. Market integration is speeding up around the world and the economy of various countries and regions are becoming more and more connected, giving rise to more economic and political communities[28] and increased inner consistency of the global economy. Meanwhile, national and regional conservatism have revived and differences in the historical traditions, political systems and religions of various groups of people have become clearer.

Second, there is the conflict between economic integration and cultural diversification. As economic globalization deepens, the integration of historical culture is also picking up its pace and responses from the traditional culture and customs of various ethnic groups have intensified, giving rise to conflicts in culture, mindset and behavior. Thus, the dispute between globalization and anti-globalization has increased.

The third conflict is between information integration and diverse thoughts. In-depth R&D of information technology and industrialized IT

---

[28] Piotr Sztompka. *The Sociology of Social Change*. Translated by Lin Juren *et al*. Beijing: Peking University Press, 2011, p. 83.

development, especially the development and application of big data, is a revolution in the IT industry and has infiltrated to various industries, creating new models of production and management. Digitalization, intelligentization, networking and global coverage has become a new massive trend and strong social force in the world, pushing mankind toward a more open and better-integrated world. In the Information Age, information spreads faster than before and nothing stays unknown to the world. Information becomes a shared resource. Meanwhile, fragmented information is everywhere. News comes from everywhere about anything, issues are more and more astounding, with more and more weird thoughts. Yet, different systems of thought have long been formed in history. Thus, the conflict between information integration and diverse thoughts becomes obvious.

The fourth conflict is between global network and the security and personal information. The Internet connects everyone to form a whole. Companies and individuals alike depend more and more heavily on the Internet. But in the meantime, personal information goes out very easily. This is a conflict. No one is against the Internet, yet no one wants to share his/her personal information with everyone. So, the problem is how to keep personal information safe while ensuring smooth Internet use.

Fifth, there is the conflict between global resource allocation and individualized consumption demands. Under the influence of the Internet and the Internet of things, global resources have actually been incorporated into one integrated system and can be allocated in a unified manner to promote the expansion of production scale. Meanwhile, market is divided into more and more refined niches, together with consumption demands. This is also a conflict.

Sixth, conflict exists between international service standards and traditional local features. In the process of global integration, international standards are a common requirement around the world. Meanwhile, people lay more and more emphasis on traditional local features and such features are indeed an indispensable part of our life. How to carry forward local traditions while advocating international standards is also a concern.

Seventh, there is conflict between social equity and the gap in wealth and income. Thanks to high and new technologies, the world is now getting smaller, and people pursue equal social status. But the reality is that the wealth gap is widening and the divide between the developed and the developing worlds, between urban and rural areas, and between the rich and the poor are all increasing. The imminent concern is how to uphold equity and justice and reduce the rich–poor gap as global integration progresses.[29]

Eighth, there is a conflict between global integration and governance diversification. As global integration deepens, the global society, economy, and politics become even more integrated. However, due to history and social conditions, conflict between the East and the West, gaps between the North and the South, and disputes between developed and developing countries are growing. The world is undergoing huge changes and adjustments, politics is moving back and forth between multiple powers, and the international strategic situation is in great disorder. This makes global governance complicated and highly uncertain. In today's world, everyone is closely related to and influenced by others, yet we still lack sound global governance. This is a conflict.

Figure 1.3 shows the conflicts in globalization today. As a natural phenomena in economic development of human society, they are the results of globalization and will be solved accordingly as globalization

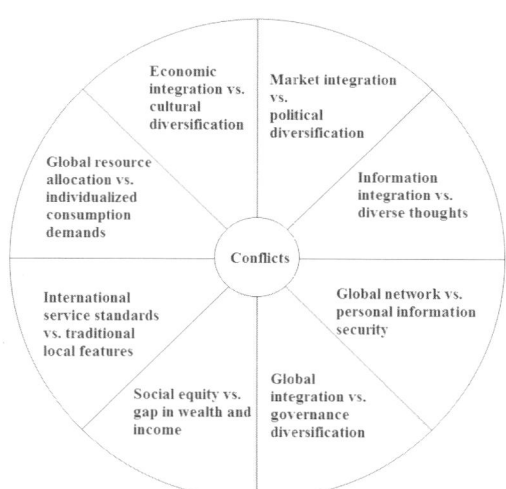

Figure 1.3. Eight Conflicts in the New Stage of Globalization

[29] Lin Yifu. *New Structural Economics*. Beijing: Peking University Press, 2012, pp. 87–89.

continues. Conflicts promote social changes in the constitution, structure, function, boundary, and environment of the human society, and changes will be dispersed at first and then intensify and spread, reshaping the human society as a whole. It is possible that human society will become an integrated community that goes beyond countries and regions. It will not be dominated by either the Western or the Eastern civilization. Instead, it will incorporate and integrate both to form a new holistic global entity.

VI. Summary: What We Learn from History

The development of human society is also subject to the law of equilibrium. Our world today is featured by integration, profound changes and great challenges, and such a situation is deep rooted in history. Human history is the history of interaction between man and nature, and the history of interpersonal interaction; human civilization is the integration of the past and the present, traditions and innovations and individual and common features; and the development of human society followed a path that goes from one source to pluralism and then to integration. Mankind has common values, pursues the truth, the good and the beautiful and endeavors for a happy life. Compared with the non-human society that sees few changes over long periods of time, the human society is unpredictable and unstable, always undergoing changes, bifurcated development, reestablishment of connections and mutual influence. Influenced by the living environment and innovative tools, the human civilization separates into different parts over time, follows different doctrines, gathers followers around different faiths and forms various groups and sects. However, all these share the same core. All ethnic, regional and national cultures formed over time constitute integral parts of the human culture. Globalization causes different cultures to collide and communicate and brings together different cultures to compete and coexist with one another. Integration, sublation and coexistence are the general trends in the development of human society.

China has an ancient civilization, and the development of the Chinese civilization, like all other human civilizations, finds its foundation in the core values shared with the entire human race. The Chinese civilization has been through ups and downs over time, withstood hardships and garnered rich connotations. Compared with many other countries, China, as a developing country, is still in the preliminary or intermediate stage of modernization. And the top priority for a developing country is reform, opening up and anti-corruption, all for the one ultimate goal of development. The development of China is comprehensive, balanced and sustainable, with both urban and rural areas given due consideration, with emphasis laid on a reasonable industrial structure, and with efforts made to maintain ecological balance and recycling of resources. China's development is a continuation from its past and long history. It has a vast market and will contribute to both the modernization of China and the human civilization as a whole. The ancient Silk Road is a hallmark of the traditional civilization. The Belt and Road Initiative proposed by China today carries forward the excellent traditions coming down from the past and opens a new horizon for the future. It shows to the whole world China's attitude in the new era to open up to the outside world, its political pursuit of peace, communication, cooperation and development, and its intention to create an international environment that facilitates the development of China.

Looking back at the history of mankind from the perspective of the law of balance, we will gain a better understanding of ourselves today, have more inner peace to face the future, and be more aware of the nature, instinct and basics of mankind. Man, by nature, is kind and gregarious, and prefers coexistence. This is the ultimate reason why we pursue peace and the ultimate way toward the survival of mankind. Man genetically tries for progress, reproduction and excellence, and tends to go all out for survival and victory. This is the fundamental motivation for human development. Therefore, it is natural that there are competitions and wars in human society. Man, by nature, is the combination of existence and consciousness. Existence is based on labor, creation and demands for life; while the consciousness is dominated by thoughts, emotions, intelligence and will power. The nature, instinct and basics of mankind are the internalization of the nature, the enlivening of materials, and the sublimation of life.[30] Man cannot live without the nature. Man has the spirit of animals,

---

[30] See Li Zehou. *The Anthropo-Historical Ontology*. Tianjin: Tianjin Academy of Social Sciences Press, 2008, pp. 38–41. In this book, Li elaborates on "the impersonation of internal nature". The "internalization of the nature" proposed in this article emphasizes the self-consciousness, lofty will and innovative spirit of mankind in its interaction with the nature.

but is different from other animals. Man is the most intelligent animal on the Earth, has its thoughts, and the ability of cognition, rational thinking and intelligence.

Looking at human civilization from the perspective of the law of balance, we will do better in cultivating our minds, bodies and behavior. To cultivate our minds, we absorb spiritual nourishment, reflect on ourselves, nurture and purify our minds, hone our health and attain perfection of both body and mind to uphold the good nature of mankind; to cultivate our bodies, we learn the skills for survival, train our bodies, improve our behavior, nurture ambitions, defend ourselves against bad habits, build our will power and enhance our abilities; to cultivate our behavior, we combine the cultivation of body and mind, put together material civilization and spiritual civilization, align our behavior to moral standards, make good use of human qualities and wisdom, and contribute to human development.

Looking at human society from the perspective of the law of balance, we will have a better understanding of people's will, livelihood and law. People's will reflects their wishes and such wishes show and determine where the society is headed. Therefore, we should establish and improve a specialized investigation and polling system so as to collect people's opinions timely and effectively. People's livelihood is how people live, is the top priority of the world, and is the motivation for social development. Therefore, the government shall take the lead, with social participation, to establish and improve a basic social security system and build a strong safety net for the people. People's law is the civil law, the manifestation of the Constitution in civil affairs, and the implementation of the civil law can be taken as a priority in achieving the rule of law in China. Listening carefully to people's voices, caring for people's livelihood and abiding by civil law are the basic conditions for a healthy development of the society.

The development of human history, starting from one ancestry with multiple sources, is a mutual impact and promotion of various civilizations. Human society has entered a stage of in-depth globalization and people around the world are more closely related to one another than ever before. Adapting to social changes and trends, respecting the history and values of mankind, keeping up with the time, incorporating different cultures while maintaining one's own characteristics, building the constructive capacity for communicating with other people and nations, avoiding going to the extremes, preventing self-inflicted setbacks, caring for life, living life serving society, being brave to explore new grounds, making continuous contributions and pursuing shared development are the basic conditions and driving forces for sustainable development amid globalization for both mankind and China.

Chapter 2

The Balance Practice in China's Modernization

What is the impact of the law of equilibrium on China's modernization? This chapter analyzes the hard course of China's modernization by using the thinking method of balance, sums up the historical experiences and lessons in China's economic and social development since the founding of New China through in-depth study, and puts forward some suggestions to improve China's development theory and policy system.

Modernization refers to the process of the transformation of civilization and international competition since the Industrial Revolution in the 18th century,[1] mainly reflected in industrialization and marketization in the economic field, legalization and democratization in the political field, and rationalization and scientization in the ideological field. In the course of global modernization, China's modernization had a rather late start and went through twists and turns. Since the founding of New China, especially since the reform and opening up, China's modernization has accelerated, and narrowed the gap between China and the developed countries. Although China has made universally acknowledged achievements, it also has many outstanding problems. No matter how hard and tortuous the process is, China is in progress. For China, the difficulties and opportunities coexist. It's no use complaining about the past; we should face reality and take on our responsibilities.

---

[1]Chuanqi He. *Modernization Science: The Principles and Methods of National Advancement.* Beijing: Science Press, 2010, p. 07.

I. Push Ahead Dauntlessly in the Last Two Centuries

Chinese people created a world-recognized agricultural civilization in the agricultural society in ancient times, and unreservedly shared their wisdom and inventions, including Confucianism, Buddhism and Taoism, paper-making, movable-type printing, gunpowder, the compass, medicine and mathematics with other countries in the world, also sharing the oriental philosophy and wisdom for the development of human civilization as well as the earliest technical support for the modern industrial civilization. But in terms of industrial civilization, China had a rather late start. In the 18th and 19th centuries, when the Industrial Revolution swept the Western Europe, China was still immersed in the dream of being a feudal empire. During the 19th century, China, as a feudal state bullied by Western powers, just passively received the modern civilization characterized by machinery industry. During the 20th century, through its struggles against imperialism, feudalism and colonialism, China actively accepted industrial civilization and was a nation seeking independence and liberation. During the nearly 200 years between the mid-19th century and the early 21st century, China's modernization can be roughly divided into three stages: the period from 1840 to 1911, marked by feudalism; the period from 1911 to 1949, marked by national capitalism; and the period from 1949 to the present day, marked by socialism. Here, we will focus on analyzing the first two periods when China was bullied by foreign powers, and many Chinese people with lofty ideals struggled bravely for the national liberation.

1. *The Period from 1840 to 1911*

During this period, the Western powers, relying on the powerful warships and cannons brought by the Industrial Revolution, blew open the door of China, and China passively received modern industrial civilization. After the first Opium War in 1840, the Western powers successively waged the Second Opium War of 1856–1860, the Sino-French War of 1883–1885, the Sino-Japanese War of 1894–1895, and the War of Invading China by the Eight-Power Allied Forces in 1900. Through these barbaric aggressive wars, the Western powers forced China to sign a series of unequal treaties to grab privileges for their activities in China, making China an object to be plundered. In China, they set up foreign concessions, opened ports by force and divided their spheres of influence, so as to control China's politics and economy. Because of the aggression and plunder of the Western powers, China was becoming increasingly weak and backward, and even disintegrated, facing a national crisis.

The invasion of the Western powers intensified the contradictions in China's society and accelerated the decay of the feudal system. The Qing Dynasty, once a world power, tried very hard to maintain its shaky rule. The contradiction between China and the imperialist powers, and the contradiction between the broad masses and the feudal forces became the major contradictions of modern Chinese society. In order to save the nation from peril, Chinese people struggled untiringly and fought bravely. Many Chinese soldiers and civilians fought bloody battles against the invasion of the Western powers, and many people with lofty ideals from all sectors rose up against foreign aggression as well as the exploitation and oppression of the domestic ruler. For example, Hong Xiuquan initiated the Taiping Rebellion in 1851; Kang Youwei, Liang Qichao, Tan Sitong and other reformists launched the Hundred Days' Reform movement in 1898; Sun Yat-sen and other revolutionaries launched the Revolution of 1911. These movements gave a heavy blow to the imperialist powers and feudal forces, and propelled the awakening of the Chinese people. Especially the Revolution of 1911, led by Sun Yat-sen, overthrew the Qing government and 2,000 years of monarchy, and opened the gate of modern civilization for China. During this period, the Westernization Movement and the Hundred Days' Reform advocated self-improvement and introduced the machinery industry to China, fostering the bureaucratic capital and national capital, meanwhile destroying natural economy structure; however, due to the heavy restriction of the old regime and old ideas as well as the heavy oppression of the imperialist powers and feudal forces, China's development was in fact at a standstill, and its productivity level decreased instead of increasing. During the same period, in terms of economic level, the US increased by more than three times, the UK increased by two times, and Japan, thanks to the Meiji Restoration, doubled its former level. Table 2.1 summarizes the level of the Chinese economy in international comparison during this period.

Table 2.1.   International Comparison of Per Capita Gross National Income from 1820 to 1913 (Unit: Geary–Khamis Dollar*)

Year	World	China	Japan	UK	USA	India
1820	667	600	669	1,707	1,257	533
1870	867	530	737	3,191	2,445	533
1913	1,510	552	1,387	4,921	5,301	673
1820/1913	2.26	0.92	2.07	2.88	4.22	1.26

Note: *Based on the consumption prices in 1990.

Source: *Angus Maddison. The World Economy: A Millennial Perspective*. Translated by Wu Xiaoying. Beijing: Peking University Press, 2003. This table was developed by the author.

2. *The Period from 1911 to 1949*

During this period, the vigorous anti-imperialist, anti-feudal and anti-colonial movements in China brought about a wave of movements for modernization. For example, the construction of the rural society allowed the national authority to extend to the rural society, and the official partition for the rural areas broke the closed rural communities; the rapid development of the national capitalist economy enabled its proportion in the whole national economy to increase from 2% in 1911 to 20% in 1949, meanwhile the proportion of the traditional feudal economy in the whole national economy fell from 90% in 1911 to 70% in 1937, and the proportion of foreign capital in the production value of China's new economic sectors decreased from 80% in 1911 to 20% in 1949.[2]

During this period, due to the development of the national capitalism and the state monopoly capitalism, a modern education system and a number of research institutions were established, which was an improvement over the Qing Dynasty, but China's development was still affected by the backward regime, the incessant fighting between warlords, the Northern Expedition, the KMT–CPC civil war and especially the Japanese invasion, which lasted 14 years and caused great damage to the Chinese economy. During this period, in terms of economic level, China increased by 20%, the US increased by 80% and the UK and Japan increased by about 40%

---

[2]Luo Rongqu. *New Discussion on Modernization*. Beijing: Peking University Press, 1993, pp. 325–315.

Table 2.2. International Comparison of Per Capita Gross National Income from 1913 to 1950 (Unit: Geary–Khamis Dollar*)

Year	World	China	Japan	UK	USA	India
1913	1,510	552	1,387	4,921	5,301	673
1950	2,114	439	1,926	6,907	9,561	619
1913/1950	1.40	0.80	1.39	1.40	1.80	0.92

Note: *Based on the consumption prices in 1990.

Source: Angus Maddison. *The World Economy: A Millennial Perspective*. Translated by Wu Xiaoying. Beijing: Peking University Press, 2003. This table was developed by the author.

respectively. Table 2.2 shows the level of the Chinese economy in international comparison during this period.

During this period, two political groups with significant influence emerged in China, constituting a balance of forces.

The Kuomintang (KMT): Based on the theory of evolution introduced by the West, the theory of natural rights and the idea of democratic republic, some Chinese progressives established the KMT. The KMT vigorously boosted the "constitutional democracy" in its early stage, but things did not progress as expected. The "constitutional democracy", which met with success in the West, deformed in China. The general election was a mere formality, and the multi-party system was just an excuse to form cliques and factions. The KMT, occupying a majority of seats in the parliament did nothing for the people's benefits. As a result, the achievement of the Revolution of 1911 was soon stolen by the Northern Warlords government led by Yuan Shikai, and China was stuck in the incessant fighting between the warlords.

The Communist Party of China (CPC): Meanwhile, other Chinese progressives turned their eyes from Europe and America to the Soviet Russia, and based on Marxism and scientific socialism they established the CPC. The victory of the October Revolution of Russia in 1917 opened up a new era in the history of mankind, and strongly attracted the attention of Chinese progressives. Advocated by the people with progressive thoughts, the propagation of Marxism and scientific socialism in China became a trend of the times. The May Fourth Movement helped scientific socialism take root, bloom and bear fruit in China. Many intellectuals with different experiences and backgrounds became followers of Marxism and scientific socialism.

The KMT and CPC jointed hands twice, and had two civil wars. The first KMT–CPC cooperation lasted three and a half years, from January 1924 until July 1927. It started with the first National Congress of KMT, where the three policies of alliance with the Soviet Union, alliance with the CPC and assisting peasants and workers were determined. This cooperation overthrew the reactionary rule of the Northern Warlords government and accelerated the progress of the Chinese revolution, but later because the fight between the two parties for the leadership increasingly intensified and the KMT right wing betrayed the revolution, the cooperation whittled away. During the anti-Japanese aggression war, the KMT and CPC established their second cooperation, and formed the Chinese united front against Japanese aggression. With the coordination of frontline battlefields and guerrilla battlefields in the enemy-occupied areas, and under the strong support of the international anti-fascist alliance, this cooperation led the Chinese people to the great victory of the war. Later due to the anti-communist policies of the Chiang Kai-shek clique, the second KMT–CPC cooperation broke up, which was the most regrettable thing in the modern history of China.

The first KMT–CPC civil war lasted from 1927 to 1937. After the April 12 Incident, which was the violent suppression of CPC organizations in Shanghai by the military forces of Chiang Kai-shek, the CPC successively launched the Nanchang Uprising, the Autumn Harvest Uprising and similar uprisings in Guangzhou and other provinces, and established the Chinese Workers' and Peasants' Red Army and some rural revolutionary bases including Jinggangshan. The Nanjing KMT Government launched five encirclement campaigns against the Red Army and their revolutionary bases.

The second KMT–CPC civil war lasted from August 1945 to September 1949, known as the War of Liberation. The war was inevitable, owing to the differentiation of social forces since the May Fourth Movement.

The KMT, led by Chiang Kai-shek, and the CPC, led by Mao Zedong, both had their own doctrines and armed forces, but they were on behalf of different classes. With their armed forces and doctrines as well as their visions of the future, they had a fierce, uncompromising struggle with each other in modern China.[3] In the KMT–CPC struggle, the communists

---

[3] Jin Yinan. *Sufferings and Glories.* Beijing: Huayi Publishing House, 2009, p. 78.

regarded Marxism as an ideological weapon to change the world instead of learning it as a simple theory. They consciously applied Marxism to the revolutionary practice. During the new democratic revolution, through the localization of Marxism in China, they put forward the ideological line of seeking truth from facts, the organizational line of maintaining close ties with the masses and the revolutionary united front by cooperating with other political forces, as well as the flexible strategy and tactics. Relying on the collective wisdom of the whole Party, they established Mao Zedong Thought, got the support of the broad masses and finally won the struggle. Influenced by the political situations and the balance law in the history, the forces of KMT and CPC changed in reverse.[4] The forces of CPC grew from weak to strong, and in October 1949, the communists established the People's Republic of China in Beijing; the forces of KMT changed from strong to weak, and in October 1949, the KMT government retreated to Taiwan after it was defeated on the mainland. Behind the two parties, there were both international powers backing them. The KMT was supported by the US, but after the KMT government retreated to Taiwan, it did not follow the intention of the US to be completely independent from the mainland. The CPC was mainly supported by the Soviet Union, but the CPC did not copy the Soviet model; instead, they "took the Soviet model as a mirror", and explored an independent development path suited to China (Figure 2.1). The CPC, supported by the Chinese people, founded

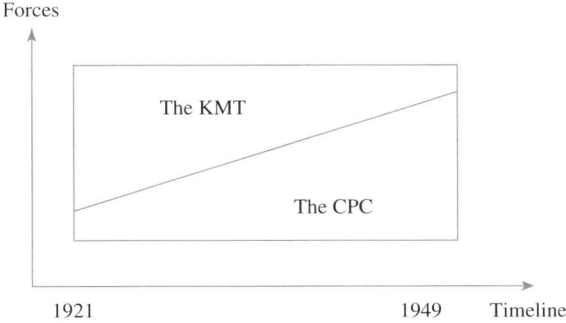

Figure 2.1.   The Changing Trend of the Forces of KMT and CPC during the Period of 1921–1949

---

[4]*Ibid.*, p. 208. Mr. Jin Yinan stated, "The failure of the KMT was not attributed to IQ, but a factor more advanced than IQ: balance".

the People's Republic of China, overthrew the "three big mountains" (imperialism, feudalism and bureaucrat-capitalism) and abolished all unequal treaties, and the Chinese people became the masters of the country.

II. The 30 Years of a Planned Economy

The 30 years after the founding of New China were a period of a planned economy. This period can be divided into four stages (see Table 2.3).

1. *The Stage of Stabilizing the Regime (1949–1956)*

The CPC took over from the KMT a broken-down country, poor and blank, in a disastrous state, so the CPC should rely on strong executive power to overcome difficulties in finance and economics, and stabilize the regime. During this time, the imperialistic countries imposed an aggressive military threat and a tight economic blockade on China, and only the Soviet Union and other socialist countries supported China, so China was influenced by the Soviet economic model at the very start. During this stage, there were three impressive major events. First, Chen Yun was commissioned by the CPC Central Committee to take charge of the national

Table 2.3.   **The Stages of the Planned Economy and Corresponding Major Events**

Stage	Major events
The first stage (1949–1956) Socialist transformation	1. Three campaigns in finance and economics 2. The war to resist US aggression and aid Korea 3. The First Five Year Plan
The second stage (1957–1966) Overall socialist construction	1. Expansion of the anti-rightist struggle 2. Great Leap Forward, Communist Wind
The third stage (1966–1976) The Cultural Revolution	1. Nationwide unrest 2. Atomic and hydrogen bombs and a man-made satellite
The fourth stage (1977–1978) Adjustment	1. Clarifying confusion and bringing things back to order

finance and economics, and he launched three campaigns[5]: stabilizing the commodity prices, planning purchase and marketing by the state, and readjusting industry and commerce. These three campaigns completely ended the hyperinflation left by the former regime, consolidated the people's regime and allowed the national economy back to the track of normal development. Second, the Chinese people, under the leadership of Mao Zedong, won a great victory in the war to resist US aggression and aid Korea, which not only defended our homeland, but also established New China's position in the international arena. Third, China launched the First Five Year Plan, and by focusing efforts on the 156 key construction projects, China established the initial foundation of socialist industrialization. From 1953 to 1956, China's gross value of industrial output increased by 19.6% per year, and its gross value of agricultural output increased by 4.8% per year; this stage witnessed faster economic development and better economic effects.[6] It was one of the stages with outstanding achievements in the history of New China's economic development.

2. *The 10-Year Stage of Socialist Construction (1957–1966)*

During this decade, due to the serious errors in the guidelines, the economic construction went through a tortuous development process. For example, in the policies, the expansion of the anti-rightist struggle in 1957, the so-called anti-rightist struggle in 1959, and the excessive struggles in the Socialist Education Movement of Urban and Rural Grassroots, caused many party members and cadres as well as a large number of talented intellectuals to suffer undue blows. In the economy, the Great Leap Forward and the movement to form rural communes initiated in 1958 ignored objective economic laws. The system of ownership of the means of production was forcedly enhanced to a higher degree of industrialization. Due to a rush for quick results in the economic development, there

---

[5] Chen Yun. *Selected Works of Chen Yun*, Vol. II. Beijing: People's Publishing House, 1995, pp. 32–33.
[6] Wang Mengkui. *Collected Works of Wang Mengkui*, Vol. V. Beijing: China Economic Publishing House, 2001, p. 333.

appeared left-leaning errors marked by unrealistic production targets, giving confused orders, tendency toward boasting and exaggeration (wind of boasting), and "communist wind" (practice of equalitarianism in the name of communism). Additionally, because of the natural calamities and the breach of contract by the Soviet Union government, China's national economy encountered serious difficulties and suffered heavy losses during the period of 1959–1961.

3. *The Cultural Revolution (1966–1976)*

The Cultural Revolution, a civil strife mistakenly waged by the leader, caused a serious disaster to the CPC and the state, and a heavy damage to the socialist democracy and rule of law. During this stage, the organizations of the CPC and the government, at all levels, was paralyzed or in abnormal operation for a long time, generating a large number of unjust, false and erroneous cases. Education, science and technology, and culture were seriously damaged, resulting in cultural gap, technology gap and talent gap. The Cultural Revolution caused the loss of national income of about RMB 500 billion, leaving the entire national economy on the brink of collapse.

4. *The Stage of Adjustment (1977–1978)*

After the Cultural Revolution, China entered an important historical stage of adjustment. The "leftist" policies resulted in a lot of contradictions and problems, and the social and economic development in urban and rural areas was far behind the times, so the CPC and the Chinese people had a strong sense of crisis and urgency. As the saying goes, "Long-term unrest requires stability and poverty gives rise to the desire for change". In December 1978, the CPC convened the Third Plenary Session of the 11th Central Committee of the CPC, opening the era of reform and opening up, which had far-reaching significance in the history since the founding of the PRC.

III. The 30 Years of a Market Economy

More than 30 years since the reform and opening up constitute a period of a market-oriented economy. This period can also be divided into four stages (see Table 2.4).

Table 2.4.   The Stages of the Market Economy and Corresponding Major Events

Stage	Major events
The first stage (1978–1984) Micro-decontrol of the market	1. Household contract responsibility system 2. Special economic zones 3. Developing the individually or privately owned economy
The second stage (1985–1991) The reform between Micro and Macro system	1. Enterprise contract and responsibility system 2. Double-track price system 3. Replacement of profit delivery by taxes
The third stage (1992–2000) The reform of the macro management system	1. Enterprise joint stock system 2. System of tax distribution 3. Commercialization of state-owned banks (SOBs)
The fourth stage (2001–2013) Comprehensive reform and opening up	1. Joining the World Trade Organization (WTO) 2. RMB exchange rate formation mechanism 3. Pricing mechanism for refined oil products
The fifth stage (2014–) A new normal	1. Negative list 2. National balance sheet 3. Free trade zones 4. Allowing private capital to establish banks 5. The Belt and Road Initiative 6. The Xiong-An special zone

1. *The Stage of Micro-Decontrol of the Market (1978–1984)*

First, the implementation of the household contract responsibility system in rural areas allowed farming households to manage agricultural production on their own initiatives, greatly arousing their enthusiasm and emancipating agricultural productivity. Based on comparable prices, from 1978 to 1985, the total value of output in agriculture, forestry, animal husbandry and fishery industry increased by 61.6%, with an average annual growth rate of 7.1%, significantly higher than the previous average annual growth rate of 2–3%, which was unprecedented in the history.[7] Secondly, the individually or privately owned economy began to develop rapidly, and the number of individually or privately owned businesses increased from 150,000 in 1978 to 10 million in 1984, which broke the domination of public ownership, and played an important role in the formation and improvement of a socialist

---

[7] Zhang Zhuoyuan. *Restart the Reform Agenda, Research on Developmental Finance* (Test Issue: No. 2).

market economy.[8] Third, special economic zones in coastal areas were established, and the government adopted a series of special policies to attract foreign investment. The special economic zones played an important leading role in China's economic development.

2. *The Stage of Moderate Reform (1985–1991)*

First, under the premise of public ownership as the mainstay, the government encouraged the development of various forms of ownership, and Chinese–foreign joint ventures and wholly foreign ventures developed rapidly. Second, the implementation of the contract system in large-scale state-owned enterprises defined their responsibility, power and interests, and through replacement of profit delivery by taxes, the enterprises could retain 40% of their profits. The enterprises transformed from government-affiliated businesses into commodity producers and business operators, relying on independent management, self-financing, self-accumulation and self-development. Third, the government decontrolled the prices of agricultural and sideline products, industrial consumables and production goods to boost the development and prosperity of commodity markets, ending the situation of buying commodities with tickets due to short supply of commodities, which had inconvenienced the Chinese people for decades.[9] Fourth, the government reformed the equalitarianism in the existing distribution system, known as the "big-pot" system, and explored the establishment of a distribution system based on labor.[10]

3. *The Stage of Reforming the Macro Management System (1992–2000)*

First, the government further defined the basic economic system in the primary stage of socialism, which called for keeping public ownership in a dominant position and having diverse forms of ownership develop side by side. Second, the government confirmed that the direction of the reform

---

[8] *Ibid.*

[9] Chen Yun. *Selected Works of Chen Yun*, Vol. III. Beijng: People's Publishing House, 1995, pp. 244–245.

[10] Jiang Ziqiang and Zhang Xukun. *The Comprehensive History of Economic Thought*, Vol. IV. Zhejiang: Zhejiang University Press, 2003, pp. 632–633.

of state-owned enterprises was to establish a modern enterprise system; corporate and shareholding systems were introduced into large- and medium-sized state-owned enterprises to replace the contract system. Third, the price reform allowed prices of most products and services to be regulated by the market, and let the market gradually play a fundamental role in resource allocation. Fourth, through reforming the financial system, the government established the financial resources adjustment system, and adopted the system of tax distribution instead of the former system where local authorities take full responsibility for their finances, enabling the central fiscal revenue to account for more than 50% of the total revenue, but the duty between central and local authorities was not very clear. Fifth, through reforming the financial system, the government strengthened the central bank's function of stabilizing currency; to boost the commercialization of SOBs, the central government took some important measures. For example, the government established four asset management companies, issued 10-year bonds guaranteed by the Ministry of Finance, and bought bad loans of SOBs with a book value of RMB 1.4 trillion, so as to create conditions for the SOB commercialization; the government also founded three policy banks to provide financial support for accelerating infrastructure construction and urbanization.

4. *The Stage of Comprehensive Reform and Opening Up (2001–2013)*

First, the government reformed the trading system. In November 2001, China joined the WTO, and the government repealed and amended the laws, regulations and policies that conflicted with general market rules, indicating that China's opening up had entered a new stage. Second, the four state-owned commercial banks were listed, and China Investment Corporation was founded. Third, to cope with the international financial crisis in 2008, the government implemented a four-trillion-yuan investment stimulus package. Fourth, the private economy gained great development, and individually and privately owned businesses accounted for 70% of the contribution to GDP, and created over 80% of the jobs. Fifth, the government further boosted reforms in various areas, such as the reform of RMB exchange rate formation mechanism, the reform of pricing mechanism for refined oil products, the reform of house property tax, the reform of the cultural administrative system, the reform of the medical and health care system and the abolition of agricultural tax. Sixth, the government comprehensively accelerated infrastructure construction and urbanization; the expressway network and high-speed railway network were formed during this stage, and China's urbanization rate was over 50%. This stage witnessed China's rapid economic development, but the problem of "unbalanced, uncoordinated and unsustainable" development was increasingly severe, and the resource and environment bottlenecks brought about by the long-term extensive economic growth were increasingly prominent, so transformation of the economic growth pattern, adjustment of the economic structure and improvement of quality became very urgent tasks. Under the impact of the international financial crisis, the Chinese economy showed a sign of weakness; although this was attributed to external impacts to some extent, it also reflected the vulnerability of the Chinese economy, which posed a serious challenge for the Chinese economy, and meanwhile provided a good opportunity for adjustment of the economic structure. Unfortunately, we did not make good use of this opportunity to adjust the economic structure and upgrade the industries by enduring pains; instead, we implemented a four-trillion-yuan stimulus package, which worsened the original serious structural imbalance, and enabled some backward technology and excess production capacity to linger on or even continue to expand. The internal growth momentum in the former economy weakened, and we needed to optimize its structure, but we gave it an injection of cardiotonic instead of eliminating its internal defects, bringing difficulties for its subsequent development.

5. *The Stage of a New Normal (2014–)*

The Chinese economy has entered a stage of changing the growth rate, adjusting the economic structure and digesting the previous stimulus policies, and this change of trend has become a new normal. The CPC central committee led by General Secretary Xi Jinping put forward the Four-Pronged Comprehensive Strategy: comprehensively building a moderately prosperous society, comprehensively deepening the reform, comprehensively implementing the rule of law, and comprehensively strengthening Party self-discipline; the Third Plenary Session of the 18th

Central Committee of the CPC made a specific plan for comprehensively deepening the reform, and the plan consists of 16 sections, including 60 reforms and 336 major measures, with an overall goal to improve and develop the socialist system with Chinese characteristics, and boost the modernization of China's governance system and governance capability. This is a six-pronged reform plan, covering the comprehensive reforms in the systems of the economy, politics, culture, society, ecological protection and Party building. The CPC central committee established a leading group for comprehensively deepening the reform, aiming to transcend the limits of authority of the departments, break down the interest groups and let the market play a decisive role in the allocation of resources. The leading group pays much attention to the major concerns of the society, such as the property right reform of stated-owned enterprises, industry access, monopoly eradication, land transfer, the household registration system, judicial independence, abolishment of the system of rehabilitation through labor, streamlining administration and delegating power to the lower levels, environmental governance, and opening to the outside world; they try to effectively solve social problems and improve people's livelihoods through reforms. In the plan for comprehensively deepening the reform, there are major institutional and mechanism innovations. For example, it focuses on establishing a unified, open market system with orderly competitions, and it proposes for the first time exploring and implementing the negative-list management mode. It proposes preparation of national and local balance sheets, allows private capital to establish small- and medium-sized banks and other financial institutions, and also allows local governments to broaden the financing channels for urban construction through multiple ways including issuing bonds. It also gives the farmers more property rights and calls for accelerating the construction of free trade zones. These major arrangements and measures prove that China's reform has entered a new stage.

IV. Analysis of the Path of China's Modernization

1. *The Arduous and Tortuous Path of China's Modernization*

China's modernization has a history of nearly two centuries. The 1st century after the Opium War witnessed the Chinese people's efforts to save the nation from extinction and achieve liberation. During the new democratic revolution (1919–1949), the CPC summed up the experience and lessons from the old democratic revolution and the new democratic revolution, and integrated the Marxism–Leninism with China's actual conditions. By localization of Marxism in China, the CPC established Mao Zedong Thought, finally overthrew the "three big mountains" (imperialism, feudalism and bureaucrat-capitalism) and founded New China.

The 2nd century, from the founding of New China to the time the PRC celebrates its centenary in 2049, has and will witness the socialist construction with Chinese characteristics under the leadership of the CPC to turn China into a moderately developed country. Over 60 years of this century have passed, and during the first 30 years, from 1949 to 1978, China's gross value of social products increased from RMB 55.7 billion to RMB 684.6 billion, with an average annual growth rate of 9%, and a 9-fold increase; the independent, relatively complete industrial system and national economic systems were established.[11] During the second 30 years and more, from 1978 to 2015, China's GDP increased from RMB 684.6 billion to RMB 676,708 trillion, with an average annual growth rate of 10%, and China's per capita GDP increased from less than RMB 1,000 to RMB 20,000; the three-step development strategy designed by Deng Xiaoping was basically realized. Especially since the beginning of the 21st century, China has made remarkable achievements in just more than 10 years. China has grown from the world's sixth largest economy to the second largest; the per capita disposable income of its urban residents and rural residents has increased respectively by 9.2% and 8.1% per year, so this is one of the periods with the fastest-growing income of residents in the history of New China. At present, China has entered a new stage of building a moderately prosperous society in all respects, and the level of China's modernization has improved significantly. Table 2.5 summarizes China's first modernization index in international comparison since the founding of New China.

As can be seen from Table 2.5, China's modernization level increased significantly from 1950 to 2010. According to China's first modernization

---

[11]The Theory Bureau of the Propaganda Department of the CPC. *The Five Centuries of World Socialism.* Beijing: Study Publishing House and Party-building Reading Material Publishing House, 2014, p. 153.

Table 2.5. China's First Modernization Index in International Comparison from 1950 to 2010

Item	China 1950	China 2010	India 1950	India 2010	Japan 1950	Japan 2010	UK 1950	UK 2010	US 1950	US 2010
Per capita GDP (USD)	36	4,240	56	1,260	112	42,050	588	38,304	1,580	47,360
Growth rate in agriculture (%)	59	10	51	18	26	1	6	1	7	1
Growth rate in service business (%)	20	43	33	55	32	72	48	78	55	79
Proportion of agricultural labor force (%)	84	40	70	51	48	4	5	1	12	2
Urbanization rate (%)	13	49	17	31	50	91	84	80	64	82
Number of doctors per 1000 people (%)	0.1	1.4	0.2	0.6	0.8	2.1	0.8	2.7	1.3	2.4
Infant mortality rate (%)	138	14	127	49	60	2	31	5	29	7
Life expectancy (Years)	48	73	32	65	63	83	69	80	68	78
Adult literacy rate (%)	43	94	98	63	98	99	96	99	98	99
Popularizing rate of college education (%)	0.1	26	1	18	6	60	2	60	18	95

Note: Based on the USD price of the year.

Source: Research Group for China's Modernization Strategy, Research Center for China's Modernization of the Chinese Academy of Sciences. *China's Modernization Report 2013*, Evaluation Data of World Modernization Level. Beijing: Peking University Press, p. 331.

index, by 2010, China had achieved a level of moderate prosperity and made great progress in the international context. This progress was the fruit of concerted efforts made by the people of all nationalities in the country in the 60 years. During the period from the Opium War to the founding of

New China, the persistent efforts made by countless revolutionary pioneers and the heroic struggles waged by them also laid a foundation for China's modernization.

However, China's level of moderate prosperity is still relatively low, incomprehensive and unbalanced. Over the past 30 years and more, since the reform and opening up, China has actually witnessed great changes; in the world, its GDP ranks the second, and the developmental level of its high-speed rail ranks the first, so does the capacity of its other industries; so China has made a serial of achievements, and its comprehensive national strength, people's living standards and international influence have reached a new high. Meanwhile, we are soberly aware that in China, there are still many problems to be solved, such as the wide gap between the rich and the poor, seriously damaged environment, serious shortage of resources and intensified social contradictions; moreover, China's economic and social development is unbalanced, uncoordinated and unsustainable. The first 30 years after the founding of New China witnessed a planned economy; due to the overuse of political means and governmental plans, there was a lack of a broad information base and a democratic decision-making mechanism, and the market, known as "an invisible hand", was seriously neglected. The second 30 years and more have witnessed a market-oriented economy; due to the overuse of market-based means and the overemphasis of GDP, China has fallen into a snare of unlimited desire, which is upheld in the Western society, while the values of Chinese civilization have been neglected.[12] Although China has made some achievements, the two economic systems are both hard to maintain. That is to say, in the socialist construction and on the major proposition of building socialism with Chinese characteristics, we have not yet got full marks so far. In addition to poor basic conditions, complex international environments and lack of experiences, the main reasons for it are incomplete ideological emancipation, inadequate theoretical construction and problems with the thinking method in development; we have neglected the function of the law of balance, and have not well learned, mastered or applied the thinking method of balance.

---

[12]Xue Guangzhou. *Mao Zedong and the Fusion of Chinese and Western Philosophies.* Beijing: People's Publishing House, 2nd Edition, 2013, pp. 16–17.

2. *China's Modernization Construction Needs Further Ideological Emancipation*

There will be no modernization without ideological emancipation, which is the law of human modernization development. Take Europe for example: the ideological emancipation movement in Europe lasted centuries and involved various fields, such as philosophy, politics, society, religion, natural science and technology, literature and art, so it was a comprehensive, in-depth and fundamental ideological revolution. Without the Renaissance during the 14th–16th centuries and the Enlightenment during the 16th–18th centuries, there would not have been the Industrial Revolution, the English Bourgeois Revolution, the French Revolution, the Scientific and Technological Revolution, or the modern civilization in European and America.

China was dominated by feudal ideology for a long time, and it created a flourishing agricultural civilization. Tradition is both a fortune and a burden, and we have to correctly handle the relationship between its inheritance and development. In the era of modern civilization based on industrial revolutions, and in the complex and volatile international environment, without in-depth ideological revolution, correct thinking method, rational and scientific ideological guidance, it is hard to correctly handle the relationship between inheritance and development, overcome the shackles of old ideas and break the confinement of old systems, and it is also hard to establish new social structures and institutions, and carry on the transfer of modern civilization centers.

China's modernization has always been accompanied by ideological emancipation. After the Opium War, the Chinese people, especially the elites, always pursued ideological change, but the real ideological emancipation movement did not happen until the Revolution of 1911. The Westernization Movement introduced Western guns, cannons and machinery industry to China, but the thoughts and ideas of the Chinese people were obsolete. The Hundred Days' Reform and the Republican Revolution, launched by a group of people with obsolete ideas, could not make great achievements. A series of ideological revolution movements intensively occurred in the 20th century. The first was the New Culture Movement around the time of the May Fourth Movement, the second was the Rectification Movement in Yan'an in 1942, and the third was the

Discussion on the Criterion of Truth in 1979. The three ideological emancipation movements, short in time and narrow in content, involving fewer people, simply served the political tasks. For example, the New Culture Movement mainly served the anti-imperialist and anti-feudal political line; the Rectification Movement in Yan'an in 1942 served the political line for the War of Resistance against Japan; the Discussion on the Criterion of Truth in 1979 served the establishment of the political line of reform and opening up. The three movements were all very important for they solved a lot of problems, but they lacked depth and breadth, and had some deviations. For example, "Down with Confucianism", and "Destroying the Old and Establishing the New", were a complete denial of tradition; there were metaphysical activities in the process of learning from Western countries, such as focusing on forms while neglecting spirit, focusing on technology while neglecting ideology, and focusing on introduction while neglecting innovation; in the economic construction, the relationship between the government and the market was not handled properly; in the construction of values and systems, the government overemphasized "Chinese characteristics" while neglecting universal values, so it is hard for our values and systems to be geared to international standards. The root of these problems lies with the lack of correct thinking methods and theoretical guidance, and in-depth ideological emancipation. Therefore, both for now and for quite a long time to come, we should continue to emphasize ideological emancipation, establish a rational and scientific theoretical system, and break the shackles of traditional ideas. China has entered a new era of modernization, but many of our systems, including political, economic, scientific and technological, and educational systems, have not adapted to the requirements of modernization, so we need ideological and theoretical innovation, and need to establish correct ideas and improve our development theory in the process of ideological emancipation.

3. *Attach Great Importance to the Function of the Law of Balance*

Our development theory requires us to attach great importance to the function of the law of balance. The law of balance dominates the paradoxical movement of matters, and in the socialist economic activities, this law functions every time and everywhere. Both the "leftist" and "rightist" policies should be corrected; overuse of political means and overemphasis of wealth will inevitably pay the price; overuse of economic plans will get more haste and less speed; overuse of the market forces will also get more kicks than halfpence. We need to summarize and analyze the causes and consequences for neglecting the function of the law of balance in building socialism with Chinese characteristics.

During the initial period of the People's Republic of China, in politics, the CPC and the Chinese people made every effort to defend the new regime, and the movement of combating and preventing revisionism had a broad mass basis; in the economy, due to the urgent requirement to combat poverty and backwardness, the Great Leap Forward in construction and production also had a broad mass basis. Due to lack of in-depth scientific research on the socialist construction, the CPC, only guided by good wishes, adopted the revolutionary means used in revolutionary times to develop the economy. In a rush for quick results, they violated the laws of economics although they saw the importance of the law of balance to some extent. For example, Mao Zedong once said, "In the socialist construction, overall balance is very important". One of his important works, *On the Ten Major Relationships*, and other guidelines put forward by him, guided the country to make great achievements such as the successful development of the atomic and hydrogen bombs and man-made satellite, but the CPC failed to fully understand and stick to the law of balance. They held that all served the politics instead of pursuing a balance between the politics and the economy; they "decided everything by one man's say" instead of pursuing a balance of democracy and centralism; they overemphasized governmental plans instead of a balance between economic plans and the market forces. Therefore, in the socialist construction, they inevitably made mistakes. For example, the agricultural cooperation movement, and the transformation of the handicraft industry and privately or individually owned business were too hasty, so the work was too rough, the transformation too fast, and the mode too simple, leaving a lot of problems; a series of "leftist" policies on urban and rural economies and class struggles caused a large number of talented intellectuals to suffer undue blows; the Great Leap Forward and the movement to form rural communes disrupted the whole economic order, and the Cultural

Revolution caused serious disaster to the CPC and the country. It can be said that during the first 30 years after the foundation of PRC, China witnessed continuous political and economic movements, including the Great Leap Forward, which aimed to realize communism earlier relying on a highly politicized planned economy and a mass movement. Its result proved that haste makes waste, so the lesson from it was profound.

After the launch of the reform and opening-up policy, facing the national economy on the verge of bankruptcy and the complex international environment, the CPC held high the banner of Mao Zedong Thoughts in the politics and conducted reform and opening up in the economy, which had a broad mass basis. All the cadres and the masses deeply mourned over the death of Mao Zedong, Zhou Enlai and other veteran revolutionaries, and meanwhile they strongly demanded that the mistakes in the Cultural Revolution be corrected and the grave situation caused by the 10 years of chaos be reversed.

Deng Xiaoping, a great statesman with plenty of courage, gathered the wisdom and ideas of the CPC members and the people in the country to develop the economy with different types of ownership, and established the guideline of taking economic construction as the central task, which greatly emancipated the productive forces. Deng Xiaoping, acutely aware of the importance of balance, placed equal emphasis on material and ethical progress. He once remarked, "If there appeared billionaires and polarization between the rich and the poor during the reform, it would show the failure of the reform". Unfortunately, during the practice, the government did not properly handle the relationship between taking economic construction as the central task and advocating fairness and justice, the relationship between the government and the market, and the relationship between domestic and international markets.

The government mainly focused on economic activities such as investment promotion, and basically let go of a lot of matters that the government really needed to deal with, such as social security and public service. In fact, the government paid too much attention to GDP while neglecting many other matters and the government was like a corporation, causing dissatisfaction of the masses. The initial cause of the 1989 student movement was anti-corruption, anti-speculative buying and selling activities of governmental agencies and government staff, and anti-inflation. During the first dozen years after the beginning of China's reform and opening up, a few old revolutionaries, still alive, could conduct discussions within the CPC to achieve a balance of opinions, so the problems were still controllable. Later, after Deng Xiaoping, Chen Yun and other old revolutionaries passed away one after another, there was no authoritative discussion within the CPC, so the problems deteriorated. The government, neglecting the fiscal balance and credit balance, issued excess currency, and it also neglected the construction of fairness and justice and the rule of law, the bearing capacity of the environment and resources, and sustainable development.

The government held that everything depended on the market, and rendered the industrialization of basic education, research, medical care and health, and housing security; it seemed that only industrialization was the feature of a market economy, which was a serious error. These problems were seen by the following leaders, who also wanted to solve them. Jiang Zemin put forward the concept of ruling by virtue, but ruling by virtue was not bound by laws and rules; replacing the principle of law with virtue was entirely contrary to the principles of a market economy. In the fierce competitions in the market, GDP and wealth remained the only criteria, so the idea of ruling by virtue did not stop the money-oriented activities. Both the government and the people chased after money, and power-for-money deals were more rampant. In a market economy, only emphasizing moral constraints and individual consciousness is not enough; the government should establish a social honesty and trust mechanism, and a norm of market behavior, supported by morals, based on credit and signing of contracts, and safeguarded by law. Hu Jintao put forward the concept of Socialist Harmonious Society, but the fact was that the society was increasingly disharmonious; when this concept was implemented by local governments, they comprehended it as maintaining stability, so maintaining stability became a top priority.

During a considerably long period, the central government controlled housing prices every year, but the housing prices increased rapidly every year. The rapid increase of housing prices erased the reform dividends accumulated in the first 12 years; the common people had hatred and fear in their hearts, but they had no alternative but to join the trend of chasing money, so as to avoid becoming destitute. The government tried to prove its ability with economic growth, so it developed the economy at all costs; as a result, the government also joined the trend of chasing money. The central government utilized high taxes and easy monetary policies to enrich the central fiscal revenue, and the local governments utilized rising housing prices to enrich local fiscal revenue. Everyone in the society chased after wealth. Corrupt officials, taking advantage of their positions, traded power for money, and the corruption made people bristle with anger; some commodity producers and business operators unscrupulously chased profits, so the market was glutted with fake and shoddy goods, to which everyone might fall victim; many manufacturers always copied the goods made in developed countries, lacking in innovation; vulgar culture was popular, and many people had no belief or ideas of their own, so they did not respect others and our traditions. Some boastful "masters" and brokers familiar with a variety of hidden rules curried favor with influential officials for personal gain, causing great harm to the society. Some people saw the seriousness of the problems, and wanted to solve them, but they had no ability to do so, so it was delayed. The new state leaders, Xi Jinping and Li Keqiang, started with anti-corruption and have achieved good results, but the problems are so serious that they cannot be solved at one stroke. It can be said that the 30 years and more since China's reform and opening up have witnessed the movement of chasing wealth, which is another Great Leap Forward. Relying on a market-oriented economy that focuses on economic development and involves the participation of the whole society, the government aims to realize an overall well-off society, but many Chinese people become indifferent and apathetic when getting rich, and there are crises in moral convictions, beliefs and trust in the whole society, which lacks the values of self-confidence, responsibility, foresight and sagacity, so the lesson from it is also profound.

Why are there so many problems? Fundamentally, it is because we have neglected the function of the law of balance, and we have no good answer to the major question about how to build socialism with Chinese characteristics. The market economy itself is not wrong, but it also needs the support of good business ethics in the process of improving consumer markets. The point is that we have not paid enough attention to moral, ethical and legal construction in the process of developing the socialist market economy; the government has not paid enough attention to its role

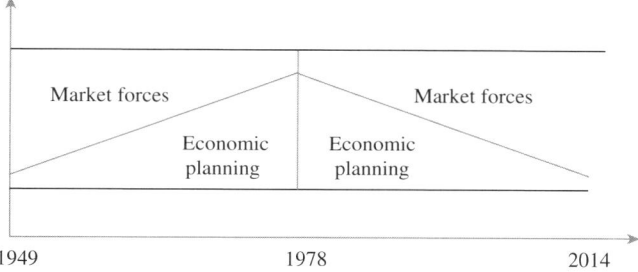

Figure 2.2. The Running Tracks of the Two Economic Systems in China

in narrowing the relatively wide gap between the rich and the poor, and improving people's livelihood. Without relying on the strength of Chinese traditional culture and universal values, and following and applying the law of social and economic development, the government, only based on good intentions and enthusiasm, has not yet solved the problems. Therefore, we should conduct an in-depth summary and research, and learn lessons from the past. Figure 2.2 depicts the running tracks of the two economic systems in China.

V. Further Improve China's Development Theory and Policy System

Mistakes during development are inevitable; making mistakes is not terrible, but repeating the mistakes of the past is. To avoid repeating the mistakes, we should follow the law of balance, and use the thinking method of balance to further improve our development theory and policy system.

1. *Improve the Theoretical System of China's Development*

Development needs a theory, which not only indicates the development direction, but also can guide the development. After the Third Plenary Session of the 11th Central Committee of the CPC, the CPC put forward the theory of a primary stage of socialism, and according to this theory, they established the CPC's basic line for the primary stage of socialism, which includes one central task and two basic points, and formulated the three-step development strategy. The theory of a primary stage of socialism, the CPC's basic line for the primary stage of socialism and the three-step development strategy constitute the theoretical system of socialism with Chinese characteristics. It now appears that it is necessary to add the ideas of giving priority to quality, innovation-driven development, environmental protection and balanced development to this system, so as to fully recognize and give full play to the important role of the balance law in the development. In addition, we also need to utilize the thinking method of balance to further improve its other content.

Firstly, the theory of a primary stage of socialism needs to emphasize the following content: in the 21st century, our country is still in the primary stage of socialism, and still a developing country; the rapid increase of GDP does not mean China is really developed, and in fact, there is still a wide gap between China and developed countries; to build a powerful socialist country and a rational, modern society governed by law, and to bring about a great rejuvenation of the Chinese nation, we still have a long way to go and a lot of hard work to do, and efforts of generations of Chinese people will be needed.

Secondly, the CPC's basic line for the primary stage of socialism, which includes one central task and two basic points, needs to further enrich its content as follows: we should take economic construction, not GDP, as the central task; we should not simply regard GDP as the only measurable indicator, to which we should add the green economy and ecological civilization; the two basic points should both be based on the rule of law, and we should attach equal importance to both of them.

Thirdly, we need to add synchronous development of the four modernizations and the "Five in One" General Layout to the three-step development strategy. Based the goal of building a socialist market economy, we should also set the goals of establishing socialist democracy under the rule of law and constructing ecological civilization in a harmonious society; we should promote the deep integration of informatization and industrialization, the positive interaction of industrialization and urbanization, and the coordination of urbanization and agricultural modernization, so as to boost the synchronous development of industrialization, informatization, urbanization and agricultural modernization; we should fully implement the "Five in One" General Layout, which consists of economic, political, cultural and social construction as well as construction of ecological civilization, boost the coordination of all the aspects of China's modernization construction, the coordination between relations of production and productive forces, and the coordination between the superstructure and the economic base, and continue to open up a path of civilized development featuring a thriving economy, an affluent life and a sound ecosystem. Thus, the new theoretical system of socialism with Chinese characteristics should include the theory of a primary stage of socialism, the CPC's basic line for the primary stage of socialism, the three-step development strategy, synchronous development of the four modernizations, the "Five in One" General Layout, and the strategy of sustainable development, as well as the ideas of giving priority to quality, innovation-driven development and balanced development. Table 2.6 outlines this system.

Based on the theoretical framework above, we can highlight some important points. First, we should both adhere to and improve the leadership of the CPC. Modern society is a society of political parties, and the modernization construction needs the leadership of modern political parties. The CPC has gathered a large number of progressives and elites, and they believe that the people make history and the elites govern the country.

Table 2.6.   The Theoretical System of Socialism with Chinese Characteristics

The theoretical system			
Development orientation	The CPC's basic line	Development strategy	Thinking methods
The theory of a primary stage of socialism (China is still in the primary stage of socialism in the 21st century)	One central task, two basic points	The three-step development strategy, synchronous development of the four modernizations, the "Five in One" General Layout, the strategy of sustainable development	Comprehensive balance, giving priority to quality, innovation-driven development

In order to better adhere to the leadership of the CPC, we should greatly improve the CPC's leadership, give place to democracy within the Party, stick to the democratic centralism within the Party, and gather wisdom within the Party. Second, we should adhere to the coordination of economic-planning and the market. In the modern market-oriented economy, economic-planning and the market are two indispensable means. A developing country should adhere to scientific and initiative economic plans, so as to exert subsequent advantages. Development finance is an important mechanism for the interaction of the government and the market, and plays a special role in overcoming the malfunction of the government and the market. Third, we should adhere to the interaction of domestic and international markets. For the economy of a large country, the domestic market is also an important part of the international market, so strengthening the construction of the domestic market holds special significance for our country. Fourth, we should adhere to the integrated development of urban and rural areas. We cannot sacrifice rural areas to develop urban areas. As the sayings goes, "With grain in our hands, there is no need to panic". The development of the agriculture, rural areas and farmers have a basic and strategic position in China's modernization process. Fifth, we should adhere to the combination of the material progress and the cultural and ideological progress. The material progress should be based on fairness and justice, and the cultural and ideological progress should be protected by law; fairness and justice and the rule of law are the key to both kinds of progress. Sixth, we should adhere to the policy of letting a hundred flowers blossom and a hundred schools of thought contend; letting a hundred flowers blossom is the guideline for boosting art development, and letting a hundred schools of thought contend is the guideline for boosting science development, but they can also be applied to other fields. Seventh, we should adhere to the combination of self-reliance and seeking foreign support; self-reliance is fundamental to realize the dream of building a powerful country, and seeking foreign support is an indispensable auxiliary condition. Without self-reliance, it is impossible for China to become a powerful country. To adhere to self-reliance, we should accelerate the implementation of the strategy of innovation-driven development; both the individuals and businesses should dare to innovate, and the whole society should foster a culture of innovation, to let the people realize their spiritual pursuit and their own value in the process of creating wealth. Eighth, we should adhere to the combination of tradition and modernity. Tradition is a valuable resource, and inheriting and carrying forward excellent Chinese traditions is of special significance in accelerating the process of China's modernization. Ninth, we should adhere to the synergetic development of the real economy and the virtual economy. In the modern economy, the real economy is the foundation of the virtual economy while the virtual economy provides important support for the real economy. Besides, in the Party building, we should pay attention to the balance of discipline and laws, and strengthen the modern culture construction of the CPC, to enhance the Party members' senses of identity, community, responsibility and mission, and to further motivate their enthusiasm, initiative and creativity.

2. *Improve the Policy System for Social Security*

To solve pressing problems, we need to further improve the policy system of social security. For example, to narrow the gap between the rich and the poor, we should take three measures. First, the government should take the improvement of social security as its important duty to relieve the common people of anxiety. The government should assume its responsibility in guaranteeing the access to basic medical services, old-age pension, housing, education and food safety. Over the past 30 years and more, in the name of reform, some responsibilities that should be shouldered by the government have been transferred to the common people; although the people's income has increased, their life is under increasing pressure. Therefore, the government should assume its responsibility and be committed to ensuring that all the people enjoy their rights to education, employment, medical and old-age care and housing. Once the people are relieved of these burdens, their yearning for a better life will be naturally activated, and they will not chase nothing else but money any more; instead, they will live a full, happy and successful life with dignity. So many years of accumulation in the Chinese economy, along with high taxes and huge state-owned assets, can play a role in improving social security. The moral bankruptcy in China is very serious, so we should strive to change this situation. We used to emphasize the liberation of productive forces, but today we need to emphasize the liberation of our-selves, to prevent ourselves from becoming slaves to money.

Second, we should improve the policy system of secondary distribu-tion to provide sustainable financial support for social security. The land finance is like draining the pond to get all the fish, and the existing finan-cial model is not sustainable because housing and land prices will not always rise. By learning from foreign experience, the government can establish multi-layered tax systems for house property, consumption and inheritance, and improve the governmental transfer payment system, the basic public service system and the basic social security system.

Third, we should reform the financial system to provide convenience for mass entrepreneurship. Firstly, we should accelerate the establishment of the food supply system for guaranteeing minimum subsistence. Based on the existing minimum subsistence guarantee system, the government should distribute food coupons and coupons for clothing and other articles of daily use to the extremely impoverished households in urban and rural areas. The sources of the fund and food should include government expenditure, and corporate and social donations, which should be managed by civil affairs departments, communities and local governments, and the coupon holders can get good from designated stores and shops. According to preliminary estimates, this system needs an annual fiscal expenditure of RMB 10 billion, and by guiding social resources to consume commodities made by enterprises and commodities in national reserves, it can benefit 20 million impoverished people and created 70 million jobs. Since the Great Depression in the 1930s, the United States has always been implementing this system. Based on the actual situation in China, the government should timely establish this system, which can not only enrich the social security system and reinforce the social safety network, but also can boost domestic consumption. Secondly, we should accelerate the establishment of a com-mon financing fund for the stable development of the stock market. The fund can be set up by policy-based financial institutions to help retail investors lacking stock-market knowledge to manage money. When there are drastic fluctuations in the stock market, the fund can be used to buy the stocks of retail investors with their consent, and when the stock market becomes relatively stable, the retail investors can buy their stock back with prices based on an appropriate ratio. This fund is of direct significance for the stable development of the stock market, and is helpful to improve the brokerage system in the capital market. Thirdly, we should accelerate the improvement of the national innovation fund system. Innovation is the best way to reduce the negative impact of the economic cycle. The government can invest RMB 20 billion to found a re-guarantee company providing further assurance for national technology entrepreneurship, and another RMB 20 billion to expand the existing guiding fund for national high-tech industry. Through the guiding fund and the re-guarantee system, the government can better solve the separation of financial investment in technology and commercial finance, and realize the demonstration and amplification effect of government funds in supporting technological innovation and industrialization, so as to promote employment and industrial upgrading as well as improve labor productivity and product quality. This fund system can also help readjust the product mix, accelerate the construction of national innovation system and improve the core competitiveness of our country. Fourth, we should accelerate the establishment of the national land fund. Based on the national level, we should analyze the whole forming process of land property, and the corresponding rights of national ownership at different levels (central, local, collective and individual) and different stages (transfer, development, construction, leasing, mortgages, etc.); by doing so, we should further confirm the property-rights boundary of each level and stage, and build a mechanism that allows the central government to restrain local governments by utilizing landownership, and a localized management mechanism that allows local governments to standardize the management and realize intensive development under the guidance of central government policies. We should also build a mechanism that allows farmers to transfer their land contract rights based on their value, and integrate the value and market-oriented operation of national landownership through land funds, so as to achieve effective control over operations related to both land and finance. We should also provide reasonable guidance for urban and rural construction and promote the self-cycling and self-growth of land assets, so as to achieve sustainable use of land, and harmonious and overall development of urban and rural economies.

VI. Summary: What is the Level of China's Modernization

During the five centuries of world modernization, capitalism and socialism have always been in conflict with each other. China's modernization is one of the greatest changes in human society, having a profound impact on the prosperity of the Chinese nation and the world civilization as well. During the course of China's modernization, through collisions and fusions of tradition and modernity, domestic and foreign cultures, advancement and backwardness, radical and conservative forces, one force always corrected another so as to promote social adjustment, dynamic equilibrium and spiral progress. For example, the people's commune system replaced the backward feudal small-scale peasant economy, and the market economy featuring the household contract responsibility system with remuneration linked to output replaced the planned economy featuring the people's commune system; in the near future, the small- and medium-sized cooperatives based on household or cooperative management will probably replace the household contract responsibility system. This tendency of one force replacing another reflects the law of balanced development. The law of paradoxical movement propels the error correction in economic activities to achieve a tendency of balanced development. In the market economy, there is economic planning, which is based on market laws; capitalism contains socialist factors, and *vise versa*.

China's modernization has witnessed remarkable achievements. For example, during the First Five Year Plan period, the achievements in industrial production and construction far exceeded the industrial accumulation of over 100 years in old China; the social vitality and infrastructure construction brought by the reform and opening up will shine through the ages. However, for a very long time, we held a thinking way of the extreme opposite, denying centrism and tending to find a solution for diverse problems; this could be proved by the simple classification of the left wing and the right wing, revolutionaries and reactionaries, the market economy and the planned economy, which were thought to be extremely contrary to each other. When the planned economy went to extremes, farmers were deprived of their possessions; they could not store up grain at home, and even the iron woks they cooked with were taken way and used for steelmaking; a series of political movements, especially the Cultural Revolution, cut-off the civilization development of our country. The market economy also once went to extremes, and during this period, almost all the people in the country chased after money, put money above everything else and tended to solve all the problems by spending money. That result was that corruption was common among officials, and the market was glutted with fake and shoddy goods; almost all the people in the country only had a desire for material welfare, and many of them worshiped things foreign and fawned on foreign countries. These phenomena caused catastrophic harm to China's sustainable development, and their devastating impact is far-reaching. The lessons learned from the past are as follows: we should abandon extreme ideas and means; either the planned economy or the market economy must not go to extremes. Ownership is only a management model of production and development; whether the ownership is good or bad, ultimately need to be selected by the production efficiency and development effects. The balanced development thinking we propose here should not only consider both redistribution and growth, but also the short term need to coordinate with the long-term development, as well as consider the balance between material development and cultural development, while ensuring the basic needs of the people, continuously improving the material, spiritual and cultural level from sustainable development. In the progress of China's modernization, following the law of balance and the idea of sustainable development is very important.

This chapter reviews the hard course of China's modernization, sums up the achievements, problems, experience and lessons in China's economic and social development, and proposes a tentative idea of improving China's development theory and policy system. According to this tentative idea, the theoretical system of socialism with Chinese characteristics should include not only the theory of a primary stage of socialism, the CPC's basic line for the primary stage of socialism and the three-step development strategy, but also the idea of balanced development; the development goals should include not only the goal of building a socialist market economy system, but also the goals of establishing socialist democracy under the rule of law and constructing ecological civilization. In China's development, making mistakes is not terrible, but repeating the mistakes of the past is; we should apply the thinking method of balance to analyze and solve problems in the development, so as to promote China's sustainable development.

Chapter 3

The Modern Theory of Economic Equilibrium*

What is the impact of the law of equilibrium on the evolution of economic theories? To further refine our theoretical system and development path, we need to take a balanced approach to analyzing and summarizing the evolution of modern economic theories. Since the mid-18th century, heated debates and competition, in theory and practice, have been focused on private ownership vs. public ownership and market competition vs. government intervention, thereby creating two camps. Generally speaking, private ownership and free market competition were the dominant drivers for growth in the late-18th and 19th centuries; government intervention and public ownership played a key role in the 20th century; mixed economies are the mainstream in the 21st century. China's socialist market economy is a mixed economy in nature, which, however, shall not be confused with mixed ownership.

I. Harbingers of Modern Economic Thoughts

The 14th and 15th centuries witnessed the fermentation of major changes in the world. In China, which was in the early Ming Dynasty, Zheng He (1405–1433) led a huge fleet of ships in undertaking seven expeditions to the Indian Ocean and visited more than 30 countries and regions bordering the western Pacific and Indian Ocean. In the Mediterranean, capitalism was burgeoning in some cities.[1] At the end of the 15th century,

---

*This chapter was published on the *International Finance News* dated April 11, 2016.

[1] *Collected Works of Marx and Engels* (Vol. 23). Shanghai: People's Publishing House, 1972, p. 784.

Christopher Columbus discovered the Americas, a new world market. As capitalist economy gradually gathered strength, the new *bourgeoisie* began to preach humanism to encourage commercial, industrial and other activities. They opposed feudal suppression, the theology of religions, abstinence which was advocated by the ruling class, and the idea of following the will of God. The dynamic Renaissance movement was started and mercantilism was proposed, along with the emergence of the unique "capitalist class", urban "citizens" and rational organizations of free labor bound by laws and regulations. The model of liberal market economy and contractual spirit came into being. All these laid the ideological and social foundation for the development of capitalism, industrial revolution and economic thought. Comparatively speaking, capitalism sprouted earlier in China, as evidenced by Sang Hongyang's proposition of centralized mintage system during the reign of Emperor Wu of Han Dynasty, the emergence of industrial and commercial guilds in the Tang Dynasty, the invention and wide application of paper money in the Northern Song Dynasty, mercantilism, urban economy and tax and welfare policies, etc. However, due to the absence of cultural movements like the Renaissance, capitalist ethos and the foundation of institutional, organizational and technological innovation, the capitalist mode of production established by small commodity producers developed very slowly and never had the chance to become the mainstream of society.

Private economy of the 19th century: In the 18th and 19th centuries, theories of capitalism and socialism emerged in Europe but the theories of private ownership and free market assumed dominance. In 1776, Adam Smith published *The Wealth of Nations*, marking the birth of economics as an independent discipline. The book offered quite a complete system of economic ideas, laying the most important foundation for economics known as the "theory of capitalist revolution". People even considered *The Wealth of Nations* as the bible about economic development, the "invisible hand" as the basic principle of market economy and Adam Smith as the father of economics. Later on, proponents of the ideas illustrated in the book kept improving and developing the classical economics produced by Adam Smith. They carried out psychological, mathematical and statistical analyses of human desires' characteristics and their influence on economic activities, placed emphasis on consumption, utility and demand, and put forward a series of views different from previous economic theories about the definition and classification of wealth, value, price, distribution, exchange, accumulation of capital, etc. They created the marginal utility theory which is also known as the "marginal revolution theory", providing strong technical support for classical economics. At the end of the 19th century, British economist Alfred Marshall (1842–1924) eclectically integrated various theories of classical economics and applied "the principle of continuity, marginal analysis and partial equilibrium analysis" to explain the laws of supply and demand and the nature of economic laws. He held that the expression of economic laws was not the certain quantitative relations between different phenomena but the general trend with probabilistic properties. Alfred Marshall established a brand new system with full competition as the prerequisite and equilibrium price as the core and founded neoclassical economics, taking the theory and practice of economics to a higher level.

As the thoughts of capitalism developed rapidly, the internal problems of the fast-growing capitalist economy became more serious and acute; labor movements emerged one after another, opposing capitalist exploitation, private ownership and the rule of the *bourgeoisie*. There was an urgent need for a revolutionary theory as guidance. That was when the theory of socialism came into being. Marx and Engels learned from the greatest thinkers of their time and proceeded from theories produced in their time. They studied and developed German classical philosophy, British classical economics and utopian socialism. By drawing upon and transforming everything valuable in the thoughts expressed over more than 2,000 years of human development, Marx established a brand new system of ideas that represent the interests and needs of the proletariat — Marxism, namely historical materialism and socialist economic theory with surplus value as the cornerstone. The book *Das Kapital* written by Marx provided the proletariat with a powerful ideological weapon. *Das Kapital,* known as the "socialist revolution theory" became the bible for the working class. However, overall, the 19th century was marked by the theories of capitalism and the booming of capitalist economy; capitalism assumed dominance in society; the classical economics founded by Adam Smith was the dominant economic theory. In particular, a *bourgeois* revolution broke out in Britain, France and Germany respectively in 1640, 1789 and 1848. As a result, the *bourgeoisie* assumed dominance in the political arena; the capitalist mode of production was further developed and consolidated; the capitalist market economy that encouraged free competition had its heyday. Just as Marx and Engels said, "the status of ruling class that the *bourgeoisie* had earned lasted less than a hundred years, but the productivity they created was greater and higher than that over previous generations".[2]

Socialized economy of the 20th century: In the 20th century, while the capitalist economy of free market continued to develop, a major economic trend characterized by socialist ideas and government-led market emerged. Two great socialist practices in human history emerged. Moreover, amid the fierce free market competition, the theory and measures of government intervention became widely recognized. A number of famous economists represented by John Maynard Keynes developed neoclassical economics from different perspectives and stressed the importance of the theory and practice of government intervention. Keynes was not like an economist in the ivory tower or a politician obsessed with his career. He conducted theoretical researches on practical issues and applied his research findings in practical work. He developed an approach to studying and solving problems from the government's perspective. On the basis of examining the first global financial crisis in the early 20th century, Keynes put forward the theory of effective demand and the view that the state should intervene in the economy. He published *The General Theory of Employment, Interest and Money*, and founded macroeconomics which is completely against the neoclassical economics, that is the "revolutionary theory of Keynesian macroeconomics". From the 1920s to the 1970s, in both capitalist and socialist countries, developed or underdeveloped, government participation in economic activities reached an unprecedented level. While the theory of government intervention gained strength, the theory and practice of socialism developed as well. In the Soviet Union, Lenin enriched and developed the theory of Marx and Engels about the two stages of communist society and held that "socialism should not be

---

[2] *Ibid.*, p. 16.

considered something still, frozen or changeless".[3] The Soviet Union established a socialist economic system based on revolutions and gave a big push for socialist planned economy with public ownership as the core. Industry, agriculture and national defense were modernized within a very short period of time. A country where wooden ploughs were used was turned into a world power with nuclear weapons, making significant contributions to the victory of the anti-Fascist war. However, the role of planning was overemphasized and even abused while that of the market was overlooked. As a result, serious problems arose from the Soviet Union's planned economy, which became an important cause of the Soviet empire's collapse. In China, the older generation of revolutionaries such as Mao Zedong put forward the economic program and thought of New Democratic Revolution according to the reality and stressed that everyone should do their part to ensure material supplies for the revolution. Upon the founding of the People's Republic of China, the general line of completing socialist transformation and the idea of developing socialist economy were put forward for the transition period — for quite a long time, efforts should be made to gradually accomplish socialist industrialization and the socialist transformation of agriculture, manufacturing and capitalist industry and commerce. The 10 relationships in building socialism were also expounded. Since the 1980s, China has gathered speed in pushing forward reform and opening up and making the transition from planned to market economy.

The CPC has held four "third plenary sessions" of great significance. The first was the Third Plenary Session of the 12th CPC Central Committee in October 1984 which adopted the decision to reform the economic system and made clear the direction, nature, task and basic policies of the reform. The second was the Third Plenary Session of the 14th CPC Central Committee in November 1993 which adopted the decision to establish socialist market economy and provided the general idea of how to do it. The third was the Third Plenary Session of the 16th CPC Central Committee in October 2003 which adopted the decision on some issues regarding the improvement of the socialist market economy. It was

---

[3] *Collected Works of Vladimir Lenin* (Vol. 31). Shanghai: People's Publishing House, 1985, p. 95.

stressed at the session that China should coordinate urban and rural development, development between regions, economic and social development, relations between man and nature, and domestic development and opening to the outside world, and leverage to a greater extent the fundamental role of the market in allocating resources so as to provide strong institutional support for building a moderately prosperous society in all respects. The fourth was the Third Plenary Session of the 18th CPC Central Committee in November 2013 which adopted the decision about major issues concerning deepening the reform comprehensively. It was noted at the session that China was entering a new stage of development requiring change of speed and model and structural adjustments and that reform was coming to a very difficult phase with many outstanding problems to solve. It was stressed that the policy of reform and opening up bears on the future of China and represents a crucial means by which China can keep up with the times so reform must be deepened comprehensively from a new starting point, the socialist system with Chinese characteristics must be improved and developed, the governance system must be modernized and a moderately prosperous society in all respects must be built, thus making China a modern socialist country that is prosperous, strong, democratic, culturally advanced and harmonious and fulfilling the Chinese dream — the great renewal of the Chinese nation. Ever since the policy of reform and opening up was introduced, China has given a big play to the role of the market as the "invisible hand", resulting in the rapid growth of commodity economy as well as extreme phenomena in some fields. There is good experience as well as bitter lessons. On the whole, the economic and political spheres of the 20th century were marked by the advocacy of government intervention or free competition but the theory of government intervention seemed to take the upper hand with its influence reaching farther and lasting longer.

II. Embracing the Era of Mixed Economies

The 21st century will be dominated by mixed economies. In a diversified world and an open era, it is dangerous to overly stress private economy, or public (collective) ownership, or government intervention in the belief that poor coordination has to bear the brunt of the blame, or free competition in the belief that reform and opening up is to become market oriented, with everything to be decided by the market. It is also harmful to overly emphasize the significance of supply or demand. Along with the development of mixed economies, traditional capitalist countries focusing on the market economy have increased spending on public services, social security and welfare and strengthened the building of the macro control mechanism, while traditional socialist countries focusing on the planned economy have encouraged the development of the private economy and made full use of the roles of the market. As the global economic system is undergoing correction today, the mixed economy, and the management system, operation mechanism and practical theories in relation to the mixed economy are set to go through profound development. Just like what Mao Zedong said in *On Contradiction*, there is a bridge leading from private assets to public assets. It is called identity in philosophy, or mutual conversion, or mutual penetration.[4]

Capitalism and socialism have been deepening, modifying and refining themselves since their emergence, sparking thinking of modern economists on the evolution of human civilization. Socioeconomic developments and their theories have confirmed that the existence of only one economic form in a society, no matter what it is, is unfavorable for optimizing allocation of resources and enhancing economic efficiency. In modern history, especially since the 20th century, tremendous changes have taken place to the mode of production and organization in both capitalist and socialist countries, and to social relations that are based on the production of material goods, while the forms of ownership have also been evolving. For example, at the macro level, most capitalist countries' economies have long begun the transformation into mixed economies or dual economies where the private economy coexist with the public economy, with the structure of ownership dramatically changed, which is characterized by stock ownership by many average families. China's private sector has also achieved significant progress over the past 30 years. At the micro level, the drawbacks of family management at private companies and non-separation of government administration from enterprise management at state-owned collective enterprises have been overcome by

---

[4] *Selected Works of Mao Zedong*, Vol. 1, People's Publishing House, 1991, pp. 299–337.

separating ownership from management power and building relevant mechanisms, further weakening the significance of pure public ownership or pure private ownership.

Alvin H. Hansen, an American economist, introduced in his book *Fiscal Policy and Business Cycles,* published in 1941, the concept of "mixed economy". Paul A. Samuelson gave a detailed description of the "mixed economy" in his book, *Economics*, believing that a mixed economy is one in which state-owned and private institutions are the fundamental controlling forces, but the government's regulation and control are more important. He also suggested that China should build a "limited centrist" economy.

In today's world, a key trend of economic thinking is interweaving; a basic framework of state governance is that "some areas must be weakened while others should be strengthened"[5]; an essential trend of mode of production is Internet and socialization; and that of economic policy, timely selection and use of the synergies of monetary policy, fiscal policy, industrial policy and environmental protection policy. In an open economy, various economic players form networks where they find multiple equilibriums as they exchange products, currencies and credit. In particular, as the global economic integration deepens, aspects of the economy have become more closely connected, as reflected in restructuring and system coordination, which are set to lead to the emergence of a mixed economy. A mixed economy is one where private and collective economies develop in parallel or integrate with each other through cross-shareholding, where market mechanism is mixed with government intervention, where traditional and new development engines work together to drive growth, where personal and macro planning are integrated, where public services and private services and consumption coexist, where the moral values of "benevolence, righteousness, humility, talent and honesty" and market requirements of "compliance, innovation, economy, greenness, sharing, and balance" both work, where multiple ownerships, economic players, ways of resource allocation,

---

[5]Francis Fukuyama. *State-Building: Governance and World Order in the 21st Century.* Translated by Huang Qiangsheng and Xumingyuan. Beijing: China Social Sciences Press, 2007, p. 5.

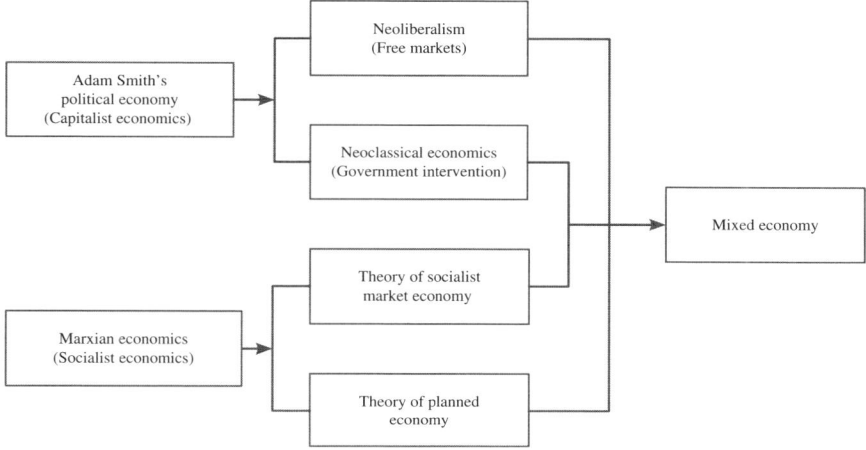

Figure 3.1.   How a Mixed Economy is Formed

market structures, and ways of distribution complement and play their roles, and where mixed operation, collaborative services and comprehensive regulation function. To put it simply, a mixed economy is a public economy in nature, in which the public sector is mixed with the private sector. This trend is a natural result of historical and social developments under the law of equilibrium to seek balanced social and economic development and is independent of man's will.

Figure 3.1 shows how a mixed economy is formed.

III. Mixed Economy is a Natural Choice of China

China's socialist market economy is a mixed economy in nature. The Decision adopted by the Third Plenary Session of the 18th CPC Central Committee highlights that the non-public sector and the public sector are both integral parts of a socialist market economy and key foundations for China's social and economic development, and proposes that a mixed-ownership economy should be developed to non-discriminatorily develop sectors of the economy and remove the barriers among different ownerships to allow every individual to participate in competition as equals.

Since the implementation of the reform and opening-up policy, Chinese private enterprises have seen explosive development from emergence to expansion, and investors of various backgrounds have set up companies through joint investment, cross shareholding and joint production that have formed diverse economic sectors, boosting the emergence and development of a mixed economy of Chinese characteristics. Private enterprises in various forms have become important employers, economic players and taxpayers, and the non-public sector has grown into a significant foundation of China's economy. In 2015, China's non-public sector contributed more than 60% to GDP, over 50% to tax revenue and 90% to employment. However, some serious problems have occurred under the free market competition, such as over consumption of resources, chaotic competition, overcapacity, serious waste, over reliance on the market to solve hot social issues, or even government functioning as a market player, driving China's socialist market economy to extremes in some aspects. Therefore, we need to learn from lessons and deeply understand the laws of social and economic development to further refine the philosophy, roadmap and policies on social and economic development.

IV. Promoting Healthy Development of China's Mixed Economy

First, the interaction mechanism between the market and the government should be refined further. Both the government and the market can function independently, but may also have failures, so they need to complement, support and collaborate with each other. The market, as an endogenous invisible force, is functioning anytime and anywhere. It is an essential requirement of the market economy that market activities are continuously constrained by market discipline in making use of the roles of the market. At the micro level, in particular, the government should not intervene in any productive or operative activities in any industries but let the market handle it. It should believe in and depend on enterprises in building more and better "Wenzhou models" to develop non-agricultural industries via household industries and specialized markets and build a development landscape of large markets with small commodities, and believe in and depend on inventors in inventing products like iPhone under a proper mechanism and continuously generating new demand. Market mechanism, however, also has failures. Even in areas where market mechanism can play a significant role, the market has its inherent weakness and drawbacks, such as spontaneity, blindness and hysteresis. China's reform and opening up has generated market demand and supply over the past 30 years, but also has led to extremes of markets in some aspects, as reflected in high demand and oversupply, which has resulted in discoordination, imbalances and unsustainability. Therefore, the government needs to directly enter or regulate in a proper way the areas that have a bearing on the overall economic layout, the national economy and people's living standards, where private capital is reluctant to access or hard to deliver a good performance, and where market failures often occur. At the macro level, to ensure stable, healthy and sustainable social and economic development, the government needs to make good macro planning, build and improve mechanisms favorable for innovation, and strictly control and regulate strategic industries and goods that are key to the national economy and people's livelihood, such as agriculture, national defense science and technology, money supply, health and security.

Second, the principle of "establishment first and elimination later" should be followed. Establishment first and elimination later, or at least attaching equal importance to establishment and elimination, is in line with the law of development. China's reform has entered the deep-water zone, with limited stones left above water to grope. Under this circumstance, establishing effective and necessary measures first and eliminating outdated and inefficient systems later will help the government firmly control the initiative of social management and the big picture to avoid collective loss and disorientation in case that outdated system is ineffective while new rules have not been in place. In the reform measures such as administration streamlining and power delegation, innovation and entrepreneurship, and the scheme of minimum standard of living, the government needs to make a further clarification that administration streamlining and power delegation does not mean the government is unnecessary but should function in other ways. For example, *ex-ante* review should be changed to end-to-end monitoring, guidance, assessment and services. In bankruptcy restructuring, establishing a new company first and announcing bankruptcy later to change the original company into a new market player is favorable for making schemes of arrangement with a bank. In pushing the mixed-ownership economy, the private sector and the public sector could be developed in parallel, with equal opportunities provided to private companies in more areas and higher levels and the investment interest security mechanism for private enterprises established. Meanwhile, cross-shareholding could be encouraged in pursuit of integrated development. By learning from the structure of collective ownership under state-owned enterprises (SOEs) in the 1990s, mixed-ownership companies participated by state-owned or state-controlled companies could be set up in a standardized, transparent and orderly manner to form new independent and sound market players in which benefits will be enjoyed depending on who invests, operates, owns, or dominates.

Third, further research should be carried out on the theories and practices of the mixed economy. A mixed economy is a new economic form with some new features such as holistic, systematic and social properties and a combination of mixed ownership at the micro level with comprehensive forms at the macro level. To accurately understand the mixed economy, we need both theoretical guidance and technical support. Both economic and moral forces function in a mixed economy, featuring a return to the nature of an economic society and highlighting the importance of putting people first, to build a market order that conforms to moral standards and the spirit of contract. A mixed economy is one where universal truths are combined with the specific national situation and one that understands the world around us in an open and broad view, and generates supply and demand in a more sensible way to improve people's life. It also features complementarity and integration, and is a natural result of the balanced development of human society, not of the negation of negation, and therefore calls for continuity, not "either-or" decisions. In a diversified world, there are various public or private economies in either public or private areas. "There are different mixed-ownership economies in the middle ground of public and private ownerships, which are objective realities that should not be neglected".[6] We should overcome the conflict between "rising demand for problem solving in reality and increasingly singular

---

[6] Li Yining (Editor-in-chief), Cheng Zhiqiang (deputy editor-in-chief). *China's Path and Mixed-ownership Economy 1st edition.* Beijing: The Commercial Press, 2014 Foreword, pp. 2.

trend of publishing academic papers" in doing research, and need to adapt to the trends of global economic development in deepening reforms, rather than simply mixing. In pushing the SOE reform to go deeper, for example, the mixed ownership should not be confused with the mixed economy, as it is just one of the ownerships, not the only ownership of a mixed economy. We encourage qualified and interested enterprises to develop the mixed ownership, not all enterprises. We still stress that enterprises should build and improve the modern corporate system based on the realities of the industry and their "internal economy" and "external economy" to reduce production costs, enhance innovation capabilities and internal vitality as well as external adaptability.

To sum up, streamlining and rethinking of the evolution trajectory of economic theories will help deepen the theories to better serve and guide practices. The development of science is always pushed by replacing earlier theories with new theories, and humans' understanding and knowledge are effective and reliable at a certain point of time but not the eternal truth. Today's capitalism is largely different from what it was in the 19th century. So is socialism. The result of the changes is that various economic sectors are inextricably interwoven with each other. The 21st century will be the one dominated by mixed economies. A mixed economy is one in which private and public sectors coexist and the government and the market interact. Regardless of changes in economic theories and forms, however, the demand for credit building, the core requirement of an economic system, has remained unchanged and even become stronger.

A mixed economy is a natural choice of China. Although a late starter in modern economic theory and practice with a weak foundation, limited experience and insufficient theoretical support, China enjoys the late-mover advantage. It is set to achieve sustainable development by further refining the philosophy, roadmap and policies on social and economic development, and boosting the integrated development of the private and public sectors, organic interactions between the government and the market, joint innovation of supply and demand, and collaborated operation of monetary policy, fiscal policy and industrial policy under the laws of social and economic development in the 21st century.

Case Study: Why Poverty Alleviation is a Tough and Time-Consuming Campaign

The central government has recently stressed that the poverty-alleviation campaign is both tough and time consuming. To deeply understand and fully implement the guidelines of the central government and finally win this campaign, we need to make an in-depth analysis of the strategy and tactics for the campaign. Strategically we should scorn the difficulties, but tactically we should pay full attention to them; this is a tough and time-consuming campaign, so we should make overall arrangements, guide social expectation properly and gain the initiative in the work.

(1) A comprehensive and in-depth understanding of the poverty-alleviation guidelines and policies of the central government. During the 13th Five Year Plan period, the campaign of lifting 70 million people out of poverty is both tough and time consuming. The 70 million poverty-stricken people are a concentrated reflection of the "Three Rural Issues". Through the efforts of the central government and local governments at various levels and the people throughout the country, there is no doubt that by 2020, the per capita income of the 70 million poverty-stricken people will surely increase to some extent, and their living standards will certainly improve; we are very confident about that. However, we have to be soberly aware that the achievement to be made in five years will probably be mainly reflected in hardware, and may represent a phased achievement. To comprehensively improve the living conditions of all the poverty-stricken people and fundamentally solve the "Three Rural Issues" is a long-term and arduous task. When we take various measures to overcome difficulties, we should be mentally prepared for a time-consuming campaign and make long-term strategic plans.

(2) A deep understanding of why the poverty-alleviation campaign is both tough and time consuming. To better understand this tough and time consuming campaign, we need to consider mainly the following five aspects: first, we all know that the last fortress is often the most difficult to capture; we are facing not only the poor living conditions of the poverty-stricken people, but also a heavy social problem caused by 40 million leftover children and 30 million elderly people in rural areas. To solve this

complex social problem, we need to have both a sense of urgency and a long-term plan. Second, the poverty alleviation work needs both hard targets and soft targets. In addition to increasing the poor people's per capita income, raising their living standards and improving rural transport, education, health and other conditions, we should also care for their mental health and spiritual outlook as well as the issues of natural environment and social management. These soft targets, from the perspective of social development, are also very important, and are not easy to improve significantly within a short time. Third, the structure of the poor people is dynamically changing, so when we make efforts to lift the existing poor people out of poverty, we need to pay much attention to the newly emerging poor population. For example, with the acceleration of China's urbanization and industrialization, many young adults and middle-aged people in rural areas have gone to cities to work, so the number of left-behind children has been increasing. According to the new standards of the United Nations, the poverty threshold is a daily per capita income of less than USD 1.25. As the level of global economic development improves, the poverty threshold and the criteria for identifying poverty are also changing. We can also see that during the development process of developed countries, there are always a proportion of low-income people, who need long-term care and help of the government and the society. Fourth, China will remain a developing country for quite a long time to come, so to adjust the industrial structure, transform the mode of production and improve the quality of development is a long-term and arduous task, and it is also a long-term and onerous task to improve the quality of poverty alleviation and increase the hematopoietic function of the poor people. We need to have a deep understanding of this.

(3) A preliminary scheme for the poverty-alleviation campaign. To carry out this tough and time-consuming campaign, we need to consider the five following aspects: first, to make medium-term and long-term plans for development-oriented poverty relief. We need not only short-term plans to overcome existing difficulties, but also medium-term and long-term plans in line with the primary stage of socialism to guide investments and gather resources for urban–rural integration development, which requires us to develop effective ways to implement the policies of promoting agriculture

through industry, urban areas helping rural areas, and agriculture and industry benefiting each other. Second, to strengthen the county economy's capability to address the Three Rural Issues. Both cities and townships contribute less to development-oriented poverty relief, so we need to regard the county economy as the supporting point for poverty alleviation, and develop the county economy by enhancing its inner impetus; during the process of developing the county economy, we should focus on organizing poverty alleviation projects and development projects for the poor, and grasp the most important link of the dynamic poverty alleviation mechanism to improve the hematopoietic function of the poverty-stricken people. Third, to establish and improve the basic security systems for the poor people, including the minimum living security system and the basic medical insurance system. We should actively explore the system of distributing food stamps to the poverty-stricken people, and provide direct assistance to them through governmental organization and social participation. This system is still effective in the United States and other developed countries. Fourth, to actively explore new ways of rural social governance based on school districts. According to the development of urban and rural areas, we should reasonably design the spatial layout, production structure and living facilities; by centralizing educational resources, we can build rural centralized schools and corresponding infrastructure systems and guide rural residents and poor people to reside near the schools so as to provide convenience for their children to attend school. Meanwhile, based on the schools, we can adjust the administrative structure of the villages, improve the rural grass-roots management system and strengthen the construction of village regulations and nongovernmental agreements; we can also improve the villagers' self-governance system and the open system of village affairs to achieve democratic management. Compared to the previous rural population migration, this measure embodies the people-oriented concept and the strategic thinking of revitalization through education and poverty alleviation through education; it can help turn some backward villages into new communities, which can become models of national rejuvenation and play an exemplary role in promoting new rounds of rural development. Fifth, to develop characteristic rural economies according to local conditions, we should help the poor people develop featured products and markets to increase their income. As long as they combine tradition and modernity, local resources and external demands, security and consumption, they will surely find a way to become well-off. We should help the poor people improve their positions in the market and the industry chain, and enhance the quality and competitiveness of the rural economy.

In short, the poverty alleviation campaign during the 13th Five Year Plan period will greatly improve the living conditions of the poorest people in China and substantially improve the income level of the poor people so that they can share the fruits of reform and opening up. We are confident that we will achieve this goal, but to fundamentally solve the poverty problem is a long-term and arduous task, so we need to make a long-term plan strategically and tactically.

Chapter 4

China's Road
to Balanced Economic Development

How to refine China's economic development strategy under the new normal? The review of history, international views and theoretical evolution in the previous chapters are to pave the way for the analysis of China's economic development in the new era. This chapter focuses on a balanced approach to studying the basics, fundamental characteristics and strategic choices of China under the new economic normal. China's economy is slowing down from a high growth rate to a middle growth rate, transforming from extensive development to intensive development, and developing from a growth economy into a mature economy, and the last change will be a critical sign of China's stepping toward a modern society. In the first 30 years since China's founding, the role of planning was abused. So has been the role of the market over the past 30 years since the reform and opening-up policy was implemented. In the coming 30 years, China's economy would become mature, and China would become bigger and stronger. In face of new tasks under the new normal, China needs new development strategies. The market-oriented and intensive economic development featuring balance, innovation and sustainability should be a strategic choice of China under the new normal.

For more than 30 years since the reform and opening-up policy was implemented, China's economy has sustained a staggering two-digit annual growth rate, which is a magic. However, this has been achieved under an extensive growth model. The extensive high growth has led to overly fast resource consumption, deteriorated eco-environment and serious structural imbalances, and is therefore not normal or sustainable. Due to the impact of the laws of development, changes in supply and demand, and internal and external factors, the maximal economic growth rate, or the natural or potential economic growth rate China can achieve when resources are optimally and fully allocated, is on a downward trend. In the coming 30 years, it is unlikely that China's gross domestic product (GDP) would sustain high growth as it has done over the past 30 years, which is a trend change and new normal.

I. China's Balance Sheet

This balance sheet (Table 4.1) can reflect the basic conditions of China.

(a) Total assets

Large economic scale: China's GDP in 2016 was about RMB 74.4 trillion, a year-on-year increase of 6.7%, calculated at comparable prices. If calculated in US dollars, it is over USD 10 trillion, making China along with the US the two largest economies in the world.

Strong political organizational capability: China's political party system is a system of multi-party cooperation and political consultation under the leadership of the Communist Party of China (CPC). This system was formed and developed in the CPC's long-term practice featuring revolutions and reforms, and it reflects the CPC's strong organizational, centralized and

Table 4.1. The Balance Sheet of China

Assets	Liabilities
(1) Large economic scale	(1) Unbalanced economic development
(2) Strong political organizational capability	(2) More unstable factors in social development
(3) A large population	(3) Severe environmental pollution
(4) Geographical basis	(4) Wide income distribution gap
(5) Traditional culture	(5) Imperfect capital market
(6) Huge production and consumption demands	(6) Enterprises lack brand awareness and creativity
(7) Achievements in infrastructure construction	(7) Citizens' legal awareness is weak
(8) Great market potential	(8) Food safety issues

political mobilization capabilities. In China, a multi-ethnic country with a large population and a vast territory, a strong leadership is needed for its modernization drive. The strong leadership of the CPC is the fundamental guarantee for the national unity and social harmony and stability. The Chinese government, based on China's political party system, has a strong ability to regulate and control.

A large population: China's population accounts for about one-fifth of the world population. A huge population provides a great demographic dividend for China, powerfully promoting its reform and opening up, and economic development.

Geographical basis: With a national territorial area of 9.6 million km^2, China has the geographical basis for becoming a large and strong economy. Diverse landforms provide a variety of conditions for China's industrial and agricultural development. China is rich in natural resources, and the reserves of many of its natural resources rank first in the world. However, because of the large population, China's per capita share of natural resources is relatively low. For example, China's per capita arable land area is only one-thirtieth of that of the US.

Traditional culture: With a long history, Chinese traditional culture is extensive and profound. It emphasizes on revealing the nature and meaning of the universe, society and life from the philosophical and scientific perspectives. It focuses on both reasoning and proving through practice, which is its unique feature.

Huge production and consumption demands: First, the rapid development of China's economy provides a huge impetus for China's consumption demands; second, China has a rural population of 700 million, which can provide huge production and consumption demands for the Chinese market in the process of urbanization; third, China has a huge production capacity, and its industrial upgrading will generate new demands. China's GDP is based on extensive production; the cost of its GDP doubles that of the US GDP. After China's industrial upgrading and updating is completed, its production efficiency and productivity will be further improved.

Impressive achievements in infrastructure construction: For example, in transportation infrastructure, from 1978 to 2016, the length of China's railway lines in service has increased from 51,700 km to 120,000 km, second only to the US; China's road mileage has increased by 300% to 4.24 million km, ranking first in the world, among of road, the highway mileage has increased to 131 km, also ranking first in the world. China's relatively leading infrastructure coverage and services have created a favorable basic environment for China's economic construction and urban development.

Great market potential: The Logistics Economy, Sharing Economy, New Household Manufacturing and New Urbanization are emerging in China and will become important new growth points. China's huge market potential is an inexhaustible resource for its economic development; in the future, there will still be a considerable proportion of residents in the rural areas, which also have great development potential. The implementation of the Free Trade Zone Development Strategy, Beijing–Tianjin–Hebei Coordinated Development Strategy and Yangtze River Economic Zone Development Strategy will generate new growth momentum.

(b) Total liabilities

Unbalanced economic development: China's GDP growth is rapid, but it has an extensive growth model, and the structural problems of the economy are outstanding; the economic development achievements have not been effectively and reasonably shared by all the citizens, resulting in the widening gap between the rich and the poor; the development of social undertakings lags behind the economic development, and there are many problems needing to be solved in the areas such as social security, housing, health care and education; the urban–rural dual economic structure has not been fundamentally changed, and the unbalanced development of urban and rural areas is very conspicuous; the unbalanced regional development is mainly reflected in the fact that western China lags behind eastern China; the resource, energy and environmental problems are outstanding, restricting China's sustainable economic development.

More unstable factors in social development: The international political situation is complicated and changeable, affecting China's international layout and development. The shock caused by deepening the reform as well as the imperfect security system and corruption problems are likely to cause psychological imbalance and social instability; in particular, the precipitous rise of housing prices is still a prominent problem, and the poor people in urban areas tend to increase; because of the losing of the moral bottom line, some people are unscrupulous and anomic.

Severe environmental pollution: First, there is a wide range of pollutants, which is spreading rapidly from the economically developed eastern and southern regions to the central, western and northern regions. Second, the degree of contamination is high. About 64% of the urban groundwater is seriously polluted. The surface water with the IV–V level quality and the inferior V level quality accounts for respectively 32% and 28%, and 48 of the 84 lakes are eutrophic. More than 10% of the China's arable land (about 150 million mu) has been polluted by heavy metals, and each year 12 million tons of grain are contaminated with heavy metals absorbed from the soil, causing direct economic losses of more than RMB 20 billion. In the urban air, the concentration of the total suspended particulates greatly exceeds the international standards; the sulfur dioxide pollution maintains a high level, and the nitrogen oxide pollution is becoming severer; there are a number of acid rain areas distributed in central, southwest, east and south China. Every year, the air pollution causes losses of up to more than RMB 400 billion. In 2013, the Chinese government invested RMB 293.2 billion in environmental protection and RMB 80 billion in curbing the pollution of Lake Taihu. Fresh air, clean water and blue sky have now become the people's wild wishes.

Wide income distribution gap: Before the reform and opening up, the Gini coefficient reflecting China's income distribution was 0.2, indicating an absolute egalitarian state, and it was 0.28 in 1986, 0.47 in 2000 and 0.469 in 2009; in recent years, the Gini coefficient witnessed a slight decline, and it was 0.481 in 2010 and 0.465 in 2016, but it still exceeds the warning line of 0.4. This implies that 1% of China's households occupy one-third of the country's wealth, which may cause a series of social problems.

Imperfect capital market: First, the development of the bond market is unbalanced. Compared to its stock market, China's bond market has always been "lame", due to its small market size, low marketization degree, unreasonable structure and fragmentation of issuance and trading. For example, in China's bond market, national bonds, financial bonds, corporate bonds account for respectively 51%, 44%, 5% and the balance of the corporate bonds is only RMB 283.1 billion, accounting for 1.35% of China's GDP. Second, the development of the stock market is unbalanced; the development of the stock market entry and exit mechanism is unbalanced, weakening securities' role in efficiently optimizing allocation of resources. Third, administrative intervention is excessive, which cannot reflect the highly market-oriented characteristics of the securities market.

Enterprises lack brand awareness and creativity: The creativity of Chinese brands is weak, so they lack market influence and cannot eventually conquer the hearts of consumers. In all kinds of competitive markets, Chinese brands are unlikely to appear as conquers, and still mainly rely on their prices to attract customers; successful brands like Haier are very rare.

Citizens' legal awareness is weak: First, traditional values and legal awareness coexist, and non-legal awareness occupies a considerable position in the people's real lives.[1] Second, the legal awareness for various branches of laws is unbalanced, which is mainly reflected in the following aspects: constitutional awareness is weak; some sectors only pay attention to the provisions of general laws, but often ignore the provisions of the Constitution and its role, and they even think that the Constitution's force is lower than that of general laws. The awareness about civil and commercial laws is weaker than that about criminal law; attaching greater importance to criminal law than to civil and commercial laws is a legacy of the past, and its influence can still be seen everywhere today. The awareness about procedural laws is weaker than that about substantive laws; many sectors pay more attention to substantive laws than to procedural laws.

Food safety issues: The foot safety issues include gutter oil, the milk containing contraceptives, the Huiyuan juice made of rotten fruits, the bacteria contained in the ice cubes used by KFC beyond the prescribed limits, illegal use of ractopamine, and tainted steamed buns. The foot safety issues are a heavy liability of China.

Excess production capacity is not included in the balance sheet because it is both an asset and a liability. Why is there excess production capacity under the system of big government? There are mainly three

---

[1] Liu Tongjun. *On the Citizen's Subject Spirit of Obeisance to Law in the Context of a Harmonious Society*, Hebei Province Law Magazine, Vol. 2, 2007.

reasons for it. First, the judgment for market demands is wrong; second, the government lacks effective regulating means, and that an institution is large does not necessarily mean its capability is strong; departmental interests and personal interests replace the national interests, and the abuse of public power contributes to excess production capacity.

It can be seen from the balance sheet that China's rapid development is abnormal, uncoordinated, unbalanced and unsustainable.

II. The Underlying Causes behind China's Economic Slowdown

China's economy is slowing down under the new normal: Its GDP growth went down from 7.7% to 6.9% in 2012–2015, and the target is set to be 6.5–7% in the 2016 Report on the Work of the Government. This economic slowdown is not accidental but has underlying causes.

First, global geopolitical uncertainties have significant impacts on economic growth. In recent years, the deepening of the global economic integration has had increasing externalities, global geopolitical uncertainties have been heightened, the gaps between the South and the North have been widened, developing countries have been further disadvantaged in global division of labor, the global economic imbalances and discoordination have become very serious, and there have been more competition than cooperation in global markets. As a result, the global situations have become more complex. These changes are set to impact the global economy including that of China. China's economic slowdown is somewhat attributable to the complex global situations. Under these circumstances, the world needs new ideas, theories and technologies: politically, global governance structure needs adjusting; economically, global economic order needs reshaping; and in facilities, global infrastructure needs improving. However, today's problems are that there are too many ideas, but none can systematically guide global governance; and there are too many theories but none is mature enough to guide global order; and there are many technical innovations but the third industrial revolution that can boost changes in production, exchange and consumption has not yet arrived. Therefore, the ongoing slowdown may not have bottomed out.

Second, the descendent phase in a long international economic cycle has significant impacts on economic growth. Unlike a business cycle, or short waves or a short cycle, a long economic cycle lasts 30–50 years on average, reflecting long-term cyclical fluctuations between booms and busts. In history, the global economy went through four long cycles of regular expansions and contractions, namely, 1790–1847, 1847–1893, 1893–1945 and 1945–1989, and is now in the fifth long wave. After a long time of upward movements in the late 20th century, the world economy has turned into the descendent phase of the new long waves since the September 11 attacks in the US, and particularly the 2007 global financial crisis. The economic geographical spaces have been narrowed, and key strategic resources that are the material basis of reproduction are running out, and despite many highlights in the new technical revolution, there is still a long way to go before the realization of large-scale industrialization and the emergence of new long waves of expansion. Since the implementation of the reform and opening-up policy, China's economy has been increasingly in sync with the world economy, and China has become a buyer's market struggling with overcapacity after the shortage economy. As a result, since the world economy is now at an interval of relatively slow growth, China's economic slowdown is inevitable.

Third, evolution of the three major industries has significant impacts on economic growth. International experiences show that, of agriculture, industry and services, changes of industry in its contributions to GDP will greatly impact economic growth. For example, our analysis of The Organization for Economic Co-operation and development (OECD) member countries finds that in 1850–2014, the three industries in these countries were at an inflection point in 1950–1960, with the share of industry dropping and that of the services industry rising, which led to fast economic slowdowns. Over the past 50 years, the share of agriculture and industry in OECD member countries has stayed around 30% while that of the services industry, 70%, which is quite different, and as these industries' contributions to GDP growth change, their GDP growth will slow down.[2] Will China hit such an inflection point? Yes, it

---

[2]Fifteen OECD member countries are selected here, including Austria, Belgium, Czech, Denmark, Finland, France, Germany, Hungary, Italy, South Korea, the Netherlands, Norway, Slovenia, Sweden and the US.

will, according to our preliminary findings. Investments in industry and relevant infrastructure have always been the driver of China's economic growth. China's accumulated investments amounted to RMB 5 trillion from November 2014 to the end of 2015, involving hundreds of projects in 12 key programs including urban rail transit, modern logistics, new-type industry, core manufacturing, health and old-age care, environmental protection, clean energy, transportation and water conservancy. With these investments, China sustained a GDP growth rate of 6.9% in 2015, but its economy was under increasing but not decreasing downward pressure. The environment and purposes of these investments were more complex than before, which, however, could not be regarded as the root causes of the poor outcome. Why could not the RMB 5 trillion pull China's economy? The underlying cause is that like what happened in developed countries, as China's economy reaches a certain level, infrastructure construction will stay stable and the role of investments in driving the economy will hit an inflection point. Although investment remains an important driver of economic growth, yet the marginal utility is diminishing. So is investment's push to GDP. Under this circumstance, overreliance on investments may not produce expected results.

Fourth, the laws of economic development have significant impacts on economic growth. Economic activities are subject to the law of diminishing marginal return and the law of economic growth rate regression. The first law states that in an economy with resources as inputs, its total outputs may be increasing but the marginal returns of per-unit resource input is diminishing. The second law states that after high growth in the early period, an economy's future growth tends to slow down. For example, the dividends of reform have a great impact in the initial period, but are set to decrease over time. Another example is that the useful space of a technology is set to narrow as time goes by. Economic growth is bound to slow down under the combined effect of these two laws. After various factors of production are invested to drive high economic growth in China, the marginal utility of resources will decline, and the economic growth rate will naturally drop, leading to slower economic growth. In addition, China's economic growth is impacted by diminishing growth of factors of inputs. For example, as the working-age population declines as a percentage of the total population, labor inputs in economic activities are bound to decline. Under the combined effect of these laws, China's economy has entered into the contraction period.

The new economic normal is a natural outcome of the laws of economic development. As the economy evolves toward the one featuring more advanced form, more complicated division of labor and more optimized structure, economic slowdown will definitely put heavier pressure on industries struggling with overcapacity, and have some impact on the economic and social structures, but the overall economy will be healthier and more favorable for sustainable and stable growth.

III. Features and Nature of the New Normal

1. *Features of the New Normal*

The new normal boasts some novel features as follows: First, novel thinking, such as Internet thinking, under which collaboration and sharing are stressed, the buyer's market and the seller's market are combined and all pillars of the social and economic system are interdependent, mutually supportive and inseparable. Second, novel speed, with economic growth slowing down into a mid and high rate. Third, novel formats, such as experienced economy, Internet, modern logistics, mobile services, big data analytics, personalized production and consumption, which will have profound impacts on our approaches to production and life. Fourth, novel structure, with the tertiary industry overtaking the secondary industry as the dominating industry to drive industry adjustments and recombination of the factors of inputs, and push services and consumption to be the new comparative advantages. Fifth, novel drivers, with economic growth being innovation-driven, rather than factor-driven or investment-driven. Sixth, novel system, with new lean and necessary regulations replacing a large number of outdated management regulations. Seventh, novel opportunity, such as rising share of the services industry, additional technology innovation forces, and eased pressure on resources and the environment, which are favorable for higher employment, better economic quality and higher economic efficiency. Eighth, novel risks, with heightened uncertainties and instability such as increased international geopolitical risk and more intense DOMESTIC SOCIAL CONFLICT, which are interconnected and easy to trigger chain reactions.

Under the new normal, China's economy is slowing down from a high growth rate to a middle and high growth rate, and driven by new growth points instead of traditional growth points, it is developing from a growth economy into a mature economy in an intensive pattern that focuses on quality and efficiency rather than the former extensive pattern that stresses size and speed, with its economic structure changed from focusing on production increase and capacity expansion to equally stressing stock adjustment and increment optimization.

China's development from a growth economy into a mature economy will be a critical sign of its stepping toward a modern society. In the first 30 years since China's founding, the role of planning was abused and People's Communes inhibited its growth. So has been the role of the market over the past 30 years since the reform and opening-up policy was implemented, and elementary education has been a slave to the market. In the coming 30 years, China's economy would become mature, and China would become bigger and stronger. These 30 years will be the key to the transformation from the primary industry to the secondary and tertiary industries, and the transformation of the secondary industry from low and mid end to high end, and of the tertiary industry from low end to mid and high end.

The transition of economic model will be very challenging as it involves industrial, corporate, and technical restructuring as well as reallocation of resources, and its success will depend on victories in several critical battles.

China's economy will slow down step by step alongside rebounds. Its real GDP may fluctuate around the natural rate of growth, higher in some years and lower in other years, and more sharply in some years due to special events. It is preliminarily estimated that its GDP growth rate will be 7.4–6.4% in 2015–2020, 6.7–5.4% in 2020–2030, 6.2–4.5% in 2030–2040 and 5.5–3.5% in 2040–2050.

2. *Nature of the New Normal*

The new normal means that China's economy is stepping into an unprecedented stage, which has novel features, concepts, formats and structure as well as various social and economic changes. But some key and basic facts will hardly change. By the end of this century, China will remain in the initial stage of the socialist market economy as a developing country; its fundamental conflict will continue to be the conflict between people's rising material and cultural demand and weak productivity; and its macroeconomy will continue to be tightened. Such tightened economy will have the following features: First, in terms of natural resources, human capital, technical conditions, management level, governance capability and development environment, China's economy always has weaknesses, especially with regard to energy, which severely constrains China's economic development. China's energy security, impacted by the geopolitical, price, technical and energy factors, is at high risk as energy supply features overreliance and slower growth than that of demand. Therefore, despite partial oversupply in a certain period of time, aggregate demand will exceed aggregate supply. Second, China's economic operation mechanism comprises the interests and differences of various economic players, with common interests as the basis of normal operation and interest gaps as the fluctuation factors. As a result, China's economy features rigid control over allocation of social and economic resources at the macro level and detention at the micro level, and will be struggling with serious structural imbalances as aggregate demand grows too quickly. Third, tightened economic operation is decided by the laws of motion of the socialist mode of production in the early stage of its market economy. China's market system needs further improvement, featuring an unhealthy mechanism to balance supply and demand, absence of a market credit system, and weak self-motivation and self-constraints of economic players in production and consumption activities, thereby deteriorating the tightened economic operation. In substance, the new normal means tightened economic operation in new situations.

3. *Economic Potential of the New Normal*

Despite various difficulties, China's economy still has a strong potential for growth. Still a developing country, China boasts room for strengthening productivity, optimistic outlooks of technology advances, and possibility of ramp-up of capital and labor force allocation efficiency, which means that it has ample room for enhancing the efficiency in exploiting and utilizing resources and total factor productivity is favorable for its growth. China is a vast country with much leeway to grow its economy. In addition to new-type urbanization, the Internet, new-type manufacturing and modern countryside, China has new growth points such as consumer economy, especially the new model of consumer economy that has strong implications for economic restructuring and sustainability.

China's new-type consumer markets boast promising outlooks. While traditional consumption such as family diet, general education and transportation continues to support economic development, the new-type socialized and Internet-based consumer economy that reflects modern lifestyle is developing rapidly, which is shown in six aspects: first, online shopping, which generated more than RMB 90 billion on the Single's Day in 2015; second, catering, which features rapid innovation and a large number of customers; third, logistics, such as SF Express, whose network has covered various aspects of society; fourth, cultural consumption, with going to matches and concerts becoming the fashion; fifth, travel, since China has been the biggest spender on international travel; sixth, experience, with rural life experience and handicrafts making experience becoming the fashion. These ways of consumption are at the early stage but have strong market potential. They are very likely to become a new growth point if further guidance is made to foster a healthy consumption culture, accelerate the formulation of rational consumption policies, develop healthy consumption products, build and improve a fiscal and taxation system to promote the development of a consumer economy, and build and refine a new market system for the consumer economy. Currently, the final consumption expenditure's contribution to China's GDP is around 50%, indicating strong potential and large development room compared with 70% in the US. It is very possible that China will achieve sustainable development if it continues with a proper growth rate and pattern.

The growth rate is important, but not so as compared with the quality of growth. GDP is a key indicator as it can more or less play a guiding role in the course of economic development, but could not reflect the whole picture, let alone the substance of the economy. If the fundamentals are normal and economic development remains sound, it is not a big problem that GDP growth slows down more sharply in a certain period. If there is no systematic financial risk, with most financial assets in good quality, it is not a big problem either that non-performing asset (NPA) ratio remains at 3–5% in 3–5 years. China is a large economy that needs to develop both new and traditional economies. Removing overcapacity cannot simply be de-capacity or capacity relocation, let alone using a single standard, pathway or model to guide the country; the supply-side reform requires a balanced approach to boosting the coordinated operation of supply and demand, motivating social and local investments and stressing effective supply's response and guidance to demand while developing effective demand like new-type consumer markets by enhancing the products, quality and cost-effectiveness through tax policy, expenditure saving and innovation to boost effective investments.

IV. Economic Roadmap under the New Normal

In the 21st century, China's economy is faced with three tasks: first, restructuring from agriculture to industry to achieve industrialization and urbanization; second, institutional transformation from planned economy to market economy, to build Chinese socialist market economy that is relatively developed; third, governance transformation from the traditional management system to the rule of law system, to build a modern country with rule of law. To accomplish these three tasks under the new normal, a roadmap toward market-oriented, balanced, innovative and sustainable economic development needs to be developed and improved.

Market-oriented development means stressing the market orientation and the decisive roles of the market, which might be a moderate adjustment to the government-led market economy. During interactions between the government and the market, the market, as the invisible hand, needs to take more responsibilities for economic activities and play greater roles, while the government, as the visible hand, is responsible for public management, such as state regulation, market regulation, social management and public services.

Balanced development means taking a holistic approach to strategic planning for comprehensive equilibrium. First, aggregate balance. Understanding demand and realities to deliver a good performance in finance, credit, materials, foreign exchange and strike a balance between them. Second, structural balance, which involves urban–rural structure, regional structure, market structure and industrial structure, in a bid to avoid structural imbalances while ensuring aggregate growth. Third, social balance, by properly handling some important relations, such as the relations between tradition and modernity, political system reform and economic system reform, government and market, supply and demand, equality and efficiency, economic development and ecological protection, urbanization and rural modernity, land title and use right, network information and industrialization, innovation and monitoring, domestic markets and international markets, collaboration and competition. Fourth, balance of mindset. The laws of equilibrium should be followed and applied to deepen conceptual identity, psychological relevance and harmony in mind, unify social will, coordinate social activities and pool wisdom to rally all sectors and promote social development.

Innovative development means transforming from being factor input-driven strategy to innovation-driven strategy, including theory innovation, system innovation, technology innovation and product innovation.

Intensive development means transforming from extensive management model to intensive management model by exploring the model of lean development, optimizing the behaviors of economic players, and saving resources to enhance the benefits of quality and seek development with Chinese characteristics.

Sustainable development requires further understanding of China's blueprint and responsibilities in terms of strategic goals. Sustainable development is the fundamental requirement, nature and motivation of human development. Our development today must be based on history and pave the way for the future. To this end, we must develop a roadmap for sustainable development using existing resources and potential capabilities. It is not advisable to exhaust resources to seek growth, not sustainable to seek extensive development that stresses size and speed, and not equitable to allocate resources in a balanced and uncoordinated way. Churning out is a waste of resources, which is a crime. We must make sustainable development the starting point and goal of our economic work.

Overall, quality will override speed, inheritance will surpass productivity, and sustainability will override efficiency in the coming 30 years. Comprehensive equilibrium should be the top priority of our economic work and the guiding principle of our economic philosophies and behaviors. So long as we follow the economic roadmap of "seeking market-oriented, balanced, innovative, intensive and sustainable development", we are likely to capture the strategic opportunities in the coming 30 years to push China's modernization into a new stage.

Case Study: How Chen Yun Applied the Balanced Way of Thinking[3]

How do we follow and apply the law of balance in the actual economic work? Chen Yun, who led the national financial and economic work for a long time, is a model of applying the law of balance. The valuable spiritual wealth left by him can help all the members of the CPC learn how to manage state affairs.

In the early days after the foundation of the People's Republic of China (PRC), the country that the CPC took over from the Kuomintang (KMT) was an awful mess. After many years of war, many things were waiting to be done; the prices were soaring and there was panic in the market. Chen Yun led three campaigns: stabilizing the commodity prices, planned purchase and marketing by the state, and readjusting industry and commerce. The three campaigns completely ended the hyperinflation left by the former regime, consolidated the people's regime, supported the movement to resist US aggression and aid Korea and, and allowed the national economy into the track of normal development. From the late 1950s to the early 1960s, Chen Yun put forward a series of economic ideas, policies and measures based on the balanced way of thinking, and did a lot of fruitful work for overcoming the negative consequences brought by the Great Leap Forward and the movement to form rural people's communes. During the period of the Great Cultural Revolution and after the starting of the reform and opening up, Chen Yun thought deeply on some issues such as introducing foreign investment, opening to the outside world, the relations between economic planning and the market, and played an important role in overcoming the severe difficulties encountered in economic and social development, focusing on macroeconomic

---

[3]"Research on Chen Yun's Theory of Comprehensive Balance", *Journal of Dongbei University of Finance and Economics*, 2015 (4), pp. 3–6.

regulation and control, and promoting the reform and opening up and healthy development. By following and applying the law of balance, Chen Yun systematically put forward a series of economic ideas, policies and measures, and made an immortal contribution to the development of new China. Chen Yun's ideological system of comprehensive balance includes not only the overall balance among finance, credit, material and foreign exchange, but also the balance of the industrial structure, regional structure, urban and rural structure, the balance between the current actual situations and long-term goals, as well as the coordination between economic planning and the market and between self-reliance and the use of foreign capital and other means to promote development (see Table 4.2).

Chen Yun's thinking method of comprehensive balance embodies the CPC's ideological line of seeking truth from facts, the development concept of historical materialism and the objective laws of economic and social development. It reveals the laws and requirements of comprehensive balance

Table 4.2.   Chen Yun's Ideological System of Comprehensive Balance

Materialism	Dialectics
Seek truth from facts (never rely on any books, never obey any authorities but only believe in the reality)	Relation between subjectivity and objectivity (exchange, comparison, iteration)
Historical materialism	Relation between existence and development
	Relation between the masses and the cadres
	Relation between the central government and the local government
	Relation between economic base and superstructure
Laws of economics	Relation between reform and development
	Relation between the government and the market
	Relation between production and consumption
	Relation between self-reliance and seeking external assistance
	Relation between speed and quality

movements in economic activities, innovatively develops the CPC's economic theory and practice and has historical and practical significance to proper handling of various complex relations in economic and social development.

Mao Zedong once profoundly pointed out, "The significance of the three campaigns launched by the CPC on the economic front is no less than that of the Huaihai Campaign". "An important lesson drawn from the Great Leap Forward is that the law of comprehensive balance is ignored". Mao Zedong also affectionately said, "In a chaotic kingdom, the king yearns for excellent generals, and in a poor family, the husband longs for a virtuous wife". "He missed Chen Yun. During Great Leap Forward period, everyone was impetuous, but Chen Yun suggested lowering the economic indicators many times; the lessons drawn from setting high targets and large quotas and exaggeration of production figures should never be forgotten and these activities should never happen again". Mao also pointed out, "Sometimes, the truth is grasped by a single person". "I don't quite know industry and commerce, but some people know them better than me, especially, Comrade Chen Yun; he knows them better than anyone else; his method is to investigate and research, and he won't speak unless he ascertains the facts through investigation".[4,5]

Deng Xiaoping also spoke highly of Chen Yun's economic thoughts. In December 1980, Chen Yun gave a report entitled "The Economic Situations and Lessons Learned" at the Working Conference of the CPC Central Committee. Deng Xiaoping then said, "I fully agree with Comrade Chen Yun's report. His report correctly summed up the experience and lessons in the economic work of the past 31 years, and is our long-term guideline for the future development".[6]

Since the founding of new China, China's economic development has gone through two major stages. During the first three decades, the economy was plan oriented, while since the reform and opening, the economy

---

[4] These Mao Zeng Dong's appraisals of Chen Yun were quoted from Mao Zedong Biography, pp. 521, 913, 925, 939 and 953.

[5] The Library of the Research Department at Party Literature Research Centre of the CPC Central Committee. Chen Yun's Life Story, Vols. I and II. Hong Kong: Phoenix Publishing & Media Group, (2007), pp. 887–888.

[6] *Selected Works of Deng Xiaoping*, Vol. II. Beijing: People's Publishing House, 1983, p. 178.

has been market oriented. Though reviewing the past and summing up experience and lessons, we have reasons to believe that in the process of China's development, if we carry on the socialist construction according to Chen Yun's thinking methods, and achieve a balance between the reform and the development and an organic interaction between the government and the market, our country will develop faster and better. Many of the defects of the economic system Chen Yun pointed out from the 1950s to the 1980s still exist, indicating that his analysis still cuts into the present-day evils; many of the reform measures he proposed have not yet been fulfilled, indicating that his proposals are still applicable to the reform and development of the present day.[7] Chen Yun was a great man who was adept in both revolution and economic construction. The thinking method of comprehensive balance Chen Yun established in his long-term practice is a magic instrument to overcome the financial and economic difficulties, and a valuable asset of China and the CPC.

Compared to the period of 1950–1980, the first two decades of the 21st century have witnessed great changes of China's economic scale and situation, but the law of comprehensive balance never changes; under the trend of global economic integration, it is increasingly difficult to distinguish the Oriental and Western stuff, but the universal value and the national interest are clear, and the law of development is permanent. The law and truth are always worth learning and always play their roles.

At present, China's economic development has entered a stage of new normal, and it is undergoing important changes in trend: a shift from high speed to medium and high speed in economic growth, a shift from medium and low end to medium and high end, a shift from traditional growth points to new growth points in development momentum and a shift from extensive growth to intensive growth in development model. Fundamentally, these shifts are the movements of comprehensive balance, reflecting the functions and requirements of the balance law, embodying the basic spirit of balance, coordination and sustainable development, and indicating that the Chinese economy is evolving into a new stage with a more advanced form, a more complex division of labor, a more reasonable

---

[7]Wang Mengkui, "Memories of Selecting and Editing Chen Yun's Works", *China Development Observation*, June 20, 2012.

structure and a more mature feature. Meanwhile, we should be soberly aware that this evolution is a long and arduous process, and since it is being carried out under the unbalanced and complex domestic and international situations, there are many variables and uncertainties as well as some intertwined and superimposed problems; China's economic operation is still "tight", and for quite a long time to come, China will still be a developing country; we should be quite clear about this positioning. With the deepening of reform, on the one hand, the deep-seated problems become prominent, and the unbalanced regional development, environmental destruction and the gap between the rich and the poor are getting severer; on the other hand, there will be some new problems such as systematic risks. Under the new circumstances, it is urgent that we need to seriously think about the following issues: how to deal with the relationships between the urban and rural areas, between the government and the market, between quality and speed, and between innovation and supervision; how to achieve overall balance and structural balance; how to reform production relations and superstructure according to the changes of the level of productive forces and economic base; how to prevent systematic risks in the process of comprehensively deepening reform; how to place equal emphasis on material and ethical progress. There are still many unexplored areas in the socialist market economy. To adapt to the new normal, cope with new challenges and grasp new opportunities, we need to conduct an in-depth study on the law of balance and learn Chen Yun's think method of comprehensive balance.

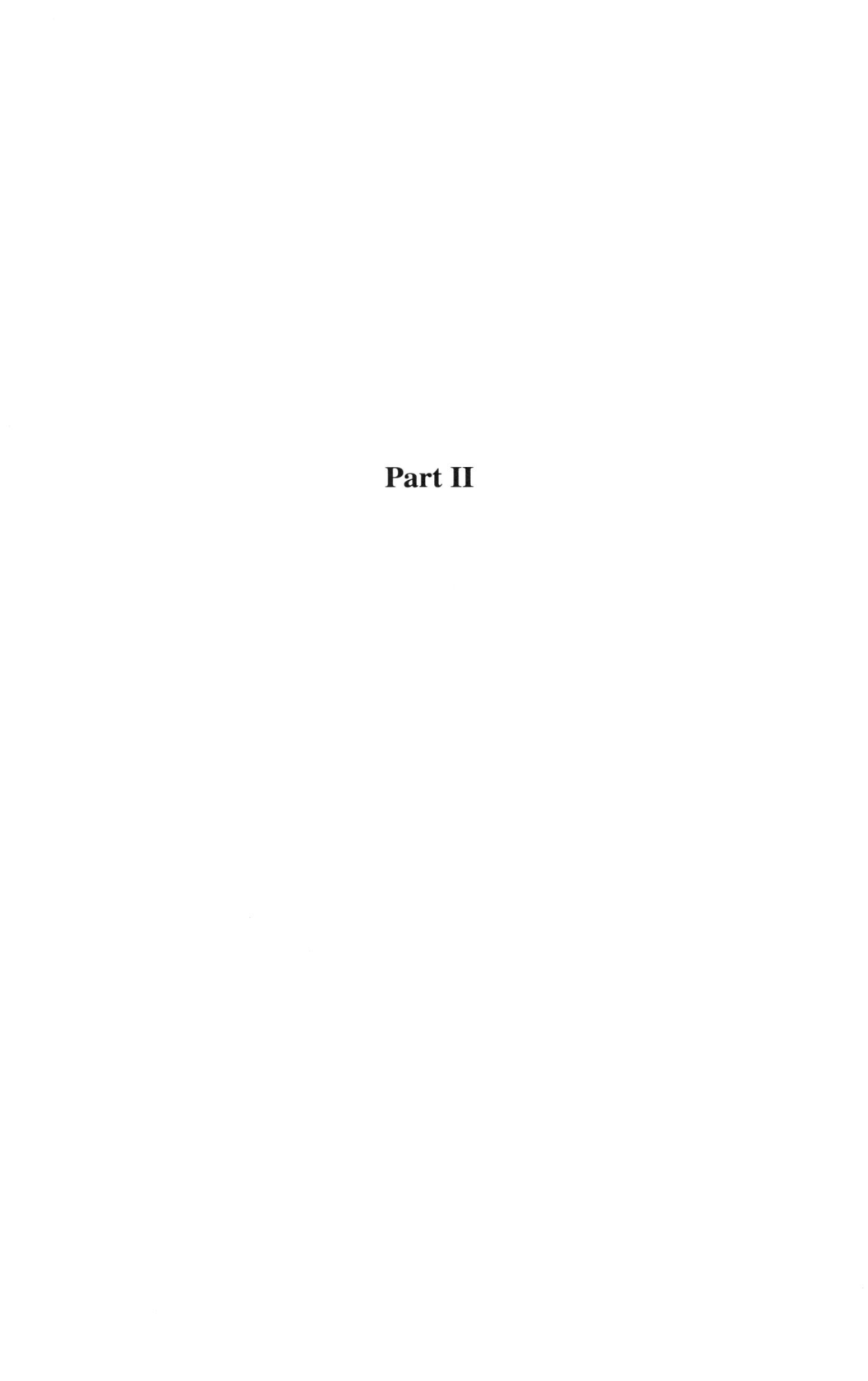

Part II

The Balance Market

The in-depth globalization of human society, under the law of equilibrium, is basically the process of building the market credit system. The "interaction between man and nature" and the "interaction between man and society" are actually human activities centering the equilibrium of market credit.

Market has the right to say. International competition is essentially wars for market places. Market economy is credit economy, and the development of market credit system is the inherent motivator that interlinks traditional and new economic drivers.

In market economy, an individual means a brand, and a brand makes a company. In the cycle of "production–consumption–production", credit is the materialized consciousness of humans and the way they do things, the core of all trades, a special asset and source of value, the strong leverage of economic growth, and is extremely valuable for individuals, businesses, nations and the society. Credit can be established and the establishment of credit can be impelled. The developing countries may foster the late-starter's strengths through the development of market credit system.

The generation and circulation of credit capital, as well as its various forms in its movement, are not fixed but variable, interactive, heterogeneous and proactive. Such attributes have large impacts on the economy. For example, the joint effect of multiple factors and variables easily causes economic bifurcation, turnaround, mutation and fluctuation. Taking its generation and circulation as a whole, credit capital is like a perpetual-motion machine that keeps stirring and amplifying a vortex that grabs all factors into the motion and creates an ever-changing market system.

In an era of Internet information-based economy and of economic transformation of China, the market sees rapidly changing conditions, tougher and more complex credit conflict, increase of inappropriate practices and uncertainties, severer disequilibrium and larger risks which call for more efforts to be made in the development of market credit, market credit regulation system and financial infrastructures.

The Internet and finance can be naturally blended. Internet finance is creating distinctive financial ecosystem but in nature is the process of credit activities and the course of agency finance. Internet finance is known of new characteristics in terms of information risk, operating risk, moral risk and process risk which call for enhanced mechanisms for financial credit evaluation, coordination and risk warning.

Along with technological, economic and social development, the core essence of credit is being expanded to include not only currency and financial assets but also intangible values such as capacity to make innovations, brand quality, market standards and formulation of rules. The intangible values of capital are the market preconditions for deepening economic globalization and readjustment to international governance structure, and are the social preconditions for innovative ideas, products, management and development. Credit, compared with currency and financial asset, is a higher-level medium of social exchanges and a sort of capital with universal significance, and reflects the equilibrium in wider social and economic collaboration. The stronger role of credit capital is likely expanding income gaps. Therefore, in the process of deepening reform, we should pay attention to the development of fundamental supporting system, and in the process of RMB internationalization, we should make good use of various values and probe into new medium of international exchanges. As financial technology becomes smarter, it is more necessary to put in place governance and other supporting mechanisms to ensure financial security.

The internationalization of RMB has evolved over the years, and presently engages in various responsibilities ranging from trade facilitation, regionalization to joining special drawing rights (SDR) and functioning as an international reserve currency aimed for global use. However, it still has a long way to go to truly becoming an important medium of exchange for international trading, investment, financial and economic activities, which in need of the time require international comprehensive strength, support as well as extensive and in-depth credit building.

The difficulty with China's private enterprise in financing is a long-standing problem, the fundamental reasons are firstly, the private enterprises have failed to realize their strategic positioning; lack of confidence and inability to understand the depth of the layout and operation have added to the woes. Secondly, it is not sufficient to only promote and help private enterprises in legal construction, credit construction, cash flow construction and project construction. The private enterprises need to venture out and thus as market players have been in a weak position. Thirdly, the existing economic management system is not flexible enough. Regulatory discipline and the law does not go hand in hand, accountability mechanism and fault-tolerant mechanism coordination is not in place thereby overlooking the uniformity among responsibility, power and play. The private firms thus picking up the sesame seeds while losing sight of a watermelon. Fourth, the new capital-injection mechanism is not perfect; lacks motivation, spirit, channels; cannot meet the actual needs of private enterprises. Fifth, the bank structure and loan conditions are not suitable for SMEs. There are not enough small- and medium-sized community banks, and the current wholesale loans and translated loans are actually for increasing the cost of corporate finance and difficulty of management. In this regard, it is necessary to firmly establish two-wheel development strategy, find out the resources, multichannels, multiforms, diversified financing service intermediaries, give full play to the role of policy-oriented funds and development of financial guidance; it is necessary to further clarify the accountability rules and define standards; to further improve the incentive measures to revise the industrial norms, which is struggling to move forward.

In the following three decades, China's financial system reform should give priority to strengthening its credit system and taking the laws, regulations, standards and deeds as the basis and dissemination of integrity as the primary purpose, establish the credit record and credit infrastructure network that cover all social members, create the incentives and punishment mechanism for credibility, and set up the human oriented and technically applicable financial regulation ideology.

Chapter 5

The Law of Equilibrium in Market Credit

What is credit equilibrium? What is the law of credit activities? The answers to the questions are the keys to market economy. This section is a preliminary study in this regard.

Credit is the basis for the interaction between man and the society. Credit is the internal driver for development. Credit can be built and credit building can be promoted. China lags behind in terms of market credit system which is critical to its mid- and long-term sustainable development. To accelerate the development of market credit system, it is necessary to gain in-depth understanding of the nature and characteristics of credit, and to reveal and follow the basic law of credit activities.

I. The Nature of Credit

According to Karl Marx, "Credit, in its most simple expression, is the confidence which, well, or ill-founded, leads a person to entrust another with a certain amount of capital, in money, or in goods computed at a value in money agreed upon, and in each case payable at the expiration of a fixed term".[1] "To move means death to trees, opportunity to man and increase to money". Money moves and credit is created. Credit activity,

---

[1] Karl Marx, *Capital* (Chinese version) (Vol. 3), 1st edition. Beijing: People's Press, 1975, pp. 410–442.

as various production activities, creates values. However, credit, in the process of creating values, features the following:

First, credit, as a spirit, concentrates a positive conscious and value of human, and is the core of all trades. "People come and go in pursue of profits".[2] Pursuing wealth is a man's inherence, and building the credit, his duty. Man's credit is built from enlightenment to maturity, i.e. the sublimation and accumulation of credit awareness. A man without credit awareness does not mean he has no credit all the time. Once he leaves the unawareness behind and gets aware of his need for and the power of credit, he will change his way of behaving and such change diversifies market trade and complexes social activities. The critical problems of the ever changing world are more or less related to credit which is the core of social relations, both vertical and transverse, between governments, enterprises or individuals.

Second, credit, as wealth, is a special asset of individuals, enterprises, regions and even states. Credit does not decrease upon use. On the contrary, it increases. Credit as a special asset has value in use and values, natural and social status. It is the wealth created by labors, enterprises and governments. Such wealth, in the form of currency or commodity and through primary distribution, partially becomes family income of employees and entrepreneurs (wage, profit, interest or rent), or enterprises' income as retained earnings, or governmental income through taxation. Such incomes must be consumed to maintain production and reproduction, and the rest is converted into credit accumulation through savings. There must be an exchange relationship or market between consumption and savings, or between the users and suppliers of credit assets. In this market, financial brokers play a special role in normal credit activities. They observe credit rules and harness their quality services to make sound allocation of credit as a financial asset, and provide essential financial guarantee for social and economic development. Otherwise, incompliant operation of financial brokers would possibly cause financial disorder. For example, the misconducts of Wall Street financial institutions in 2008, who over relaxed consumptive housing credit, triggered global financial turbulence.

---

[2] Sima Qian (145 BC–), *Historical Records — Biography of Merchants*.

Third, credit, as a conduct, is a powerful leverage that promotes social and economic development. Credit is a priceless resource of individuals, enterprises, states and the society, and a powerful leverage to solve problems. Credit is essential for individuals and businesses. Modern economy by nature is credit economy, while various trades are made centering credit. Modern economy features social credit relations represented by debt and creditor's right, i.e. social credit. Without social credit relation, social resources cannot reasonably flow or be used. Good social credit means all members of the society have strong credit awareness, established credit instruments and standard credit order. The credit of individuals and enterprises are the foundation of social credit, and business credit and government credit, the guarantee. Modern finance is made up of individual credit, enterprise credit, business credit and government credit. The elementary objective of modern financial activities is to organically combine all credits, build a sound credit market, make reasonable allocation of limited resources, and support social and economic development with credit services.

II. The Characteristics of Credit

Credit activities are characterized by sociality, morality and subjectivity.

Sociality reflects the social need of credit and the far-reaching impact of credit on social development. Credit is closely related to people's practical interests, involves all social factors, and is reflected by public involvement, credit exchange and balance proportion.

Public involvement is the concrete representation of the sociality. Such representation can be positive or negative, because the public participates in credit activities consciously or unconsciously, or even passively. Conscious and active credit activities are positive and build up social credit. Unconscious credit activities are an integral part of social credit which, if appropriately guided and managed, can be converted into favorable forces for social credit building. Passive credit activities, or "be credited", are usually used by unlawful credit activities which bring about destructive effect on society. Therefore, the government should properly guide public involvement, at the same time, normalize credit activities to shape a credit supporting network, accelerate the development of public financial system, ensure the equal right to financing, and mobilize social resources to protect financial security.

Credit exchange is the result of the sociality. Credit exchange is the necessary requirement and result when social development reaches a certain level. This way, level and coverage of credit exchange develops along with social progress. In modern economic activities, credit as an important component of social resources has become the instrumental element of finance and financial trades. Such element flows. That is to say, it works as a special exchange carrier among resources and shows different liquidity in different exchanges. In balance of payments (BOP), for example, such liquidity of credit activities reflects the ability of a country to make up deficits, and reflects necessary foreign exchange reserve. Another example, in the banking system, liquidity reflects the bank's ability to satisfy customers' cash requests, and reflects cash holding and short-term asset of the bank. It is safe to say that liquidity from credit exchanges is the fundamental characteristic of the sociality. Gaining good knowledge of the characteristic helps elevate the efficiency of credit.

Balanced proportion is the basic requirement of the sociality that reflects the proportion between sectors. An appropriate proportion should be maintained between consumption and savings, commodity credit and currency credit. Any imbalance would obstruct credit activities. Ye Ziqi of late Ming and early Qing Dynasty said in his book that, "It is like a pond where water enters from the inlet and exits from the outlet. If the entry equals to the exit, the pond is full of vitality". That is to say, the balance in credit activities is dynamic and should be considered from the perspective of time and spatial conversion. Analysis on the balance proportion helps make appropriate planning for credit building, make reasonable use of resources, carry out financial and economic activities in a proportionate and step-by-step manner, and create effective, healthy and balanced financial ecosystem.

Morality: Credit is the use and embodiment of morality in the process of commodity exchange. Morality is sourced from human relations and exhibits wider social values along with social progresses. Morality from a general perspective is a result and foundation of social development. After the industrial revolution, morality has been widely applied to large-scale exchange of commodities, hence has evolved into the essential precondition and foundation of social development and the social framework and guarantee for commodity exchange, driving the exchange of commodities between human from material-based to currency-based and then to credit-based exchanges. Credit transaction is based on credit morality which requires credit subjects to conform to specific ethic rules and social expectations (including integrity, justice, equality and compliance) in terms of trade activity and methodology. This is the core and basis of financial and other laws, regulations and rules. In this sense, credit integrity has become the soul of market economy. Those who neglect or violate the bottom line of morality would eventually destruct their market reputation, which is hardly reversible. Therefore, morality is another important property of financial activities that reflects the economic attribute and internal value of market subjects, and is directly related to their health growth and sustained competitiveness. Meanwhile, morality is the most important precondition for other social activities, the necessary philosophy of market economy, the basis for harmonized social progress, and the essence of human civilization. The civilized human society is indeed built upon credit relations and will be upgraded as credit relations deepen.

Subjectivity: It refers to appeals for interest and the social subjectivity of trades and dialogues. Social subject by nature means credit subjects or interest subjects. Social subject is the fundamental element of social marketization, which translates all transactions into a common language and through credit converts all economic activities into homogeneous and comparable behaviors. The acts of credit subjects, e.g. activities of individuals, organizations or countries, reflect their distinctive appeals for interests. Credit subjectivity has a conglomerating structure. Credit subjects are symbiotic. Individual credit at various levels needs to form a solid combination to provide socialized credit activities with credit resources. Credit subjects could not stand along. They coexist in certain credit ecosystem. In particular, in the market economy, the interests of economic activities take the form of credit, and credit relations can be seen in all aspects of economic growth. Therefore, credit economy needs specialized credit management institutions (e.g. credit management company, guarantee company, etc.) to offer services, prevent risks and safeguard market order at low cost. Credit economy also needs special financial institutions, such as development financial institutions, to, relying on state credit, implementing state mid- and long-term development strategies in areas where market credit is not yet matured or hardly has any effects.

III. Basic Laws for Credit Activities

We also found out that credit activities observe three basic laws:

First, credit activities are the basic social activities that reflect realistic existences and create new existences. Realistic existences are the foundation of credit activities, while credit activities are not only reflections of but possibly the distillation and extension of realistic activities. Taking real economy and virtual economy for example, the former is visible and the latter, a new existence in contrast with the former, is an independent credit activity which comes from yet operates independently from the former. This can be explained by the many financial products that are getting further and further away from the reality. However, real economy and virtual economy in nature are the different sides of a coin. One side is carved with concrete materials and the other by the sign of its value. Real economy is the basis and virtual economy reflects the specific credit form of the former which, if overstated, would give a distorted effect — economic bubble. China should refuse and control economic bubble but not restrict credit activities from creating new existence, otherwise the economy will lack catalyst and vigor. The reason is that, economic growth calls for deepened finance to foster constant improvement of productivity, the real economy needs to make leap-forward development through the expansion effect of virtual economy which also facilitates restructuring of real economy. Virtual economy as a new existence reflects the independent movement of the value of capital. Such independent movement can be divided into four levels according to the virtualization degree of financial assets. The first level is credit currency. Currency is transformed from gold standard to paper notes, creating preconditions for the birth and development of virtual economy. Modern banks have the function to create credit and issue currency, and credit currency is relatively the independent movement of banks. The second level is that the real economy takes the forms of property right, bills and securities, i.e. monetization of tangible assets in the forms of stock, bond, securities and foundation. Such currencies are traded and come with virtual relation with real economy. The third level is securitization of financial assets. Banks securitize credit assets (housing and mobile mortgage loans, credit card loan, etc.)

after "repacking", or add in securities of higher credit level and make them as collaterals to issue new securities, in order to improve asset liquidity. This approach makes the virtual capital even more virtual. The fourth level is absolute virtual economy independent from real economy. A large number of financial derivatives are produced and used to form highly virtual capital, such as financial option, financial futures, forward agreement and financial swap. Neither buyers nor sellers directly possess the amount of securities or currency specified in the contract, and they do not have to actually buy or sell them in trade settlement. Index futures and option trade, in particular, are absolute virtual capital, or just betting on index rise or fall without any material equivalent. The progress of virtual economy, i.e. capitalization of idle currency, socialization of interest-bearing capital, marketization of marketable securities, globalization of financial markets, integration of international finance, etc. give birth to series of new existences. This is the characteristic and trend of modern finance, and should not be neglected or negated by China. Nevertheless, if virtual economy is too far from real economy, it would be too virtual. We should have a clear understanding about this issue and make innovations to serve real economy.

Second, in the process of creating new existences, credit activities may have both positive and negative effects. We also take virtual economic activities as an example. The independent movement of virtual capital value constitutes another economic form in contrast with real economy. Virtual economy emerges and develops as an instrument to disperse and avoid risks in economic cooperation, standardize trade and investment, promote specialized labor division, and hence reduce trade cost, improve scale benefits. That is to say, virtual economy plays an irreplaceable role in optimizing resource allocation and accelerating asset restructuring. Unfortunately, it seems that virtual economy is out of control. A large amount of actual and expected financial wealth is in rapid virtual movement and inflation without physical materials, which threaten the economic and social stability and to which we have to be highly alert. Judging from the development of virtual economy, credit currency, capitalized credit, capital securitization and priced intangible assets are the access channels of real economy to virtual economy, expectation for future profit is the soul of virtual economy, electronics, network and data are the driving force for rapid development of virtual economy. Investors expect to pass down their "relay batons" to the newcomers and seek huge risk return from speculation, that is, the inherent driver for the rapid development of virtual economy. In the capital market, if investors share the same expectation and have not formed mainstream contrary expectations in the process of investment operation, there would be newcomers to take the "batons". We should recognize the existence and understand the characteristics of virtual economy before making positive use of it. If we fail to realize the possible negative roles of virtual economy or to put it under macro control, we would have to face potential economic bubble and economic crisis. Virtual economy generates huge profit and at the same time risks — economic bubble, which accompanies virtual economy all the time. Economic bubble is a part of the market value of virtual capital that exceeds the corresponding valid value of actual capital, and is mainly caused by overspeculation. Speculators focus on the profit margin through buying and selling. They do not care about, do not understand or are not interested in the status of real economy on which virtual economy relies. That is, why the price of capital easily deviates from its value and causes bubble. Surely, although real economy takes physical form, it also suffers bubbles, including overcapacity and excessive products resulted from low-level repetitive construction, and the part of bank loans, stocks and securities of enterprises that exceeds their actual capital. Economic bubble includes bubbles brought by capital market speculation and shrunken actual capital. The major reasons are problematic economic structure and overspeculation in capital market, which explain the burst of Japanese bubble economy, Mexican and Brazilian financial crisis, East Asian financial turmoil and the 2008 global financial crisis in the past three decades, though they varies a lot. An important message from them is, the preconditions for sound economic operation are the proper relations between real and virtual economies, and between financial innovation and regulation. The relations should be dealt with through effective infrastructures, strategic industries, public projects and financial innovation.

Third, credit can be built, and building of credit can be propelled. The development process of market economy is characterized by the conflict between credit observance and credit violation. Though most traders abide by credit rules, there are always frauds in the market. The European and American countries spent a century and huge cost to build a sound market order. China is transforming from planned economy to socialist market economy, and sees frequent infringements, tax cheats and economic frauds. Credit will definitely overwhelm anti-credit obstruction, and establish its dominant role in market economy through normal operation. According to in-depth analysis, this is because the laws of values, supply and demand, competition and balance forms synergy during the development of market economy and create an objective market balance mechanism for the survival of the fittest. This is because the social and moral appeals of credit activities generate a natural effect that facilitates social balance and credit building. Of course, the "invisible" force is not enough. The superstructure should take necessary measures and create effective mechanisms to accelerate the progress. Credit activities can be developed and enhanced, just as material productivity. The combination of external factors (e.g. governments' advantages in organization and coordination) and basic functions of market is sufficient to expedite building of market credit system. The government plays an irreplaceable role in the developed Western market economy system, which seems to be the result of spontaneous evolution. The following-up developing countries, including China, do not have the time and potential for, nor the financial power to support the expensive natural evolution. They have to proactively develop credit, give play to initiative, seek the possibilities for conversion from immature to mature credit, and boost market credit system in the infrastructural and weak aspects of social economy, in order to keep pace with times.

Along with the development of modern economy, the values and attributes of commodities become differentiated, diversified and complicated. Such a tendency intensifies the credit-related problems in the modern socioeconomic scenario. This is inevitable during the transformation from traditional to modern economy which should be coped with through modernized conversion and interpretation of traditional morality and credit. In the process of economic globalization, integrity is a universal principle applying to all nations and eras. Integrity and loyalty should be observed in both China and rest of the world, in the past or in modern times. The West and China stress different aspects of integrity. China focuses on individuality, namely the inherent quality of individuals. While the West stresses on conducts and process, namely the spirit of contract. Particularly since the industrial revolution, the developed Western countries have established modern market credit system and code of market conducts based on morality, credit, deed and law, implying sound integrity mechanism. In China, however, the integrity mechanism is under "remote regulation", relying on moral restrictions and individual consciousness. Therefore, China needs to review and enrich its integrity. In the contemporary world, China should not only stress moral restriction, but also accelerate the development of integrity management mechanism.

Case Study: How to Assess the Infrastructure of China's Financial Market

Financial infrastructure is the basic condition and guarantee for the healthy development of financial markets, and determines a country's national economic security and national security. Financial infrastructure includes hardware and software; its hardware includes institutions, systems and databases, and its software mainly refers to the relevant systems and mechanisms. The international community has a wealth of experience in this area and continuous improvement measures. In April 2012, the Committee on Payment and Settlement Systems (CPSS) and the International Organization of Securities Commissions (IOSCO) jointly issued the Principles for Financial Market Infrastructures (PFMIS), which mainly involves Payment Systems (PS), Central Securities Depository (CSD), Securities Settlement Systems (SSS), Central Counter Party (CCP) and Transaction Database (TD). The PFMIS consists of 24 principles, including legal basis, governance structure and market data disclosure. These principles are based on international experience and lessons.

China's modern financial markets started late and had poor foundation, but in the process of global economic integration, they have followed up closely based on its strong adaptability and enjoyed rapid development. After more than 30 years of construction, the infrastructure system, operation mechanism and system framework for supporting the development of China's financial markets have been initially formed, and to some extent, have ensured the issuance, registration, trusteeship, transaction, pricing, settlement and risk control of securities. China's financial markets mainly include the exchange market and the interbank market; the exchange market is subdivided into markets for stocks, funds, bonds, futures, bills and gold. In the exchange market, Shanghai and Shenzhen Stock Exchanges were established in 1990; the interbank market, the interbank lending market, Foreign Exchange Center and China Government Securities Depository Trust & Clearing Co. Ltd. were respectively founded in 1984, 1994 and 1996, and the Shanghai–Hong Kong Stock Connect was established in 2014. In line with the hardware, the financial market system also made some progress. For example, the Law of the People's Bank of China, Regulations on the Foreign Exchange System of the PRC and Measures on the Administration of Inter-bank Bond Transactions were respectively promulgated in 1995, 1996 and 2000. In January 2016, China Securities Regulatory Commission (CSRC) started the trial of the "Circuit Breaker" mechanism. In July 2017, the People's Bank of China guides the construction of "network payment clearing platform of non-bank payment agencies". These are all landmark events in the infrastructure construction of China's financial markets, indicating that China has made some achievements in the development of financial infrastructure.

However, China's financial infrastructure construction is still arduous; there is still a lot of work to be done in terms of depth, breadth, efficiency and risk prevention, and compared to the international financial markets, China's financial markets still have some deep-seated problems, such as the imbalance between the bond market and the stock market, the uncoordinated structure, the low proportion of direct financing, the prominent contradiction between separate construction and centralized management, and high time-difference risk. To properly address these deep-seated problems, the infrastructure construction of the financial markets needs to be further strengthened, and this requires that some measures should be taken to strengthen the construction of a comprehensive regulatory system, financial channels, and an assessment and pre-warning system.

Chapter 6

Change in Credit Conditions and Perfection of Policies

What are the preconditions for credit balance? What are the changes in credit conditions in the new era? How should policies be readjusted? Those questions should be addressed in current economic works and in mid and long development. In this section, we adopt balanced way of thinking to analyze the changes in Chinese monetary conditions, financial conditions and credit condition as well as the perfection of policies. In response to the new circumstances and changes in market credit conditions, it is necessary to, from the ideological perspective, gain new understanding of the financial market to reflect our determined ideology, from the strategic perspective, to define and refine the tasks and objectives of financial system reform to reflect our strategic planning, and from the perspective of policy and technical methodology, improve and diversify operating kits to take the initiative and enhance risk control ability.

Modern market economy is based on credit which, in the "production–exchange–consumption" cycle, brings in "oxygen", discharges "waste", exports health and safeguards the balance. However, credit plays its role under certain preconditions. So, what are the preconditions and how will the preconditions change in the new circumstances? We will make analysis on monetary conditions, financial conditions and credit conditions, respectively.

I. Monetary Conditions and Policies[1]

Monetary condition refers to the preconditions that something is deemed as currency. In this chapter, it refers to the preconditions for the conversion of money to capital. An object with the functions of "trade medium, unit of measurement, value storage and means of payment" is money. Money is essential for economic and social development. However, money has to create values before being converted into capital, which is also an important precondition for the development of market economy.

According to *Capital* (Karl Marx), "In themselves, money and commodities are no more capital than are the means of production and of subsistence. They want transforming into capital. But this transformation itself can only take place under certain circumstances that centre in this, viz., that two very different kinds of commodity-possessors must come face to face and into contact; on the one hand, the owners of money, means of production, means of subsistence, who are eager to increase the sum of values they possess, by buying other people's labor-power; on the other hand, free laborers, the sellers of their own labor-power, and therefore the sellers of labor".[2] Marx made it clear that, the most important precondition for the conversion of money to capital is available labor force. Without labor, money is an average commodity or an IOU. Money cannot become a special commodity nor be used and converted into capital unless labors are departed from their ownership of labor and the direct producers converted into employees. Therefore, from the perspective of ownership, social division of labor is the precondition of money (Table 6.1). The division of labor enables labors to own products, and surplus products allow

Table 6.1.   Money Condition

Precondition for the conversion of money into capital	
Available labor force	Labors are departed from their ownership
	Producer converted into employee

---

[1] This section is published in *Journal of Dongbei University of Finance & Economics,* 2016, Vol. 2, pp. 3–12.

[2] Karl Marx, *Capital* (Vol. 1, Part 2) 1st edition. Beijing: The People's Publishing House, 1976, pp. 782–783.

exchanges and loans. Looking back into the history of primitive accumulation, establishing the market of employable labors, or creating the preconditions for capital relationship, was a crucial revolution, particularly in the early stage of capitalism. In modern and contemporary economy, labor market has been well developed and labors are entitled to their rights to work. It is easy to convert money into capital. Nevertheless, primitive accumulation and creation are complicated, through hard working, resource abundance, or even violence, cheating and theft. "As a matter of fact, the methods of primitive accumulation are anything but idyllic".

Along with the development of information economy and knowledge economy, on the one hand, new changes have taken place in modern social environment and social development conditions. For instance, labor is converting to human capital, technical factors impose great impact on labors, and laborers vary largely in terms of technique, efficiency and ability to create wealth. On the other hand, primitive accumulation of wealth is further complicated. For example, labor capital takes the form of stock, i.e. individuals, families and professional brokers take shares. It has become an important phenomenon of market economy. Human capital is available through networks, allowing laborers to look for jobs and create capital relationships throughout the world. Moreover, e-money and online payment boost financial popularization.[3] In this sense, it is impossible for monetary authorities to maintain market order and control the total circulation of social capital merely through traditional monetary administration tools and measures. They have to make innovations in policies and improve monetary conditions. Possible options are, for example, considering the impact of Internet finance on traditional currency and of faster circulation of Internet finance on the multiplying function of base money, in addition to traditional concerns. We will address this issue in the following sections.

II. Financial Conditions and Policies

Financial condition refers to the preconditions for the entitlement to valuable things. In this chapter, it refers to the preconditions for the

---

[3] Robert J. Shiller. *The New Financial Order: Risk in the 21st Century*. Beijing: CITIS Press, 2014, p. 237.

conversion of financial assets into financial capital. It reflects not only the claim for financial assets, but also the value-creating effect of financial assets. This is an important property of modern financial market development.

Financial asset is a financial claim. As long as you are entitled to claim for something valuable, you own the financial assets, which means you can claim for assignment.[4] At present, financial assets include five categories: currencies, stocks, bonds, foreign exchange and securities loan. Of them, currency brings bank deposit interests, stock brings dividends, foreign exchange has exchange rates, bond and securities loans also have interests. All the assets can be transformed into capital when conditions are matured. The matured conditions are: first, potential of appreciation of the asset; second, entitlement to claim for the assets, i.e. for the total value of the appreciated financial asset; and third, liquidity. Liquidity reflects the encashment ability of various assets as required, which determines how they are accepted in the market. This is a special precondition for transforming financial assets to financial capital. We have to consider its liquidity when analyzing a financial asset, in addition to its appreciation potential and total value. The market concerns the liquidity of financial assets. The investors would accept lower interest rate of high-liquidity investment options, given the rest of the conditions remain the same, and *vice versa*. For poor-liquidity bonds, investors expect higher return to compensate the loss in liquidity (see Table 6.2).

In the current financial market, the five categories differ in encashment ability due to different difficulties in market transaction. If a sector

Table 6.2.   Finance Condition

Precondition for the conversion of financial asset into financial capital	
Liquidity	Appropriate operation of supplier and user
	Appropriate action of financial intermediaries
	Appropriate correlation between virtual and real economy

[4] R. Glenn Hubbard and Anthony P. O'Brien. *Money, Banking and the Financial System.* Translated by Liu Qi, Lan Botao and Wang Yin, 2nd edition. Beijing: Tsinghua University Press, 2014, p. 4.

has all its financial assets in the form of money, it boasts high asset liquidity. If all the assets of a sector are hard to convert into case (e.g. long-term bonds), it suffers poor asset liquidity. In international payment, liquidity represents the ability of a country or a region to cover external payment balance and its necessary foreign exchange reserve. In the commercial banking system, liquidity explains the ability of the bank to satisfy depositors' claim for cash, and its cash and short-term assets in possession. In the securities market, liquidity reflects the market values of marketable products. The more stocks issued, the more market makers; the wider the stock holding, the higher the stock liquidity in the market. In the monetary market, liquidity shows the total volume of money and short-term financial assets in economic activities. Excessively loose monetary policy makes it hard to distinguish base money from other current assets, and easy to lead to excessive liquidity. Therefore, in terms of financial administration, various countries/regions should, depending on their financial market and financial asset, develop their own macro policies, the general goals of which are economic growth, price stability, higher employment rate and trade balance.[5]

Along with the development of technology, non-liquid assets have liquidity, and all forms of assets can be used as money in settlement or as capital in investment. As the financial market develops, more and more financial derivatives come into being, financial assets have more complex structures, financial liquidity are harder to control, and more and more assets have capital properties. Therefore, it is necessary to make in-depth study on and appropriate improvements of financial conditions.

We learn from history that, the smooth circulation of financial assets in financial system is decided by three preconditions. First and fundamentally, it calls for appropriate operation of the suppliers and users of financial asset. That is to say, the financial asset suppliers should provide adequate financial assets. It depends on various factors: whether enterprises could improve economic efficiency and convert more profits into retained earnings, whether governments at all levels could control the

---

[5] Lixing Zou. *Financial Empire: American Development and Enlightenment,* 1st edition. Changsha: Hunan University Press, 2009, p. 129.

situation and maintain budgetary balance and appropriate surplus, whether individuals make proper arrangement of income and expense and have savings after meeting their demands. If they do so, it means adequate supply of financial assets and benefits low-cost, high-quality fund supply. Meanwhile, the users of financial assets should make good use of the funds and maximize investment returns. It also depends on various factors: whether enterprises, for the purpose of expanding production, could choose appropriate projects that benefit industrial upgrading and new products for marketing, whether the governments could spend more money on public infrastructures that benefit the knowledge and innovation system, and whether individuals could spend money on improving their physical, mental, living and cultural qualities. If they do so, the financial assets are essential and in effective use and the expansion of reproduction is in a sound cycle.

Second, financial intermediaries should abide by financial regulations. They are responsible for providing good services to help reasonable allocation of financial assets and maintaining financial order. If they are reluctant to loan fund users, it would possibly cause financial stringency and economic downturn. However, if they keep lending when the economy is overheated, the economic situation would worsen and lead to calamity.

Third, it calls for appropriate correlation between virtual and real economies, which are closely interlaced but vary in attributes. Virtual economy pursues high profits at the cost of high risk, with real economy as the carrier. Real economy, by means of industrial assets, provides visible and physical products and services. If virtual economy deviates too far away from the real economy, it would likely cause financial crisis. In late 2007 and 2008, the US suffered subordinated bond catastrophe, one of the major reasons for the ridiculously overheated financial derivatives. The history tells that, poor control over financial innovations and financial conditions are the prelude of financial crisis. To avoid the negative possibilities, it is necessary to improve financial conditions through, for example, first, raising the access bar of mortgage loans, second, keeping the price of financial derivatives below the base value, and third, requiring financial enterprises to make available credit risk reserves (no less than 5% in the US).

China's total financial assets are worth three times the GDP, or over RMB 120 trillion, and is continuously increasing. How to make good use of the money and create more wealth, and how to safeguard financial assets against the complex and ever-changing circumstance and to create appropriate values? The financial sector has to work out proper answers to the questions through not only creating conditions to encourage financial innovation, improving the financial market, diversifying financial products, providing more investment products, but also making good control over the conditions, standardizing the market, enhancing the regulatory system and mastering the laws and characteristics of financial assets. For example, it is necessary to, according to the new trends of liquidity, monitor "normal, pre-warning and abnormal" market situations, or integrate the Alipay accounts of individual companies to create a consolidated public safe payment system based on a careful study on the development trends of e-money and e-cash. This will also be elaborated in the following section of the chapter.

III. Credit Conditions and Policies

Credit condition refers to the basis of mutual trust in trades. In this chapter, it refers to the precondition for the conversion of credit into capital. It reflects not only the significance of integrity in credit trade, but also the special role of credit in creating new values. It is a fundamental condition for the modern economic and social development. As capital, credit causes great change in country's view of wealth, as wealth, it causes great change in country's economic strength, and as a strength, causes great change in country's comprehensive competitiveness.

Credit is a special and rare resource. It is special because, first, utilization brings more credit, on the contrary to average rare resources that get fewer; second, it is closely related to personality, political wisdom, friendship and affection; third, it is the basis of sound social development — without credit, social development would be expensive and plans hard to implement due to high cost (or unpleasantness); fourth, credit exists freely and forms network, without compulsory obligations. Credit is intangible, unlike labor force or commodities that can be measured by time or quantity. It is communication based on mutual trust. You deposit your salary in a bank because you trust it. The banks earn credit and you get interest, which is a special award for your trust in the bank.

The coverage of credit is far wider than financial assets. It includes, acts as and replaces money and financial assets. Credit creates values in wider and deeper levels. Credit value takes both tangible form such as means of production and living and virtual form such as brand, trademark and reputation. Credit value, in addition to value maintenance, trade, deposit and payment functions, also increases values such as real estate, copyright and patent. In the US, intangible assets such as marketing skill and reputation are increasingly important for giant banks' acquisition acts. About 70% payments are made in the form of credit.[6]

Therefore, in the conversion of credit into capital, "reliability and standardization" are special conditions in addition to monetary and financial conditions.

The reliability reflects the possibility of credit to smoothly fulfill its functions in normal conditions. It has three essential factors: quality, safety and compensation. Quality is required in the process of design and generation of credit. Safety is the lifespan and the faultlessness when credit is used. Compensation is the guarantee for credit. Therefore, the reliability based on quality, safety and compensation is essential for the generation and development of credit and for the conversion of credit into capital. For instance, land is a potential credit but not necessarily the capital. "Land can only be converted into capital through sales or mortgage under the condition of well-established property protection system and contract protection system".[7] Furthermore, credit, either in the form of product or regulation, agreement or otherwise, should be reliable and trustworthy. Credit reliability reflects how much the virtual value of credit is converted into physical value, and how the society recognizes the value of credit. Credit reliability can be surveyed, analyzed and rated by such important indicators as credit development level. The three top appraisers: Standard and Poor's, Moody, and Fitch, from time to time issue credit rating reports that determine the wealth cost of countries, regions and companies.

---

[6] *Ibid.*, pp. 219–220.
[7] Chen Zhi Wu. *Logic of Finance*, 1st edition. Beijing: China International Culture Press Limited, 2009, p. 21.

The standardization reflects the stability of credit value in the market, and reflects the standard of credit products, credit instruments and credit forms in normal context. Despite of diverse credit products, credit instruments and credit forms, credit should be normalized and standardized in each process ranging from design and operation. Standardization helps in credit training and dissemination, enhance credit stability and fair value, establishes professional managerial system, benefits agency business, and expands specialized credit businesses. Therefore, standards are important for credit development and conversion of credit into capital. In nature, modern economy is aggressive and calls for standard credit activities to expand production, improve techniques, update equipment and sell products. Domestic debt relationship or international economic partnership rely on standardized credit. Standardization is the foundation for modernization. As the economic globalization deepens, credit standardization will play an even more important role in creating capital and wealth (see Table 6.3).

In the new era, credit reliability and standardization are getting more and more important, while some negative factors keep on increasing. First, market uncertainty may impair credit reliability. Second, market asymmetry increases, for example, between virtual and real economies, between urban and rural finance and asymmetric agency — larger gap between the interests of agents and clients. Third, market fluctuation gets tougher. In particular, the recent financial market both at home and abroad saw the largest abnormal fluctuation which imposes new pressures upon the reliability and standardization of credit. Fourth, conventional money is weakened. For example, the exchange and storage functions of conventional money are weakened. Young people prefer credit cards to cash. Fifth, the market culture is being readjusted. China is experiencing a revolution of consumption credit and Internet finance which is reforming the wealth management methods of families, especially young families. Sixth, there

Table 6.3.   Credit Condition

Precondition for the conversion of credit into capital	
Reliability	Quality, Safety, Compensation
Standardization	Stability, Standard

is a clear trend that financial capital is transforming into social capital, which sets new challenges for the traditional credit standardization.

In addition, changes in international finance tendency have significant impact on China's credit conditions and policies. For instance, new financial institution gains rapid development driven by new economy. The New Development Bank (NDB) and Asian Infrastructure Investment Bank (AIIB) have emerged as important supplements to modern international financial system. The developing countries hold more shares in the International Monetary Fund (IMF) equity structure of the World Bank. The international regulatory regulations are readjusted after the financial crisis. The important changes, on the one hand, benefit the interaction between the two markets and improvement of China's credit system and policies; on the other hand, throw heavy pressure on China's credit conditions and market development.

China suffers severe credit defects. In the recent six decades, especially the three decades since the reform and opening up, China presented remarkable achievements in modernization. The ancient country has shaken off its long backwardness and poverty and is exhibiting admirable development potential and the vigor of Chinese people. However, during the 30-year market-oriented development, China also recorded extreme market economy with severe credit defects, including poor social integrity and credit, negligence of contract and trustworthiness, business fraud, counterfeiting, tax fraudulence, academic misconduct, etc. The reasons are: first, the market values are improperly extended to politics, education, science, religion, culture and other social sectors, where individual profits are the exclusive goal. Second, importance is blindly attached to impractically high development speed and leap forward, while the development balance is neglected. Third, the governments give priorities to investment attraction and project development rather than overall planning and balanced development. Fourth, it is the capital, technology and market rather than the institutions — the spirit of credit-centered market economy — that are introduced.

IV. Four Credit Development Campaigns

In view of the new trends and market credit conditions, we have to succeed in four campaigns to readjust and perfect policies, enhance credit development, take the initiatives and improve risk control.

Campaign 1: Strengthening strategic guidance

We should gain in-depth understanding of the importance of market credit system: Market credit system reflects the development level, environment and condition of a country/region. Countries with developed market economy have sound market credit system. China, in the early stage of socialist market economy, is in urgent need of the support from market credit system. It is hard to achieve smooth economic transformation and development in a country with poor market credit. In the credit-centered modern economy, market credit system is the regulator that coordinates capital supply and demand, and is the essential foundation for harmonious social development.

We should highlight the essential role of credit development: The core of financial business is to satisfy the society with products and services, and create market value and credit in the process of finding demands (business opportunities) and meeting demands. No matter how financial and market pattern changes or how up-to-date the financial products are, finance remains to be credit-centered. It is necessary to stress that credit development is the foundation for market development, and stress that integrity is the guideline for conducts.

We should define the tasks and goals of deepened financial structure reform: In the following three decades and a longer period of time, we should set credit system development as the core of financial structure reform, in order to, based on laws, regulations, standards and contracts, build the credit record and credit infrastructures covering all social members, set up the credit incentive/punishment mechanism, promote and foster the spirit of contract, create integrity culture, disseminate traditional credit morality, and upgrade social integrity awareness and the level of credit. Regarding contracts, we need passion and capacity, motivation and pressure, and strong sense of contract. Respect to contracts is the epitome of the rule of law. We should create a sound climate for the execution of contracts and disseminate strict code of conducts. The climate includes severe punishments that make violation of contracts unaffordable, social criticism that may violation unacceptable, frequent audit and inspection that may keep contract parties warned, feasible provisions that allow contract parties to perform, and definite contract value and importance that build up contract parties' sense of responsibility.

We should stick to market-oriented finance, which is essential for China's financial theory, financial market concept and the development of finance. First, we should reform the market concept and managerial rules to highlight high quality and high price, high risk and high return. Financial institutions will not simply prevent risks, they should make good use of risk as an important means of profit. Second, we should reform the market structure and share. Along with interest rate liberation and market segmentation, the market will be divided by cost and risk. Driven by risk and return, cost and profit, the capital market will be divided by high, medium and low financing cost and risks. Large numbers of financial instruments and methods, such as bonds, securities, guarantees, insurance, intermediary services, will be created to apply to market segments and satisfy various customers. Third, we should reform the way of market competition. Interbank competition will transform from simple competition in terms of loan condition, loan procedures and general services to overall competition in terms of interest rate-centered management level, business level, cost accounting, customer relationship and operating strategy. Fourth, we should reform customers' understanding of financing and liability. Interest rate liberation observes the principle of credit rating-based pricing which will effectively improve customers' credit awareness, reform enterprises' ideas and practices that the more the liability, the higher the leverage, and set new and higher requirements on banks' level of credit rating and especially companies' level of financial analysis. Fifth, we should reform customers' idea on financing risk management. Banks give different interest rates even for preferred industries, i.e. interest rate rises as loan increases. For enterprises, it means the cost of external finance varies: the more they rely on external finance, the higher the cost is. Exceeding the rational financing cost bar will bring heavy burden. The financing risk control philosophy is commonly accepted by banks and enterprises, which helps enterprises constrain blind financing impetus, improve self constraint and rationality.

Campaign 2: Putting in place sound institution and mechanism for financial reform

We should create multi-dimensional and diverse financial ecology: The currently plain financial system hardly meets the demand for market diversification and segmentation and China will have to cope with the harsh competition in global market. We need to build multi-dimensional financial structures to adapt to the open market. Multi-dimensional and diverse financial system, regarding the structured economic interests, calls for banking institutions at various levels and with various functions that meet the financial demands of customers from various levels and involve the entire process of social and economic activities to swiftly detect potential market risks, especially in the early stages of risks. It is necessary that a large number of prosperous small and medium-sized banks form a multi-level banking system with strong market competitiveness and adapt to the structure of the primary, secondary and tertiary industries. In this sense, we need to coordinate the relationship between "decentralization" and "centralization", and between "subtraction" and "addition". Chinese banking sector adopts the "centralization" strategy, resulting in over-centralization of human and financial resources in the Big Four. The numbers of outlets and employees of ABC are 50,000 and 500,000, ICBC 28,000 and 420,000, CCB 24,000 and 310,000, and BOC 13,000 and 200,000. Over-centralization impairs optimized allocation of social resources. Therefore, the strategy for regulatory system reform should be "decentralization" and then "re-centralization". Decentralization means that the state-owned giant banks appropriately break down some small and medium-sized banks targeted at local, regional or transregional customers, and attract social funds to set up more banks through market operation, so as to fully tap market competition. Decentralization can be made through subtraction or addition. The subtraction means banks are streamlined (cut down some institutions and lay off employees) for better corporate competitiveness. This may not be practical and may possibly result in heavier employment burdens to the society. The addition means to divide the banks into new banks and encourage private regional banks, in order to create new jobs. The Big Four, though large in size, can hardly satisfy the financial needs of all market segments, regions and customers. To make effective use of the RMB trillions financial resource of the country, we need to extend our focus, through market competition or organizational division. In China's financial ecology, it is necessary to have six or seven internationally competitive banks, including policy and development financial institutions, and thousands of small and medium-sized banks. Only in this way could Chinese economy be guaranteed with strong financial support.

We should establish and perfect long-effective management mechanism: In the process of social and economic development, human perception is restricted despite of the developing technology. Neither can we fully understand the objective reality, nor can we fully control it. Regulators are part of the market and could not go beyond the market when identifying risks. Therefore, it is necessary to raise the awareness of risk prevention, establish and perfect the long-effective risk management mechanism. Risk compensation should be a key concern throughout the process of credit approval and management. In terms of organizational structures, credit policy formulation, execution and supervision should be separated to ensure the adequacy and effectiveness of constraint mechanism. Loan granting and post-lending management should be integrated to gradually optimize asset structure and elevate overall return of assets. Practical efforts should be made in fundamental works and priority given to the basic works of risk management, allowing the frontline to be informed of customer dynamics and make technical analysis of credit risks. We should actively explore credit innovation and credit portfolios, syndicate loan business, asset securitization and credit risk management tools. We should set up mechanisms for examination, certification and continuous learning of credit risk managers to provide HR guarantee. Only through the long-effective banking and credit risk management mechanisms could we ensure the healthy and sustainable development of Chinese banking industry.

We should strengthen financial services: Finance as a medium changes along with the development of market economy. We made profits through increasing sales of products in 1950s, through increasing market shares by customer-centered integrated marketing in 1980s, and through increasing customer satisfaction by quality- and value-centered integrated corporate activities in 1990s. While in the contemporary times, profits should be gained through customers' lifelong value by value-centered integrated supply chain, which should give concerns to the following:

The first is the structure of financial services and marketing, including organizational core, value communication, delivered value and performance evaluation. The key points of organizational core are assessment of market environment, analysis of competition status, and decision on development scale and capability. As for value communication, it is to understand customers' value, demands, value flow and satisfaction. Those of delivered value are market segmentation, differentiated advantages, meeting customer needs and providing integrated supply and services. Those of performance evaluation are evaluation standards and the expressions of customers' value.

The second is the different service strategies for new and old customers, i.e. attracting new customers and maintaining old ones. New customers are attracted through increasing market share and expanding market scale, while old customers are maintained through better satisfaction, especially through good knowledge about competitors and timely improvement of services.

The third is the establishment of customer asset management system, which is an important element of financial service strategy, including determining the key characteristics of customer service quality, designing satisfactory and understandable customer reply language, and setting up a complete set of customer asset and resource models.

It is necessary for China's banking sector to make in-depth study on the marketing and service system, the financial service and marketing strategy, standards, methodology and performance evaluation rules adaptive to the current condition of China, define market segments and target markets, and the marketing strategy to fulfill market objectives.

Campaign 3: Building mechanism for communicating macro control policies

The objective system of monetary policies includes ultimate objectives, intermediary objectives and operating objectives. The ultimate objectives, in addition to the common concerns over "economic growth, job opportunity and price stability", should include balanced development, which should also be the objectives of national macroeconomic policies. The intermediary objectives means that the central bank should aim to achieve the ultimate objectives, including for observable and re-adjustable indicators as well as disseminative financial variables such as interest rate, growth rate of money supply and total social financing. The central bank also sets up operating objectives, namely the short-term and direct objectives of monetary policies, such as base money and short-term interest rate. From the perspective of the law of balance, the above objectives are subject to three factors: maturity of the monetary policy dissemination mechanism, the acceptance of monetary policies by market players, and external pressure and level of gaming. Before the reform and opening up, China's monetary policies were disseminated from the people's Bank of China (PBOC) to its branches and then to enterprises, i.e. simply and directly to the ultimate objectives. In the early stage of the reform and opening up when the central bank system was established and financial institutions gained growth, the monetary policy dissemination system, from the central bank to financial institutions and then to enterprises, were initially set up, but the monetary market was not included. Since 1990s, the financial macro-control gradually transformed, the monetary market developed, as a result, the dissemination was made from the central bank to the monetary market then to financial institutions and finally to enterprises, or from policies to operating objectives then to intermediary objectives and finally to the ultimate objectives. In recent years, the central bank made significant reforms to the monetary policy control and dissemination system including, for example, the loan limit was canceled to lay institutional foundation for the transformation of intermediary objectives to control over the amount of money supply, the deposit reserve system was reformed to make use of fund centralization and credit restructuring, the rediscount mechanism was reformed to link up rediscount interest rate with re-lending interest rate, operation of open market was expanded to highlight the initiative, flexibility, time-efficiency and justice of monetary policies, and interest rate including that of foreign currency was liberated to allow financial institutions to determine foreign currency loan interest rate based on the interest rate of the international market. In general, the policy dissemination mechanism made some progress, but some critical weaknesses exist. First, the interest rate is yet to have access to the intermediary and operating objectives, and macro-control lacks flexible measures. Second, the monetary market infrastructures need to be reinforced to bring more market players, solve market separation and push forward the bill market. Third, Chinese banks do not have in-place modern bank operation mechanism and it is hard to increase the vertical and cross-coverage of financial services. Fourth, the micro basis (enterprises and residents) of monetary policy dissemination should be improved. The above factors and practical conditions impair the effectiveness of China's macro-control policies. Along with the application of advanced technologies, such as Internet banking, the monetary circulation is accelerated, the efficiency of commodity trade improved, fund coverage expanded, and, however, the uncertainty and risk of financial activities are increased. Market players differ in appetite towards monetary policies, making it harder to implement the policies. External pressure and competition expose domestic problems to the world and make monetary policies more complicated. Chinese monetary policies are also subject to financial variables including interest rate, inflation, exchange rate and credit distribution policies. Against world economic restructuring and globalization, Chinese monetary policies are facing unprecedented tough challenges which call for proper economic philosophy and strategy to meet the demand of development and public expectation. At present, the monetary policies should "effectively deliver policy information, define policy orientation, so as to effectively guide market expectation and mitigate the pessimism".[8]

Campaign 4: Improving and enriching policy and technical tool kits

For example, in terms of monetary policy, in addition to traditional policy tools — legal reserve, central bank discount rate, central bank fund ratio and open market operation, we may consider innovations such as bond portfolio, use policy tools at proper time and use monetary policy combinations. Control over money supply alone is not enough to prevent asset bubble because increase or decrease of supply is stiff. We may also consider using securitization to regulate the volume and increment in the capital market, and control the scale by adjusting liquidity, such as directional lending, favorable policy, control channel among others.

In recent years, the stock market exhibits abnormal fluctuations: rapid ups and severe downs. The reasons could be: more speculations instead of investment, more individual investors instead of agglomerations, more passion instead of rationality, more amateurs instead of specialists, and more desires instead of market development. A large part of the individual

---

[8] Liu Yuanchun. "Restructuring China's Monetary Policy", *Caijing Magazine*, 2015, Vol. 10, pp. 20–30.

investors are students, self-employed, jobless urban people and secret investors. Thank God, there are not many farmers. The speculators are domestic and foreign floating capital, institutions seeking for quick profit, and students and citizens dreaming of sudden wealth. Chinese stock market is the playground of individuals, or herds, making it distinct from institution-dominating mature markets. The individuals are unprecedentedly free in the stock market, which to some extent encourages gambling and irrationality. There are few institutional investors although they have the right to say and always win the game. The conclusion is that, in an immature market, 70% of the investors are speculators and the rest 30% rely on confidence. The stock market is a casino that calls for close attention but is not necessarily worrisome. Do not be easily influenced by public opinion. When speculators are being punished by the market, we should learn from the lessons, control fund flow and over-the-counter (OTC) fund distribution, slow down initial public offering (IPO), control financing scale, control position transfer rate and scale, give special concern over the fluctuation between position and disposition, credit violation exchange, synergic security issuance and possible systematic risks resulted from centralized securities positions.

There must be definite policies regarding the large and "must-standing" financial institutions. Examples are: control over the leverage ratio; necessary constraint on deposit savings from proprietary trading; raising the risk rating of bank-held securities; firewall between markets. Though complete segregation management is not practical, there should be inherent mechanisms to separate proprietary trading in different markets. In face of complex changes in the situation and credit conditions, it is necessary to make overall concern and stick to the combinations between, first, determined strategic goals and tactics: flexibility and effectiveness based on firm strategies; second, virtual and real economies, i.e. problems in the capital market should be solved in the sphere of real economy. In addition to establishing stock market stability foundation, perfecting pertinent policies and rigid control over capital inward and outward flows, we should make innovation in the parent and affiliated foundations, and at the same time set up national land foundation, create labor protection and food supply system to protect the vulnerable groups with basic supports, prevent harsh fluctuation in the stock market from impairing the reform and opening up; third, short-term and long-term policies: we may hold back before making large leaps forward. The short-term policies may conflict with long-term strategy. For the purpose of economic stability, it is necessary to increase money base and liquidity, provided that liquidity is under appropriate control to prevent inflation; fourth, domestic and international markets: we should make domestic market as the basis to implement "internal demand and supply" strategies, and seize the opportunities brought by the international market and geopolitics to expand market for mutual benefits.

In a word, along with technological, economic and social development, the content of value keeps expanding to cover, in addition to money and financial assets, such intangible values as innovation ability, brand quality, market rules and formulation of regulations which are the market basis for the in-depth economic globalization and international governance restructuring, the social basis for China's modernization and sustainable development, and the guarantee for ideology, product, management and development innovations. Therefore, credit, compared with money and financial products, is a higher-level social exchange medium that reflects wider social and economic cooperation.

Chinese economy has entered into the new normality. Slowed-down economic growth, restructuring and transformation are new and long-term trends instead of temporary abnormality. During the long-term transformation, some economic development conditions change (e.g. financial, monetary and credit conditions) and some remain unchanged (e.g. the essence of market economy based on credit development do not change and becomes more important). In face of the changes and challenges, we should not only amend pertinent policies but also exhibit ideological determination. For example, in term of monetary policy, we should be prudent with such proposals as "critical turning", "policy restructuring" and "relax monetary policy via price". In China, steady appreciation of money builds up public confidence. It is critical for state stability and joining international community. Liquidity varies in different sectors in China, while in general is sufficient. The economic problems are not exclusively about nor can be solved solely through monetary policy. Should China have excess liquidity, it could not be absorbed by the international market, as the developed countries did. In this sense, it is steady monetary policies that are needed in China.

China is in a critical stage of deepening economic reform and fixing socialist market economy, a stage with strategic opportunities for accelerating transformation of development mode, a key stage of economic and social transformation, a stage of upgrading its open economic level. Speeding up social credit system development is an important approach for the standardization of market economic order, improvement of market credit climate, lowering trade cost and preventing economic risks, an essential precondition for optimizing resource allocation, expanding domestic demands and promoting industrial structure upgrading, an effective measure for elevating social integrity, boosting social trustworthiness and minimizing social conflicts, a necessary precondition for building international brand and reputation, reducing foreign trade cost and upgrading China's soft strengths and international influence, and an urgent need for establishing an objective, just, reasonable and balanced international credit system to adapt to the new trend and pattern of globalization.

In early 2016, RMB was depreciated to a historical low and the stock market experienced abnormal fluctuation (two suspensions), causing panic in the market. The reasons are, internationally, regional politic turmoil, pressure on RMB due to USD appreciation, global market downturn, and domestically, the superimposed effect of the "three phases", conflict from market immaturity, poor management experience and low investor confidence. The countermeasures could be the following. First, timely communication and negotiation between the Chinese and the US governments or competent authorities. The two countries are competitors but share common interest in market stability and need interaction against globalization. Second, priority should be given to domestic market to enhance credit development, strengthen innovation, develop "river basin economy", build brands of consumer goods and tap urban and rural potentials. Third, we should appropriately readjust interest rate to lower down financing cost. Fourth, we should enhance policy study, test and dissemination, and decide appropriate time, method and extent of tests. Fifth, we may consider appropriately postponing the "sale restriction" period, and choosing appropriate time to dispose of stock index futures to allow slow "bull-bear" swap, coordinate interests and demands, and achieve "soft landing". In all, the global politic situation is controllable and predictable, the international environment is free from fundamental change, and China enters the new normality as scheduled. The recent problems in exchange rate and stock market are regular problems caused by historical deposit and lack of experiences. If timely handled, they would affect the general situation. The fundamental problem is market immaturity, the key to which is enhanced market credit regulations.

Chapter 7

Market Uncertainty and Financial Risk

What are the uncertainties in the economic development in the new age? What are the changes in market activities especially under the impact of information network technology? How to identify and control financial risks? This section deals with the potential uncertainties and financial risks when Chinese economy enters into the new normality and proposes countermeasures.

I. International Market Uncertainty[1]

Uncertainty is the essence of the real world. The real world is never certain, follows no tactics and not subject to the human will. The particularity of reality is common existence and the difference in reality keeps changing. Everything of the real world and its environment cannot be fully determined. We cannot completely and accurately estimate changes in the real world. Change of things is uncertain, and such uncertainty is normal. There are natural and social uncertainties. The former reflects the essence of natural environment, and the latter is closely related to social, political and economic regulations, policies and status, and reflects the essence of social development. In market economy, there are not only inherent uncertainties rooted in the market system, but also external uncertainties resulting from the outside. The increasing uncertainties in the world have significant impact on market economy.

In the following section, we talk about the uncertainties in the international market.

---

[1] This section is published in *International Finance News*, February 2, 2015, p. 19.

First, the historical and cultural particularities are highlighted: As economic globalization deepens, in the accelerated integration of histories and cultures, regions and ethnic groups witness their traditions rapidly change, evolve, react and produce effects. Cultural differences, ideological and behavioral conflicts get increasingly remarkable, and the contest between globalization and anti-globalization gets tougher. In the new century, national and regional conservatism has risen, and nations and ethnic groups increasingly disagree in terms of tradition, religion, culture and politics. The terrorist attack on Charlie Hebdo of France is an unfortunate proof that, in the contemporary world of cultural diversity, free speech in the real world indeed has a boundary due to cultural differences. Historical, cultural and religious antagonisms are causing more and more conflicts which worsen the uncertainties in global development.

Second, uncertainty increases in global governance: The existing global governance structure is based mainly on the political pattern of Europe and America, and is playing and will continue to play an important role. Meanwhile, it is focused on promoting good global governance and balanced development, for it cannot fully reflect the changes in global pattern and the development of emerging economies. In the recent decade, the developed countries have transferred 3.13% voting weights to the developing countries in the World Bank and International Monetary Fund (IMF), increasing the voting right of the latter to 47%. Unfortunately, the reform of global governance did not present satisfactory progress. The disagreement and contradiction between the east and the west, the north and the south and the developing and developed countries have evolved to deep-level conflicts. Great shifts in world pattern, competition in geopolitics, severe conflicts, terrible disorder in international strategic situation make global governance even complex and uncertain.

Third, international events occur frequently: For example, Ebola panicked the world and impaired global trade, investment and tourism. Slump of oil price, from USD 110 per barrel in early 2014 to USD 21 per barrel in December 2015, intensified market turmoil. Ukrainian crisis dragged Russia, USA and EU to the "new Cold War" and triggered new changes in international geographic pattern. The two air crashes of Malaysia Airlines that threatened the international safety are possibly related to politic and social factors. ISIS and international terrorists are causing global turmoil. Ferguson incident frightened the entire USA and destructed Barak Obama's mid-term election. The worldwide incidents worsened international chaos and market fluctuation, and dragged down the sliding world economy.

Fourth, market economy shows increasing discrepancy that can be seen from the following. First, innovations in technology, product and market change production and consumption modes. Enterprises are more inclined to "flexibly specialized" production and provision of personalized commodities and services in refined market segments. The consumers exhibit more changeable appetite, making consumption demands harder to predict, diversifying shopping modes, highlighting comfortable and functional consumption environment, and complicating consumption culture. Second, wide application of network and information technologies triggers incompliant behaviors of market players. Along with the miraculous rise of information and online shopping, network information foster average market participants' synchronous tendency, herd effect, gambling acts and animal spirits that challenge human psychology and make them even more active, outward, complex, extremist and uncertain. Third, physical and biological revolution may further change our way of life and values, bringing more consumption appetite, technical options of manufacturers and demand uncertainties, and making it harder to predict market prospect. Fourth, the large gaps between developed and emerging markets and between the emerging markets add more discrepancies to market economy.

II. China's Market Uncertainty

China has to cope with even more complicated uncertainties in its development. The economic and social transformation of China from planned to market-oriented development is far more complicated than the natural evolution from the infant to developed market economy. In particular, as affected by objective uncertainties and uncertainties in international context, Chinese economy, in the process of entering the new normality, exhibits more complicated uncertainties in the following aspect.

First, uncertainty increases in policy effect: It is mainly because economic fluctuation is harder to predict. The precondition for macro policy is timely and accurate judgment and prediction of economic fluctuation which, however, is harder. It is theoretically difficult to make predictions. Except traditional economic cycle theory that believes economic fluctuation has rules to follow, all other theories, such as practical economic cycle theory, monetarism, neoclassicism and Keynesianism, hold that economic fluctuation to a large extent is impossible to predict. The real world is unprecedentedly complex, making it even harder to analyze the macro situation. Taking Chinese real estate sector for example, the inherent driver is weak and more vulnerable to regulations, the markets vary in different regions and cities in terms of prosperity and structure, which, plus the changes in population flow, force the government to promulgate differentiated control policies and set diverse objectives, making it harder to determine and balance regulation strengths and to make sure of policies and their effects.

Second, the influence of emergencies is growing: Emergencies are characterized by uncertain social nature and sudden non-programmed decisions. Under the influence of information and network, in the process of economic and social transition, emergencies, especially mass events, in great degree are unexpected, information on life, property and safety losses spread faster, wider, and are more likely to trigger a chain reaction. It is difficult for competent authorities to make clear and objective response to rapidly changing events, and cannot take countermeasures step by step. This series of phenomena easily become the focus of the world. Network emergencies, often dominated by public opinions, easily produce "the butterfly effect" and "group polarization" phenomenon, and find it difficult to predict the development and possible social impacts.

Third, systemic risk accumulation overlay: The risk of economic downturn, restructuring, institutional constraints, regional and local debt, market credit risk and liquidity risk and so on, are interactive, interlacing, overlapping, and increases the uncertainty of the economic and social development in China. For example, the risks of bank deposit and loan structures, local government financing vehicles and related-party in the real estate market are indeed not promising enough. According to the estimates, the gap between the average terms of deposit and loan in Chinese banking sector had been extended from 1.87 years in 2006 to 2.88 years in 2013. It is obvious that money goes to long-term investments, and maturity mismatch is further aggravated. In long-term investment, 86% of local government finance vehicle (LGFV) are related to the real estate market. The asset-liability ratio of real estate enterprises as a whole rises every year to as high as 75% in 2015. The effective measures in the real estate sector are deepened reform of the land markets, real estate registration and real estate tax system, which, however, might have impact on government borrowers and the revenue of local governments, and will in particular cool off the property market in the third- or fourth-tier cities. These problems, if intertwined and superposed, would disrupt reform and development as well as macro-control.

Fourth, critical thinking is becoming the normality: China needs to simultaneously deal with the slowdown in economic growth, make difficult structural adjustments, and absorb the effects of previous economic stimulus policies. It has entered the deep-water zone of reform, facing a lot of difficulties such as how to prevent systematic risks in the process of overall and deepening reform and development, how to organically integrate real and virtual economy and effectively fill the fund gap of small- and medium-sized enterprises (SMEs), how to break up the urban–rural and industrial barriers and achieve free circulation of production elements, how to lift the financial access bar under the precondition of preventing financial risks, how to reform the rural land system, explore into various effective rural land collective ownership, how to ensure the soft landing of real estate industry and avoid comprehensive systemic risk caused by sharp drop of quantity and price, how to establish market value compensation mechanism for public goods and social resources, and achieve social justice on the premise of increased efficiency. There won't be perfect solutions. What is more, in the process of resolving these issues, the deep-seated problems and contradictions of domestic economic system may also be exposed that would challenge the risk-bearing capacity of the general public. After more than 30 years of reform and opening up, the risk-bearing capacity of the public has greatly improved. Meanwhile, the judging capacity of the mass and their requirement for the quality of the reform and opening up and the development has also greatly improved. Reflections on history, hatred for corruption, anger against injustice, and criticism on incompetence will become normality.

How should we cope with the increase of such uncertainties?

First, we should enhance study on the uncertainties: In the early 20th century, German scientist Werner Heisenberg discovered the uncertainty in particle motion. It is believed that the macro world is dominated by the law of uncertainty regardless of how it changes. Factors of a thing cannot be certain at the same time. The certainty of one factor implies the uncertainty of others. If we are sure that the economic growth will slow down, the other factors such as employment rate and export volume may become more uncertain. We need to address all the possibilities in all circumstances, analyze the non-correspondence of the objective world, probe into the law of uncertainty, effectively prevent uncertainties, minimize the negative effect, overcome the difficulties brought by uncertain decisions, formulate plans for harsh internal and external changes, build up policy reserve, make appropriate readjustments, coordinate the policies, and make sound management of development prospects.

Second, we should enhance emergency education: Human history is a book of struggles against challenges and crisis. In recent years, we recorded frequent natural disasters and emergent public events and increased uncertainties. We should, in addition to improving the government's capacity, foster public awareness. It is necessary to provide family-based emergency response training, enhance emergency bases to prevent and mitigate natural disasters and emergency events as much as possible, so as to protect life and property, safeguard state security and social stability.

Third, we should strengthen the emergency response mechanism: We should first strengthen the development of emergency theory. We should ensure the preparedness in order to deal with changeable and unchangeable, typical and non-typical, symmetrical and asymmetrical, innovation and regulation, further improve the national master emergency plans, establish the emergency organization system, operational mechanism, safeguarding measures and supervisory management adapted to the new normality, and guide the prevention and treatment of all types of emergencies. Second, we should strengthen the platform for data integration and analysis. We should make full use of advanced technologies, such as big data, to closely track the trajectories of uncertain factors analysis, make better use of asymmetric means, innovative approaches and new perspectives to judge the situation, actively and courageously to break the "temporary" preparedness and "relative" steadiness, and minimize the major uncertainty factor. Third, we should strengthen the early warning mechanism, including early warning for natural disasters, social unrest, risk in the financial system, the network emergencies, and raise public sensitivity to all kinds of risks. Fourth, we should strengthen the emergency reserve. We should further establish and improve national emergency center and material reserve system, improve the efficiency of disaster relief, material allocation and information integration. We should further enrich and improve fiscal, monetary, industrial and social policy toolkits, and prepare not only quantitative, flexible and traditional regulatory tools, but also innovative policy instruments that are readily available. Fifth, we should enhance the emergency insurance fund system. We should establish and improve government-led, social-participating, market-oriented emergency insurance system, strengthen the innovation in financial tools, and build multi-dimensional emergency financing and insurance mechanisms.

III. Change in Market Behavior[2]

As uncertainty increases, market players get more active and exhibit new characteristics as follows:

Synchronous tendency: Financial institutions and market players operate similarly in financial activities. They would swarm forward to buy or sell. Synchronization is the instinct of market players who tend to converge. An individual that differs from his surrounding in terms of attitude or behavior would try all the means to converge into it. The issue is, market

---

[2]This section is published in *International Finance News*, January 5, 2015, p. 19.

trade modes and technical routes increasingly intend to converge. Many traders' technical routes and programmed trade modes are designed by a small number of engineers. Such models are quite standard, process similar information and are widely applied to the financial sector. However, under certain circumstances, they generate similar oriented results at the same time, i.e. the results are highly correlated, thus easily lead to highly similar investment, worsen securities price fluctuation and trigger systematic risks. During the financial crisis in 2008, the programmed models used by financial institutions aggravated changes of the values of securities assets and liabilities and, through fair value pricing tools, amplified the fluctuation of profits from securities against investors' prospects. The result was a terrible chain of "price down–asset write-down-panic undersell-price down". The investors collectively, spontaneously, synchronously and blindly escaped. In this sense, we should stress the impact of technical changes on market synchronization when studying financial situation.

Herd behavior: This is typically "going with the stream". Average market players are uncertain about the market. They observe those around them or industrial leaders to get information, make decisions and take actions. This is the typical herd mentality. People consciously or unconsciously follow the majority and negate the ideas of their own. In the market, they consciously or unconsciously follow other investors, despite of not having obtained first-hand information or formed their own prospects or ideas. They just rush ahead and encourage each other, and are collectively zealous or panicked. Investors blindly trust the three top-rating institutions who provide the majority of credit rating information and are highly correlated, resulting in strong overlapping force. Market players refer to their rating as business instructions and the basis for internal deliberation, without making careful study on the inherent risks of investment portfolios. Such dependence and laziness lead to the herd effect. Good rating cheers them up, while poor rating throws them in deep pessimism. In an era where online information rapidly spreads in particular, the herd effect ferments and adds irrationality to the market.

Gambling act: People would like to "gamble" in specific circumstances. Without sufficient knowledge about necessary information, they hope to bet on luck. This is especially the case in current Chinese market. Chinese people would not like to run a company unless they are short of money. It is the contrary in the West, where people would not do that unless they have enough fund. Many Chinese companies are built on debts, even usury, that is why they do not hesitate to chase sudden wealth to repay debt and develop. It is hard to believe that those entrepreneurs, under such pressure, would play in the market with credit and integrity. They are more like gamblers. Uncertain environment gives few opportunities to win the game, and the bet raises. Gambling causes numerous financial crisis and social problems, as well as high liability, excessive profit and high risk in Chinese financial market. In addition to quick money, some starting-up entrepreneurs hope to win social recognition and control social environment, the possible results of which are on the one hand, they build up their social responsibility and on the other, they destruct social and natural environments, e.g. expansion that causes severe pollution and waste of resources. *Animal spirit* is a sort of irrational wisdom that makes confidence-based heavy investment. Mood, psychology and confidence have widespread impact on market behavior. Confidence has a considerable effect on the real world. With confidence, we buy in; without it, we sell out. According to George A. Akerlof and Robert Shiller, winners of Nobel Prize for Economics, market players have animal spirits, whose investment acts cannot be explained simply by theory or rationality, but natural instinct. Animal spirits have positive side that encourages people to shoulder risks and boosts innovation; while the other side leads to irrational anxiety and panic, market turbulence and financial crisis. The two sides are closely related to the reasonability and fairness of market information. The grassroots build confidence upon social equity. Currently our information is complex and asymmetric. The irrational anxiety and panic caused by the rapid dissemination of online information seem to aggravate. Taking the financial crisis in 2008 for example, some listed companies, though having sufficient cash, were driven by the animal spirits, i.e. they panicked due to some signs from the market, and hurried to escape. All of them demanded debt repayment. The collective irrationality burnt down the cashes and threw the healthy companies into real trouble.

IV. Motivation of Market Behaviors

Knowledge decides behavior. Human brain is an information processing system. Information is imported and exported, psychology is created, and behaviors are thus decided. As Internet information technology is widely applied, the material society is transformed into an information society where uncertainties increase, things change both quantitatively and qualitatively, and human mental fitness is challenged. People get more mentally active, open, complex and extreme. In addition to common adventure and herd, new psychological attributes have larger impact on behavior.

First, "fragment psychology" is a fragmented, "jumping" way of thinking. Your brain has access to large amount of fragmented information and your mind jumps from here to there, causing weird associations and distinctive sparkles, disabling you to make overall and systematic consideration. We find more and more fragments, miscellaneous news, and hot issues. Things jump in and fly away. You feel out of date if you sleep few hours more, or abandoned if you shut down your cell phone. It is like the partition fragmentation. Files are stored in different areas of the disk. If the physical memory is not enough to run a program, the operating system (OS) generates temporary files and massive fragments in the hard disk. When the fragments are too many, OS has to search on and on. This impairs system performance, or even shortens its lifespan. Human brain is like a disk, and psychology is like the system performance. Excessive information will add pressure on psychological mechanism and harm your health. Mentally fragmented people are always distracted. They know a lot but specializes in nothing. Such fragmentation can also be seen in organizational acts. Credit information are fragmented and not effectively integrated by any institution, severely impairing the value of credit information.

Second, "mianzi" (face or prestige) is a pursuit of reputation and ostentation. *Mianzi* is an important social psychology, especially in China. Chinese, the noble or the average, in the past or at present, are fond of it. Lin Yutang said that *mianzi* is the goddess that rules China, and is more powerful than fate and benefaction. *Mianzi* as a cultural psychology means ambition. Appropriate *mianzi* helps us to control ourselves, protect our dignity and keep a sense of shame. Neglecting *mianzi* leads to shamelessness, while attaching too much importance to it leads to vanity. Some people tend to take too much care of *mianzi* regardless of the cost. They spend too much on unpractical needs. For example, many families buy cars to show off rather than for practical needs. Some leaders care only about fame and merit rather than local needs and preferences. They dare to use their power to implement "image" projects.

Third, "resentment" is repelling others' goodness that you do not have. This is a process of envy — others made it but you did not, jealousy — consciously or unconsciously unhappy and ashamed, and resentment — strong counter-emotion and resistance. This is a special feeling rooted in human's competitive psychology. It comes from, on the one hand, poor mental compatibility or narrow mindedness, on the other, the worry that someone would harm your interests. It is also an instinct that, seeing others' success, you may be possessed by bitterness, jealousy, hatred, suspicion, disappointment, shame and sorrow. In the information era characterized by increased opportunities and challenges and gaps between individuals, resentment aggravates and threatens the society.

Fourth, "petty gain" is a part of human sub-consciousness. People feel content with petty gains, and the degree of content depends on the degree of difficulty in getting the gains. The more you get, the more you want. In the era of network information, especially in the "Single's Day sales", buyers get crazy. Many vendors use their psychology well to bring in new products, expand existing market shares and suppress competitors in the forms of "cut-throat price", among others. Many netizens love "Single's Day sales" because it saves money. In the "festival", more and more e-commerce websites, in addition to taobao.com, launch sales campaigns. For most consumers, the "50% off" is indeed a rare chance. The producer–consumer relation is getting flat and lowering down trade cost, significantly benefiting the consumers. However, if you concentrate on frequent shopping in a single day, you might be in a hurry and miss out on saving money. "Impulse shopping" spends your money and returns with many items that you do not need. It is hardly a regular social behavior. In addition to market innovation, we have to consider how to guarantee market quality, eliminate payment risk, provide good after-sell services, improve vendors' reputation, protection consumers' mental and physical health, and promote sound social development.

V. Identification of Financial Risks

How to balance credit earnings and credit risks? How to control market risks? What are the characteristics of market risks in the information network era? This section is dedicated to the identification of financial market risks and analysis of typical financial risks in the development of market credit system and countermeasures.

1. *Internet Finance Risk*

Internet finance does not simply mean that financial institutions use the Internet for swift network operation. It refers to new investment and financing operation platforms based on Internet technology (search engine and social networks). Though Internet finance aims to constitute flat, effective and transparent intermediate and create distinctive financial ecology, it in nature is the process of credit activities and financial intermediation. The Internet and finance share common genes. The Internet is open to everyone, and the Internet and mobile terminal allow access to and analysis of big data. This is important for realizing the financial functions that are difficult in traditional transactions.

China's Internet finance consists of third-party payment and wealth management, peer-to-peer (P2P) online loan, public financing, e-commerce petty loan and virtual money, which are innovations of Internet companies on the basis of channels and user experiences. Facing the Internet finance, technical innovators in the finance sector do not have adequate understanding of risks and from time to time launch "profit guaranteed, zero risk" advertisements which are revoked by regulators. After years of development, Internet finance is widely applied to traditional financial industries. Internet companies and financial companies merge and supplement each other, and compete within the financial sphere, i.e. from channel and users to product innovation and risk management.

Product innovation and risk management is a new theoretical and practical issue in the Internet finance sector. We take P2P as the analysis object. Theoretically, all financial arrangements are subject to transaction cost, except in P2P market where the marginal cost tends to be zero, thanks to the power of Internet. P2P market may have the best credit resource allocation efficiency because those who need loans can be rapidly targeted, and petty, urgent and transregional loans can be provided.

Besides, the marginal income ratio of such loans are pretty high, which is impossible in traditional banking sector. As Internet technology develops, Internet search engines become faster, allowing P2P to continuously expand to adequate and valid markets, and enhancing nationwide credit allocation function to realize public-benefiting finance. Traditional finance hardly satisfies the rights to loan and investment, which is possible through P2P online loan, evidencing the favorable effect of P2P on social progress.

Internet finance accelerates reshaping of financial modes and social revolution, and brings about new challenges on economic and social development. Internet finance exhibits new characteristics in terms of information risk, operating risk, moral risk and procedural risk. For example, big data needs time to accumulate Internet credit gathering features "first come first go", third-party collection/payment platforms operate like financial institutions but actual are not, thus becoming regulatory "blind spots". Specifically, Chinese P2P suffers from incomplete credit system and less-developed data base, which directly constrain P2P online loan in terms of credit valuation, loan pricing and risk management. In practical operations, many information are not open to free access and many P2P platforms have to make offline due diligence, which increases transaction cost and loan interest rate. P2P actually issues small-value bonds, thus is direct financing. Some P2P websites have to use principal guarantee, risk reserve, professional loan granting and debt right transfer, access to or establish capital pool, which are quite close to regulation redlines. To improve their security reputation, many of them set up their own guarantee institutions or use their affiliates as third-party guarantor, which easily cause huge related risks. When a certain number of borrowers break contracts and threaten capital security of the platforms, the guarantee, which is essential for investors, must be endangered. Therefore, such guarantee of high related risks cannot shelter the investors and possibly cause systematic financial risks. According to the report by *Economic Daily* dated November 2, 2015, as of mid-October 2015, there were 2,935 P2P platforms in China, 42 of them stayed above basic earning power (BEP), and the rest, over 90%, were running at loss.[3]

---

[3] Refer to Huang Xiaofen, "Enhancing Risk Prevention against the Rapid Growth of P2P". *Economic Daily*, November 2, 2015.

How to solve this? To cope with Internet finance risks, we may have to stress the following:

First, third-party collection/payment platforms should be standardized. Third-party collection/payment platforms are the soul of Internet finance activities. We should legally define third-party collection/payment platforms as financial institutions. Though, theoretically, such platforms do not directly involve in lending, they are doing businesses similar to bank settlement and securities companies. In this sense, it is necessary to perfect the rules and policy standards: P2P platforms cannot directly involve in lending, nor can they bear credit risks or liquidity risks; they should disclose their information, shareholders, transaction process and managers, and keep record of transactions.

Second, regulatory mechanism should be enhanced. This includes credit-evaluation mechanism, coordination mechanism and risk-warning mechanism. Risk-warning mechanism is particularly an urgent task. In an era of Internet economy, increased mutation of some tiny factors in economic activities easily leads risks to accumulate and evolve into market cyclone and systematic risks which is a social risk similar to such sudden, fuzzy, rapidly disseminating and destructive natural disasters as hurricane, typhoon and tornado. To get timely understanding of such disasters, it is necessary to build effective financial information platforms and far-reaching, multi-dimensional monitoring system to cover social and economic activities at all micro aspects, measure economic processes, provide detailed basic financial operating data, effectively identify financial cyclone, precisely measure financial risks, and form financial risk warning mechanism.

2. *Complicated Related Guarantee Risk*

With the rapid expansion of bank loans, more and more affiliated enterprises are getting involved. The business relationships are more complex, risk of related party transactions has become prominent. In China, listed companies are guaranteed in the following forms: mutual guarantee between associated companies, mutual guarantee between potentially associated companies, and serial guarantee among unassociated companies.

Guarantee for large shareholders: Due to the dominant shareholder and corporate governance deficiencies, controlling shareholders, in fundraising and use, credit guarantee, often use their controlling stake to seek personal gains at the expense of the legitimate interests of listed companies. The big shareholders of ST Houwang, namely that parent company Houwang Group made the listed company to guarantee its huge loans, ultimately emptied the listed companies and had to announce bankruptcy of the parent and listed companies.

Guarantee for subsidiaries, i.e. a listed company guarantees its or its former subsidiary. For example, Shenzhou provided joint liability guarantee for the RMB 30 million loan of its subsidiary Taiyuan Dongsheng Coking Gas Co., Ltd. From Taiyuan Branch of China Minsheng Banking Corp. Ltd., Shanghai Port Container Co., Ltd., guaranteed the RMB 40 million circulating fund of its subsidiary Shanghai Pudong Container Logistics Co., Ltd. Such guarantee is popular and should be attached great importance.

Indirect and serial guarantee: Driven by the demand for funds, listed companies guarantee each other and form interlacing guarantee chains. From the perspective of related-party transaction, the listed companies (1) guarantee each other indirectly or in a chain, (2) guarantee each other in the capacity of potential related parties. If a single enterprise of the chain cracks, the others are affected. Therefore the listed companies cooperate with professional guarantors who make the guarantee market even more complex.

Causes of risks: Listed companies guarantee for big shareholders and improper guarantee for related companies trigger series of risks. The former, in particular, is widely concerned and doubted because it makes the listed companies act as ATM machines for big shareholders. The reasons are: first, improper equity structure. Chinese listed companies usually have a dominant shareholder who may use its power to transfer interests from listed companies via guarantee. In addition, the complex interrelationship between listed companies and local governments forces listed companies to relieve local enterprises with the guaranteed part, thus impair their own interests. The second reason is the improper governance structure. Most companies launch an initial public offering (IPO) in the form of "list the principal part and transform the enterprise to parent companies", resulting in interlaced relationship between the listed and the parent companies. State-controlled companies have their senior management appointed by the government, thus the controlled company and the listed company do not have substantial equity, resulting in lack of operating and management incentives. Moreover, the senior management of many listed companies do not have good knowledge on guarantee risks and make decisions blindly or casually, which increase guarantee risk. The third reason is that some financial institutions have lost their credit function and operate as pawnshops. They are reluctant to assume risks, and will not grant loan without mortgage or guarantee. Listed companies have good credit, therefore financial institutions would like to accept them as guarantor or mortgager, without caring about the related party relation between the listed companies and the debtor. The fourth is poor market credit. Legal imperfection causes lack of credit mechanism, easily leading to guarantees that derivate from the will of listed companies or fraudulent guarantee that impair the interests of listed companies who have to choose mutual guarantee for fundraising. That is why related guarantee is popular. Frequent related guarantee implies that, the regulated tries everything to detour those regulatory policies that neglect their proper demands or do not fit the market.

In a word, the disordered guarantee by listed companies is complicated, but fall into the following. It is believed that listed companies have strong power and good credit. Commercial banks prefer guarantee by listed companies, some of them even imply the borrowers to provide such guarantee. Some listed companies would like to provide guarantee in order to charge commission. Some local governments force listed companies to guarantee. What is more, the regulatory system is imperfect. Some listed companies' guaranteed amount exceeds 100% of their asset, which should be addressed by the regulator.

3. *Loan-Cheat Risk*

This is an important source of financial risks. Enterprises frequently cheat loans of large amount and threaten the security of financial assets. They usually take the following forms:

False borrower: First, they register a shell company. Second, they counterfeit signature or company seal, convene illegal shareholders' meeting or Board meeting, counterfeit resolutions. Third, they make false financial statements, report untrue profit and expense. For example, the senior management of Guangxi Liuzhou Liyu Group made false financial statements and false contract to cheat RMB 200 million in loan from the Agricultural Development Bank of China. Fourth, they conceal their capacity of reaching contracts, exaggerate repayment or guarantee capacity. Fifth, they have complex equity structure and relate-party transaction, and use related parties to cheat loan.

False transaction: First, they counterfeit contracts, invoice, bill of lading and guarantee, one example is the financing fraud in Qingdao and Tianjin ports. Second, they counterfeit bills, and documents under letter of credit (L/C). For example, Tianjin Nande Economic Group counterfeited the export contract with an Australian company and got thirty-three 180-day time L/C from Bank of China (BOC) through Hubei Qinggong, with a total value of USD 80.13 million, and BOC lost USD 35.49 million. Third, they counterfeit payables and receivables with other companies, counterfeit invoice and payment voucher.

False project: First, they counterfeit approval. Second, they counterfeit capital contribution voucher. Third, they counterfeit loan purpose, and fraud construction scale and investment amount.

False guarantee: First, they counterfeit land and property ownership certificates with untruly high value. Guangdong Nanhai Huaguang Company counterfeited the certificate of 1,574 mu of land, overestimated the collateral eight times for excessive guarantee, and cheated RMB 7.421 billion loan from Industrial and Commercial Bank of China (ICBC). Second, they counterfeit deposit receipt to cheat loans, for example, RMB 10.1 billion from Qilu Bank. Third, they use cross-guarantee by related parties or mutual guarantee by non-related companies to amplify financing leverage. For example, Shanxi Shenglian used frequent related-party transactions and related guarantee and got RMB 30 billion loan, despite of its poor power to follow the agreements.

Evading repayment obligations: First, they conceal or transfer assets. Shenyang Gaokai Company set up new companies and transferred equity to transfer quality assets and leave the debts with the borrower, making it incapable to repay. Second, they withdraw or transfer funds. Zhejiang Zhijun Holding Group cheated loans via a number of related companies and then withdrew its contribution or transfer funds through related-party transactions, leaving the borrowers incapable of repaying. Third, they use false suit or bankruptcy. In Tonghua Maoxiang case, the company applied for bankruptcy to evade repayment obligation. Fourth, they use unlawful merger, association, joint venture or auction or revocation of business license.

The above examples exhibit the defects in banking risks control system. First, the system has leakage and is vulnerable to attack by financial frauds in all forms. Under the pressure of performance, some personnel lift the operating bar and grant loans in violation of rules. Second, the banking sector has not established systematic prevention mechanism, and should improve their identification capacity through field survey, website inquiry, investigation with competent authorities and written materials. Third, the information mechanism has to be perfected. It is hard to prevent frauds if the social credit system is ineffective and customer information and changes are not effectively monitored. The most important thing is, we should establish and improve the whole-process prevention mechanism and build up identification capacity, set up the case reporting mechanism, enhance legal education and business training, and have knowledge about frauds.

4. *Ineffective Fund Circulation in Financial Channels*[4]

Chinese banks, foundations, trust and securities institutions share mutual channel to launch businesses which break monopoly and allow business overlaps. On the one hand, this improves fund circulation and efficiency. On the other, we should be aware that such circulation in mainly found in monetary and capital markets, not in the real economy. For example, some listed companies issue RMB 1.4 billion bonds to which a single fund

---

[4]This section is published in *Study Times*, May 30, 2016.

company subscribes RMB 19.6 billion. It is estimated that the total bond subscription is 1,000 times that of the issuance, and stock subscription is 2,000–4,000 times that of the issuance. This abnormality indicates that the monetary increment, as a result of People's Bank of China (PBOC) reduction of interest rate and reserve, circulates mainly in monetary and capital markets, not effectively accessing the real economy, leaving the real economy, especially micro and small enterprises (MSEs) with financing difficulty. It is worse that, monetary increment, together with monetary volume, circulate in monetary and capital markets and build up the tie between monetary policy and the two markets, weaken that between monetary policy and real economy, defecting the driving force of policy for economic growth and adding potential systematic risks. Such potential systematic risks are created in the following process. PBOC monetary increment flows to financial institutions, such as commercial banks, who, for fund safety and profit, timely operate the money in three ways: first, direct inter-bank short-term money transaction, such as repurchase and inter-bank borrowing; second, access to capital market through trust, fund and securities companies; at last, the rest of the money enters real economy through credit loan. In the process, the majority of money is in monetary and capital markets. This circulation aggravates. Many fund companies launch unlicensed securities business. Such capital circulation weakens the function of monetary policies and easily cause bubble economy, trigger potential systematic risks.

How to solve it? In the short term, we may take three measures. First, we may increase the proportion of direct financing, such as more enterprise bonds, to increase government-led special debt financing within financial framework. Second, we should strictly control and lower down the return rate of capital market, and maintain proper loan interest rate, thus not increasing the cost the market loan, and encourage financial institutions to grant loans. Third, we should increase the innovation and issuance of open financial bond varieties to provide diverse investment options and reinforce real economy. In the mid and long term, first, we should enhance the infrastructure in monetary and capital markets, including early warning and evaluation systems, to timely master fund flow, and we may set up transaction centers in the West and North of China to parallel Shenzhen and Shanghai Stock Exchanges. Second, we should speed up financial reform, enhance mixed financial management system in line with financial development rules, to minimize or prevent regulatory blind spots. Third, we should strengthen credit laws and regulations, enhance valuation and certification of business quality and channel quality, standardize credit subject's activities, and improve market demand–supply mechanism and risk prevention mechanism. Fourth, design of financial system and economic restructuring should observe the law of overall market development, to combine short-term effect and long-term normalization, combine international experiences and the practical conditions of China, prudently establish local exchanges, and probe into feasible policy measures. In this way, we could not only give play to the initiatives of the central and local governments and greatly lift the proportion of direct financing, but also effectively control risks, and promote the synergic development of virtual and real economies. In a word, we should be careful about financial channel businesses to play its role in increasing liquidity and prevent it from generating and disseminating systematic risks.

VI. Regulatory Philosophy and Mechanisms with Chinese Characteristics

What is the spirit of the Basel regulation? How to look at the balance of credit income and credit risk from a global perspective? How to improve the financial regulation system with Chinese characteristics? Financial innovation requires appropriate financial regulation which means "to keep things under control while protecting innovation and development capacity". These issues are to be addressed in this section.

The Basel Committee on Banking Supervision (the Committee) proposed the Basel Concordat in September 1975 and the Basel Accord in 1988, which defined the Tiers 1, 2, 3 capital of banks, established a set of universally applicable Corrective and preventive Action Requirements (CAR) standards that measure banks' in- and out-of-balance sheet risks by weights, and identified the basic model and factors of capital regulation, ushering in a new era of standardized banking regulation.

The main spirit of the Basel regulatory framework is to look at the *balance of credit income and credit risk* from a global perspective. The

Basel Committee and the Basel Accords have made important contributions to the building of banking supervision system. They have inspired banks to continue improving risk management, and also brought certain risk of regulatory policies. The US-led Basel I, EU-led Basel II, as well as crisis-driven Basel III are based primarily on effective market theory and risk management practice of large international banks. In the process economic globalization, the Basel regulatory framework generally benefits economic stability. However, it should be noted that capital-based risk management framework is flawed and is not totally applicable to developing countries.

First, the regulatory logic is contradictory. Risk coverage is expanded and various methodologies are given to measure risk weights. Such measures are intended to make more economical use of capital and objectively reduce the weights of risks. For example, compared with low-level measures, high-level measures easily lead to risk weights in favor of commercial banks, that is why they throw money on developing new measures. This is contradictory to strict regulation.

Second, it relies too much on technology. Technically calculated risks do not necessarily reflect objective existences, nor can technical progress eliminate risk potential. The Basel model becomes increasingly complicated, thus easily causing model risks, computation errors and misleading statements.

Third, it aggravates market competition. There are more and more risk management options, and more and more severe market competition, causing market turmoil. Meanwhile, excessive capital substitution effect causes arbitrage irrespective of regulation, thus amplifying asset risks.

It should be noted that, the new rules bring more and more complex risk management measures, which provide more options for banks and regulators based on the business status and risk management level and help banks improve risk management, but make universal applicability even more difficult. In particular, countries and regions vary largely in terms of internal risk management basis and technical level, especially between developed and developing countries, thus possibly aggravating the North–South gap.

The new regulatory framework brings more challenges to the Chinese banking industry.

The first is regulatory competition. China stays in separated industries and separated regulation, the fundamental problem of which is market segmentation. In a segmented market, regulatory competition in nature is competition among authorities. Neither does such a regulatory mode fit the development of a unified market, nor is it consistent with the centralized regulation requirement of the new rules.

The second is CAR. China has strict regulation requirements on bank capital, e.g. the minimum CAR and core CAR are set respectively at 11.5% and 8.40% for big state-owned banks, 10% and 7.20% for joint-stock banks, 10% and 6.92% for city commercial banks. Nevertheless, the Chinese banking sector is of typical capital consumption operation, with a large part of income from interest rate differential. Rapid business development calls for high capital supplementation. CAR has long been a major challenge to the Chinese banking sector, which will be even more critical according to the new CAR computation specified by the new Accord.

The third is economic growth. Research made the Bank for International Settlements indicates that, in the transitional period, one percentage point increase of the overall CAR of a country's banking sector will drag down its GDP growth rate 0.32% below the benchmark and the impact will last for 4.5 years; the implementation of liquidity indicators will drag down its GDP growth rate 0.08% below the benchmark. China's economic growth is in a shift — the growth rate drops from 8% to 6–7%. Thus, Basel III may be a factor that hinders Chinese economic growth.

The fourth is profitability. The stipulated core Tier 1 capital proportion will theoretically lower banks' operating leverage from 50 down to 15 times, impairing their fundraising and lending capacity and, possibly, profitability. Against the new trend, Chinese commercial banks should develop new profitability and accelerate transformation of operating mode from excessive reliance on interest rate differential to diverse operation, from extensive development to refined operation, i.e. customer, product, industry and region segmentation. It is also important to note that China is under a heavy pressure to develop capital measurement model. According to the Basel Accords, the forecasted loss of core capital risk measurement model is based on consideration of default possibility, default loss rate and default exposure in a certain period to come. Many advanced banks, to make the "future computation", make use of modern mathematical statistics based on, for example, historical internal credit rating, market neutral principle, insurance actuarial and equity pricing theory. For the Chinese banking sector, such measures are in their early stages and not normalized, for example, the infrastructures such as the default database is almost a blank. Credit rating of borrowing companies is mostly used for selecting customers and early warning, and has not formed a systematic and in-depth measurement management framework. The Chinese banking sector is actively applying such technologies but has a long way to go to create a standardized capital measurement model, due to the varied foundation and technical level.

Economic globalization has both common and individual features. Market diversity is also an important feature of international economic development. Financial regulation should observe international guidance and, at the same time, should give full consideration to national conditions and requirements. In drawing upon international experience, China may consider the following measures.

The first is to establish the regulatory philosophy with Chinese characteristics. (1) People-oriented: We should, besides prioritizing the core effect of capital, enhance the development of market players and market credit which are the essences of risk prevention and the basis of regulation. Capital does not move and people have ideas. Capital is valueless unless it is used by men. Regulation should focus on people. (2) Technical applicability: Technical and computation progress is essential for higher regulatory effectiveness and standardized management flow. However, technology does not eliminate risks, and complicated technology increases operating cost and risk. According to the development and trend of Chinese financial market, China should stress on traditional banking businesses, and the first power rather than the second or third power of financial derivatives. China should simplify the models and emphasize technical practicability and applicability. (3) Regulatory coordination: We should give full play to the strength of political organizations, establish the regulatory coordination mechanism, highlight coordination and prevent transmarket risks. (4) Policy systematization: Regulatory policies are an integral part of public policies. Bank risk regulation should involve both internality and externality of banks and the banking industry. We should systematically design financial policies, industrial policies and financial market infrastructure plans.

Second, we should establish the rules on full economic cycle risk evaluation and dynamic capital reserve and provision. China's economy will stay in the state of growing for a long time. The CAR calculated as per the risks during the period may underestimate all the risks in face of Chinese banking sector during the entire economic cycle. Therefore, it is necessary to make "complete cycle" computation of such regulator indicators as CAR, and take such measures as pressure test to make proper evaluation of the impact of cyclical effect on core regulatory indicators, establish rules on forward-looking dynamic capital reserve and provision that requires banks to make more provision in economic prosperity and less provision in economic depression.

Third, we should further improve the capital management plan and capital in cash supplementation mechanism. Chinese businesses usually do not supplement capital in cash until loan and asset increase brings regulatory pressure. The supplementation, passive and lagging behind, is often in a process of: loan increase — asset increase — CAR decrease — seeking capital in cash — application to regulator — preparation upon approval — implementation plan — acquiring capital in cash — trying to match CAR and asset. In 2009, CBRC issued the *Notice on Soliciting Opinions Concerning the Guidance for the Implementation of the New Capital Accord* and the *Notice on the Mechanism for Capital Supplementation of Commercial Banks* that require commercial banks to sharply lift CAR standards. According to Basel III, international capital regulation is aiming to become dynamic, reasonable and long-effective. Chinese banks should make proper plans for capital supplementation, intensify innovation in this regard, enhance research on such products as trust hybrid securities portfolio (priority share and mixed securities), and develop feasible supplementation instruments.

Fourth, we should build long-term risk management mechanism. (1) Establishing real-time capital monitoring system: We should accelerate IT-based capital management, analyze daily changes in risk assets, calculate CAR changes in a timely manner, fix the increase of risk assets, estimate short-, mid- and long-term CAR, plan beforehand capital supplementation on this basis, and adopt in-depth, real-time and refined capital management. (2) Establishing and perfecting risk warning analysis mechanism: We should establish the monitoring, evaluating and warning system that covers all business risks including credit risk, market risk and operating risk to make early risk warning, timely risk identification, and, through measurement models, accurate risk quantification and scientific risk pricing for timely prevention and mitigation.

Case Study: What is the Chongqing Land Quota

The Chongqing Land Quota represents a very influential land reform. In 2008, the Chongqing municipal government actively explored the organic combination of increasing urban construction land and developing rural collective construction land. After confirming and quantifying the quotas for the rural collective construction land that will participate in the program, the government publicly auctions the land in the form of bills at the Chongqing Rural Land Exchange, a case worth studying.

1. *Case Background*

Extensive use of rural collective construction land. With the deepening of urbanization, new industrialization and agricultural industrialization, more and more farmers go to cities and towns, so the value realization form of rural land needs to be explored urgently. Since Chongqing became a municipality directly under the Central Government, the proportion of its rural population has dropped from 69% to 48%, but the rural residential land has only decreased by 48 km².

The imbalance between supply and demand of its urban construction land is outstanding. In recent years, the total area of the construction land applied for use by Chongqing's city and county-level departments has remained at 3.73×104 h m², 4.5 times Chongqing's quotas for new construction land planned by the Central Government each year, the imbalance between supply and demand of its urban construction is prominent.

The rapid development of urbanization provides a possibility for rural construction land reclamation and quota transfer. At present, Chongqing's rural per capita residential land is nearly 240 m², while its urban per capita land is only 80 m², so if a rural resident becomes an urban resident, 160 m² of construction land will be saved.

2. *Reform Process*

Design ideas of Land Quota: "The goal is to protect arable land and realize the property value of rural land; to achieve this goal, the government will establish the market-oriented incentive mechanism for reclamation and guide the farmers to reclaim the unused and abandoned rural construction land; some of the reclaimed land will be used for guaranteeing the rural development, and the rest of it will be publicly traded in the form of land quotas, which can be used in Chongqing's urban and rural construction planning". Its specific implementation includes the following four steps:

First, reclamation. The premise is that the farmers have enough residential land, and they voluntarily apply for the reclamation of their unused or insufficiently used residential land. The municipal government and district governments are responsible for the approval of the application for reclamation and the acceptance of the reclamation results. Upon the acceptance, the property certificates are the legal basis of the construction land, and the scope of land reclamation is controlled based on the data of the second land survey; the area of the effective construction land is confirmed by the measured results, and the technical standards for reclamation are strictly observed (soil layer thickness: 40 cm; slope: less than 15°).

Second, the transaction. Once the reclamation passes the acceptance, land quotas are issued. The land quotas are property certificates that can be traded publicly. With land quotas, the property owners — farmers or village collectives — can sell their land at the Chongqing Rural Land Exchange. According to the regulations of the Rural Land Exchange, legal persons, natural persons with independent civil capacity and other organizations in the urban and rural areas can all bid for land quotas. The transaction price of land quotas is determined by the market and the information is disclosed publicly.

Third, the implementation system. The Chongqing municipal government has established a system for planning land quotas and classifying increase–decrease of linking land quotas to guarantee land demands. The system specifies that new profit-oriented land in Chongqing's districts and counties must be traded with land quotas, and land quotas should not be applied to the land outside urban construction planning to ensure that they serve the urban construction.

Fourth, the distribution. After the cost of land reclamation is deducted from the transaction price of land quotas, the remaining revenue belongs to the farmers and the village collectives. Around 85% of the residential land revenue belongs to the farmers, and 15% of it belongs to the village collectives; the Chongqing Rural Land Exchange will transfer the revenue belonging to the farmers to their bank accounts. The land quota revenue for the reclamation of the construction land for rural collective use should be given to the village collectives. A minimum transaction price is also set to ensure that the farmers' land quota revenue will not be less than RMB 120,000 per *mu*, and the village collectives' land quota revenue will not be less than RMB 21,000. The reclaimed arable land belongs to the village collectives.

Anyone who wins the bidding for a land quota, whether it is legal or by inheritance, has the right to develop a plot of land designated by the land quota at a proper location within the scope of urban construction planning. The net proceeds of developing the plot of land belong to the land quota holder, who shall pay taxes to the government. Therefore, the land increment revenue brought by converting rural construction land into urban use is divided among the farmers, village collectives, developers and the government based on the rights defined by the land quota system. It is not difficult to understand that Chongqing's land quota system is derived from the "urban and rural construction land increase-decrease linking system", and the economic function that the two systems want to achieve is also similar; both of them is designed to "move" the unused, abandoned and insufficiently used rural construction land to urban locations with higher land prices through reclamation, in order to release the potential of land appreciation. However, the goal of the land quota system is not achieved through administrative means, but realized through the mechanism of supply and demand and competition in an open land market.

3. *Characteristics of the Case*

First, the transaction scope expands. Chongqing's land quota transaction transcends the administrative boundaries required for the traditional "increase–decrease linking land quota" trial, and includes the outer suburbs into the "increase–decrease linking land quota" trial.

Second, to achieve price unification. According to the land quota system, the linking land quotas in different districts are auctioned together, and their auction proceeds are divided according to the land areas under them. The prices of land quotas have nothing to do with the land's locations and differential rents, and are determined only through auction to achieve price unification.

Third, the land transaction in the form of bills. The land quota system turns linking land quotas into bills. Through this system, the land is transferred from immovable property into exchangeable commercial paper, enabling solidified land resources to become floating assets. Besides, the land is auctioned uniformly in the form of land quotas, which is conducive to selling the linking land quotas in both inner and outer suburbs with a unified standard.

Fourth, land occupation following reclamation reduces the "linking" risk. Under the land quota system, the arable land quotas are increased after the reclaimed rural construction land passes the acceptance, and then auctioned at the Land Exchange; this is an operation mode of land occupation following reclamation.

The land quota system was designed by the Chongqing municipal government to provide convenience for urban land use, but it is undeniable that it is an attempt to reform China's rural collective construction land system. If we combine the land quota system with the urbanization policy that encourages farmers to reside in town, it may help farmers dispose their residential land when they decide to leave rural areas, because it is designed to benefit the government and the farmers. The land quota system is also conducive to the reform of China's household registration system and rural land system, so as to promote the coordinated urbanization.

Chapter 8

Interaction between the Government and the Market

What is the interaction between the government and the market? How to create the mechanism for the balance between the government and the market? This section focuses on the changes in government functions, and stresses that development finance is an important bridge between the government and the market. The role of government in market depends on its authority and strength. As the development environment changes, the government and the market need to interact, and government authority and strength need to be readjusted accordingly.

I. Dynamic Process of Government Function Adjustment[1]

In the early stage of capitalism, rapid economic growth was attributable to market promotion which, however, was inadequate to maintain national security and stability, providing the government with opportunities to expand its authority in specific times and specific conditions. World War I (WWI) told people how government concentrated social resources to cope with emergencies. During the war, for the provision of military materials and mobilization of defense market laws could not be observed, the government was considered more powerful than the market. During emergencies, it was understandable that the government will surpasses its set authority boundary. This laid the foundation for the government to expand its social, political and cultural functions. This way the Roosevelt

---

[1]This section is published in *Hongqi Wengao*, 2012, Vol. 23, pp. 12–15.

government was not challenged by the society when it intervened into the market during the economic crisis in the 1930s. After the crisis, the government expanded its functions in the name of market correction. Economic crisis and war changed the relation between government and market, and changed public's opinion on government intervention, giving the government more opportunities. After WWII, Western governments not only maintained market order but also took active measures to overcome market defects and improve market conditions, implying they had transformed from guards to dynamic governments.

The governments of developing and socialist countries, in particular, stress "catching up" development strategies. They use centralized and planned ways to enhance the foundation of the national economy. In the process of economic development, the government not only acts a navigator, but also directly organizes and participates in the entire national economy, hence becoming "active government" or even "government economy", with its functions covering all social and economic aspects. The "active government" features excessive intervention and monopoly, leading to extreme government functions. Things turn into their opposites when they reach the extreme. Since 1970s and 1980s, as such countries transformed from planned to market economy, the government authority reduced and gave priority to building effective governments.

From the government transformation, we learn that: they transform from guard, dynamic, active to service governments as shown in Table 8.1.

Government function transforms from external safeguarding to internal regulation and social service, or from small to medium and active

Table 8.1.   Types of Government Functions

Type	Function	Characteristic
Guard	Small functions: maintain market order, work as guardian	Market economy
Dynamic	Medium function: overcome market defect, market restructuring	Mixed economy
Active	Comprehensive function: overall aspects	Planned economy
Service	Active function: coordination, promotion, allocation, public service	Knowledge and public economy

functions. Small function provides pure public objects: national defense, law, property right, macroeconomic management and basic medical service; medium function, on this basis, stresses basic education, environmental protection, monopoly regulation and information integrity; comprehensive function covers all aspects; active function, in addition to those of small and medium function, focuses on coordination, promotion, allocation, reallocation, public service, protection of the vulnerable and social equity. The transformation is the shift of government organizational idea, structure, institution, function and methodology from a type to another, in order to meet political, economic and social development needs and adapt to and keep balance with social evolution. As a whole, government transformation does not simply mean fixing current administrative system, nor general readjustment to government functions, but establishing a systematically improved modern government adapted to economic and social transformations. Functional transformation is a continuous process to meet the needs of social and economic development. Government administrative function, as the superstructure, should fit and benefit social and economic development.

Public management is an important government function in this new century: In the 21st century, as economic globalization deepens, international economic order is experiencing profound changes and world economic structure great readjustment. Governments of all countries are reviewing their role in economic development. Though countries differ in development level and environment, the international community has reached a consensus that government must perform its critical functions and play a proper role in economic and social development. One of the important functions is public management, including economic control, market regulation, social management and public service. This is a special public power[2] which, in the new century, should be seen from: first, macro control to maintain macroeconomic stability; second, establishing market rules to maintain economic order; third, adjusting social distribution and organizing social security; fourth, enhancing international cooperation to cope with global issues. Government in the 21st century should highlight

---

[2] *A Collection of Works of K. Marx and F. Engels* (Vol. 4). Beijing: People's Press, 1995, p. 94.

strategy, science, sustainability and responsibility, and should be an advocator, planner, guide, public servant and judge.

Government should advocate scientific social and economic development, adhere to eco-social coordination to administer the management and stress social responsibilities in economic development. It should also enhance education on social cost, social benefit, social management and social protection. For this purpose, government personnel should build up their awareness and responsibilities for eco-social coordination.

Government should make social and economic development plans. Before executing economic intervention, it should analyze the budget in line with cost efficiency principle. Only if intervention brings more benefits than the cost can we say it is effective.

Government should actively guide and standardize activities, lower down economic and social development costs and promote overall economic and social progress and social security through formulation and execution of laws and regulation, policies and development planning.

Government as a public servant should use its legislative, judicial and administrative powers to serve and help economic and social development.

Government as the judge of economic and social activities should define the boundaries between the powers of economic organizations and individuals, and between the ownership of natural resources to internally minimize external cost and effectively control the cost of economic and social development.

From the perspective of international experience and trend of social transformation of the government, public service function is becoming the core function of the Government. Therefore, the core task of the transformation of the government is to deal with the relation between government and market, overcome their disadvantages and give play to their advantages, realize sound interaction between government and market. This is the precondition and institutional basis of modern public service system.

II. Transformation of the Functions of Chinese Government

Transformation of government functions is crucial to deepening the reform. Transformation of government functions is the inner demand of economic and social development, and is decided by the objective law of social development. Of the three-decade reform and opening up, the first decade was top-down reform driven by the government and it has been bottom-up reform since 1989. With the deepening of reform and opening up, the Chinese government made two evolutionary function transformations: The first was the transformation from political leader to economic development government, and the second, from economic development government in market development to public service government in market maturity. This means, the government gradually shifted its priority to establishing and perfecting basic public service system, improving market environment and promoting fair competition, or it handed over efficiency to the market. The market provides efficiency and government safeguards justice. This has become the basic value of Chinese market and government. For this purpose, China is in urgent need to set up the mechanism for the conversion between government and market powers and realize sound interaction between government transformation and social and economic transformation. Government transformation and social and economic transformation have triggered profound changes in Chinese economic development mode, management mode, social structure and social life. The success of government transformation decides its social and economic transformation, affects coordinated economic, social and political development, and enables the rise of the nation and modernization.

It should be noted that the transformation of Chinese government is quite difficult. Compared with governments of market economies, Chinese government shares commonness in terms of the relation between government and market, but features different start point and track. Market economies have higher start point and stress comprehensive functions with large flexibility. While Chinese government has low start point, and takes the overall power and manages the industry. It easily leads to overcontrol and market abnormality, overstressing the role of government or of the market. While creating economic wonders, China's economic and social development suffers imbalance, disharmony and improper transformation which is caused by excessively rapid or slow transformation. Many policies are transformed, the laws or ideology of the system did not. Credit imbalance is particularly seen in market credit system development, i.e. the imbalance between backward credit system and the financing demand out of the rapid economic and social development. The imbalances are closely related to government administration system, management measures and methods and public administrative capacity, and to government's macroeconomic control capacity. To solve this, we need to set up effective government–market interaction mechanism and dissemination channels.

We should enhance the government–market interaction mechanism: The interaction is a process where government and market mutually support and mutually promote internal and external powers. The key is to properly define the scope and strength of government and market authorities. That is to say, the government should further shrink its function and give power to the market and micro subjects, and enhance its power within its authority. Government authority refers to the scope of government powers, and the strength to the capacity of government to develop and execute law and policies. Referring to World Bank's 1999 World Development Report and Francis Fukuyama's (USA) *State-Building: Governance and World Order in the 21st century*, and based on Chinese development, I worked out the scope of government authority (*x*-coordinate) and the government capacity (*y*-coordinate) as shown in Figure 8.1.

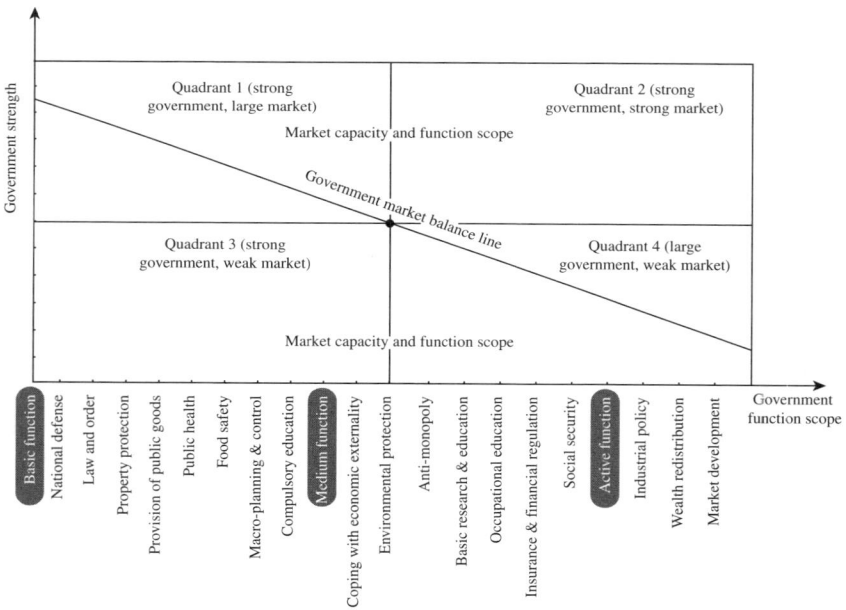

Figure 8.1.   Government–Market Interaction Mechanism

The line from top left to bottom right represents the balance of government and market authority scope and strength, indicating the strength of the two vary largely in different authority scopes. Any scope, more or less, is the combination of the two powers. In some fields like national defense, and law and order, government power should prevail to reflect state will and public interests. While in other sectors like social security, the market should prevail to exhibit its value, and the government should do nothing more than providing guidance and supplementation. As market economy deepens, the government still performs active functions, but cannot take the place of enterprises and other market players. For example, in terms of industrial policy, as long as an industry does not violate the macro planning, it should be the industry or the enterprises that decide the development and structural readjustment, and product update.

The figure can be divided into four quadrants. The first on upper left explains that social and economic development rely on market power, and government function scope is narrow but the power is strong, especially in national defense and law enforcement, i.e. large market and strong government, such as the US. The second quadrant on the upper right shows that government function scope is large and government plays important roles in many sectors. In contrast, market function scope is small but market plays significant roles, i.e. strong government, strong market, such as France and Japan. Quadrant 3 on the lower left is weak government, weak market, such as the developing countries. Quadrant 4 at bottom right is large government, weak market (China), but in some sectors the market is given too much freedom. China's deepening reform and setting up of the government–market interaction mechanism should go along the parallel line from Quadrant 4 to Quadrant 1, i.e. gradually narrowing the scope of government functions, and give power to the market to play a leading and decisive role in more sectors. At the same time, we should further identify the sectors where the government should enhance its authority and play a leading and decisive role.

In conclusion, the Chinese government's functions in economic sector, in the following years, should include the following. First, strategic planning function. The government should formulate industrial and urbanization strategies and accelerate system development in favor of economic development. System arrangement is a public object and an important task of the government. Governments at all levels should perfect local mid- and long-term strategies, and formulate supporting rules and regulations, to create favorable economic environment and safeguard social and economic development. Second, market system design and organization. China's market system is not perfect and economic subjects should be fostered. Government is the designer and leader of market reform and should make use of its organizing and coordinating advantages, work as a market builder and cultivator, enhance guidance, supervision, training of industrial organizations, in order to promote market economy. Third, macroeconomic control. Government function features planned economy, i.e. government attaches too much importance to specific projects and neglects its due obligations. Overcapacity of some industries is somehow due to functional disorder of the government. Therefore, the government should lighten its microeconomic control and enhance macro management and service functions. Fourth, basic social security. The government assumes large responsibilities for eco-protection, food safety, protecting the vulnerable and providing public products, thus needing larger control. Fifth, the government should give play to its special credit to support market credit system. Sixth, China as a large developing country will play an increasingly important role in the international community and accordingly shoulder more responsibilities. It is urgently needed to enhance the government's capacity of formulating and executing laws, of macroplanning and effective management, of corruption and malpractice prevention, of environmental protection, of basic security, of combating counterfeiting, and of coping with external competition.

We should strengthen policy communication: Domestic and foreign experiences show that effective government–market interaction mechanism must have a valid communication channel. The government cannot directly address various imbalances, but needs an effective channel between the government and the market, usually market intermediaries, industry associations, financial institutions, and especially development finance institutions. In market development process, government has national resources, including various resources for the development and implementation of laws and regulations. Effective government role is necessary for sustainable economic and social development, boosts market economy and individual activities, and improves market efficiency. Market intermediaries, industry associations, financial institutions, and especially development finance institutions rapidly cover those fields that used to be subject to the government, and drive up economic development. China should make good use of the capacity and resources which, combined with state credit, form the third power that benefits microsystems and financial infrastructures, boosts market development, and fulfills government objectives. In the past, the government decided the development, and at present, it is the government and the market that boost the coordinated and sustainable social and economic development through market development.

Communication channel is necessary due to market imperfections and flaws of the government. Every market more or less has defects, making government intervention in the market economy possible. But the government is not a panacea. Government defects are seen in many aspects. For example, lack of information or information asymmetry leads to errors in decision making; public institutions or government officials work for the interests of particular sectors or individuals, not always public interests; various interested groups affect government decision, and seek non-directly productive profit; the government sector has remaining legislative and discretionary powers, so abuse of power is possible; the decision-making, implementation, supervision and feedback mechanisms of government administration are not well coordinated and thus not very effective; distortion of government performance evaluation does not help develop incentive mechanisms; long-standing "government body culture" makes supervision and constraint mechanisms ineffective. Defects of the government force people to look for markets and complement the government's way. In the current market conditions, we should take the initiative to develop the market and institutions. In developing market economy, Western countries experienced a long process and paid a very high price. China should not take the old road or wait for the natural formation of the market. China can and must give play to late-mover advantage, combine factors such as the government, finance, and businesses, and build the credit system in an active, efficient and systematic way, for leapfrogging development.

III. China Development Bank (CDB) is an Important Bridge between the Government and the Market

Development finance is an important product of China's financial reform. In 1994, in response to the guidelines of the Third Plenary Session of the 14th CPC Central Committee, in order to more effectively focus on funding key national projects, the State Council decided to set up the China Development Bank (CDB). Unlike traditional policy banks, CDB strives to serve national strategies and policies in a market-based way. Based on constant innovation, it supports projects in key fields with medium and long-term investment, helps to improve people's living standards through inclusive finance, and facilitates the implementation of the "going global" strategy through international cooperation. It has created a model of development finance with Chinese characteristics, and plays an irreplaceable role in China's economic and social development and financial system.

Based on continued reform and innovation, CDB has strived to find the effective way of serving national strategies with finance, and transformed underdeveloped fields that are beyond the reach of the government and the interest of businesses into commercially sustainable sectors with mature markets. CDB stresses medium and long-term investment in key sectors and major projects. It puts over 80% of its loans into coal, electricity, oil, transport, agriculture, forestry, water, communications and public facilities, 60% into central and western China and old industrial bases in northeast China to promote urbanization and sound interaction between industrialization and urbanization and facilitate economic restructuring and change of growth model. In terms of people's livelihood, CDB seeks to establish an inclusive financial system where everyone has equal access to financing. CDB makes global business planning. It has successfully implemented China–Russia oil project, China–Pakistan oil project, China–Turkmenistan natural gas project and other major projects to safeguard national energy and resource security, and helped Chinese enterprises to explore international markets. It facilitated the establishment of Shanghai Cooperation Organization (SCO) Interbank Consortium, China–ASEAN Banking Consortium and BRICS Interbank Cooperation Mechanism which are important financial platforms China can leverage to boost regional economic cooperation and increase its voice and influence in international affairs.

CDB itself has also achieved great development and gradually shaped a business model featuring "one axis with two branches". It has created and maintained leading market performance. As of the end of 2014, CDB's total assets exceeded RMB 10 trillion, and net profit of the year registered RMB 88.4 billion, with a non-performing loan (NPL) ratio of 0.63%, less than 1% for 39 consecutive quarters. It leads the industry in terms of asset quality, and has made breakthroughs one after another in business performance. It has become China's largest bank in terms of medium and long-term loans, bonds, international investment and financing cooperation, and development finance, providing prominent support for the national economy.

In April 2015, the State Council's *Reply to Approving the Plan of China Development Bank for Deepening Reform* (the Reply) expressly states that the Bank should adhere to the role of development-oriented financial institution. The role is significant in five aspects. First, it points to the strategic importance of development finance under the new normal. The background to CDB's reform and development is the same as that to China's reform and opening up and economic development. Development finance ensures the healthy development of economic transformation. Second, it fully recognizes the practice and theory of development finance with Chinese characteristics. CDB as an institution of development finance, under the leadership of the central government and the State Council, with the support from all sectors and government at all levels, and with wide public participation, makes innovations in financial model, serves national strategies, and has found a way to integrate advanced international financial principles with China's realities, playing an indispensable role in the efforts to build socialism with Chinese characteristics. Third, it gives full expression to the need of financial diversity. Development finance represents a new model of finance that combines role of policy banks to serve the country and commercial banks' market awareness. It is an integral part of China's financial ecology and financial system. Fourth, it is conducive to the creation of China's development economics. The Reply not only put forward the orientation of development finance, but also clarified the concept of development finance, providing important theoretical and policy basis for China's development economics. Fifth, it marks that development finance has entered a new phase. CDB's role as a development-oriented financial institution goes with the launch of Asian Infrastructure Investment Bank (AIIB), showing the growing influence of development finance in China.

The Reply further defines development finance and underlines its characteristics in terms of purpose, credit, operating mode and management principles. First, it serves national strategies. In addition to financially supporting mid- and long-term strategic projects, it provides financing services for projects at the center of government attention and helps to coordinate immediate economic growth with mid- and long-term development. Second, it acts as the bridge between the government and the market, i.e. to give full pay to the roles of the two and organically combine government-led and market-based means. Third, it should develop innovative financial models: as the debtor, it should make use of state credit and be supported by other sides to raise funds via financial bond issuance and liquidize existing assets in the financial market; as the asset holder, it should boost the development of market credit system, foster market players, activate the vitality of and the inherent driver for economic development to organically integrate state credit and institutional credit. Fourth, it should adhere to planning-led development, grasp and make use of the inherent law of planning-led development to make sound fund arrangement, give priority to risk control, stick to market operation, raise the awareness of the operation, make overall use of "investment, loan, debt, leasing and securities", build specialized brands, intensify refined management, and highlight risk prevention and control, in order to achieve sound interaction between serving national strategies and development of the bank. Fifth, it should make overall planning for people's livelihood-related business such as inclusive finance, affordable housing, student loans and SMEs, thus supporting both country-focused and people-centered projects. Sixth, it should stress Party building, bank operation and development, give full play to the strength of political organizations, establish and perfect the management system combining business risk control and cadre supervision mechanisms, and organically integrate Party building and bank business. CDB as a national institution of development finance will exhibit its special strategic focus under the new normal and contribute to China's sustainable economic development.

In the 21st century, China faces the task of making the transition from agriculture to industry, and from planned to market economy. To speed up China's development, it is very important to give play to the government's organizational strength and the role of development finance, integrate the government, finance, businesses and other resources, actively develop the market, credit and systems, and promote transformations in a market-based way. Development finance should act as an important bridge between the government and the market and help to translate the power of the government into market forces.

Case Study: What is a Local Government Financing Vehicle

1. *General Information*

Local government financing vehicle (LGFV) is a new thing in the process of China's reform. It is an important way of interaction between the government and the market. In the 1990s and the first decade of the 21st century when there was no market financing system, local governments, to accelerate urban infrastructural development, allocated land, equity, fees, national debts and other assets to set up companies that meet the capital and cash flow eligibility for financing, and gave financial subsidies as commitment on repayment. The companies made direct or indirect financing, and received funds from various channels. The funds were put into municipal development, public utilities and other profitable or less profitable projects. LGFV helped local governments to raise funds from the market, and greatly benefited government-led infrastructure development and urbanization.

At the end of 2008, the Chinese government announced the RMB 4 trillion stimulus policy, which brought rare opportunities for local governments to increase financing and investment. Many projects of local governments involved RMB 30 trillion or more investment. LGFV gained rapid expansion all across the country. As of the end of 2009, there were 8,221 such companies, 4,907 of which were county-level companies. A large part of the debts of local governments were borrowed in those years. As of the end of 2015, government debts amounted to RMB 36 trillion or more, accounting for 60% of GDP, 1.78 times the central government's budgetary revenue for 2015. Of the debts, local governments were obligated to RMB 16 trillion, 1.23 times their budgetary revenue for 2015, or 1.84 times if contingent liabilities are considered.[3] By the end of 2016, China's social debt situation (debt / GDP) reached 259%.

How to look at the debt growth of LGFVs? Under China's economic management system, LGFV debt, after all, is the budgetary deficit of the central government. The Ministry of Finance is the ultimate debtor. The central government does not allow local governments to go bankrupt or close down and will take over the debts. Therefore, it is essential to look at the debt of China. According to the "National Balance Sheet of China" taskforce of the Chinese Academy of Social Sciences, China's total debt is lower than most of the developed economies, staying at a moderate level. Moreover, the debt of CDB (statistics as per capital construction loans), mainly for infrastructure development, made up 69% of the total, implying that the debts are controllable. In the 21st century, China stays at a developing stage so it is helpful to appropriately lift budgetary deficit and draw on the future profits in order to foster infrastructure development and the real economy.

However, the fast-growing debt of China has some negative impacts. First, government debt growth increases macro-economic risks. To pay the debt, local governments and the companies need more funds, which stimulates excessive investment and brings heavy burden to Chinese economy. Second, banks face more systematic risks. In the short term, tightened investment and credit policies push local governments to shadow banks, causing rapid expansion of the latter and systematic risks. Some underfinanced projects have to be abandoned, leaving banks with bad debts. In the mid- and long term, LGFV investments are characterized by long payback period, low or even zero profit, i.e. the project itself has poor solvency. The repayment is mainly made by the government which largely relies on land-related income and is subject to the real estate market. Such circumstances may be long-standing potential risks threatening the credit quality of the banking system. Third, the real estate bubble

---

[3] Refer to *The Beijing News* (December 22, 2015): The Report on Standard Debt Management by Local Governments was reviewed in groups at the session of the Standing Committee of the National People's Congress (NPC) and Chen Zhu, Vice Chairman of the NPC Standing Committee, warned about some local governments' *de facto* "bankruptcy" in the future.

expands. Many local governments rely on land-related income to repay the debts, and thus have strong motivation to raise the price of land which significantly pumps up real estate bubbles. Moreover, the land finance is overdrawing the profits or the financial resources of the next five decades. Collectively owned farmlands are acquired in the name of public interests at low cost and by force from farmers, which deprive their rights to survival and development, and endanger social stability. Fourth, land transfer fee, as a major source of repayment, goes down, local debt risk is increased. The Vice Chairman of the Standing Committee of Liaoning Provincial People's Congress said, "Years ago the land transfer fee was about RMB 200 billion per year, and now it is RMB 70 or 80 billion, down by two-thirds, indicating solvency risk".[4]

LGFV is an innovative financing channel created by CDB before its transformation into a corporation in 2008. It represents a strategic thinking to make comprehensive use of resources to hedge risks. It was less risky then because CDB issued bonds in financial market with state credit as guarantee, featuring long term, low cost, long loan term and low interest rate. However, a large number of commercial banks rushed in, "short-term saving and long-term loan by commercial banks, or short-term borrowing and long-term investment by LGFVs, resulting in risk increase". This issue should be carefully considered and we should be on high alert.

2. *Deep-Seated Causes*

The governments lack decision-making mechanism and policies that respect economic rules.

First, there is no democratic and scientific mechanism for major economic matters. The Chinese government is very efficient in political mobilization and decision making, which is a strength that benefits economic development. Nevertheless, if it is used excessively or improperly, this may impair sustainable economic and social development during which funds control should be neither too tight nor too relaxing. We have to respect economic rules and local conditions, make appropriate resource allocation, arrange fund input according to project progress, and steadily march forward.

---

[4] *Ibid.*

Second, LGFVs lack self-control. Most of infrastructure projects are exclusively invested in by local governments which, out of hunger for investments, increase investment budgets year by year, forcing LGFVs to raise funds irrespective of cost and risk. Thus, financing actually is out of control, out of limit and incapable to shoulder risk liabilities. It is the governments that take the place of companies to choose and select projects.

Third, LGFVs are not transparent and subject to external constraint. Seen from the entire society, local governments' investment and financing activities exhibit great externality, involving the practical interests of financial sector (creditor) and ordinary people (beneficiary), and should be subject to social supervision and constraint, while currently there is hardly any strong external constraint. Firstly, there is no soft constraint from banks. The financing vehicles confuse local government credit with corporate credit. In case of a solvency problem, it is next to impossible for banks to recover loans because the loans are out of lender's supervision. Secondly, there is no NPC constraint. As of today, government-invested projects are yet to be included into the NPC public budget regulatory system.

Fourth, local governments are out of external and internal constraints. Firstly, the leaders are in service for a short term (1.7 years on average), and thus can hardly take over the debts. Secondly, the budget law specifies that local governments are prohibited from borrowing or guaranteeing. But the fact is that they did borrow money and made guarantee. However, no one was blamable because everyone does so. The right to borrow and the obligation to repay are not bounded, so no one cares about repayment. Thirdly, the central financial authority and local administration are not clearly identified. Local leaders would like to contribute to local region, but had few choices other than borrowing.

3. *Countermeasures*

In solving the problem, the national policy should neither be "a sudden stop" nor "let it be". The former may ignite some long-term problems, causing "hard landing" of credit, investment and economy. The latter will allow potential risks to accumulate. An appropriate option could be gradually standardizing LGFVs and leading them to the market. Existing debts and investments should be sorted out and standardized, and increments should be subject to rigorous control. Specifically, LGFVs should be categorized: standard ones should be supported for further development; average ones be merged and reorganized; infrastructure projects that meet long-term local needs, irrespective of advanced investment and low profit, should be protected because it would be more expensive to abandon them; investment in energy-intensive, highly polluting and overcapacity industries should be suspended. Private capital may be introduced to vehicles with good cash flow to enhance its Corrective and preventive Action Requirements (CAR); vehicles for charity projects should seek or enhance secondary repayment sources. Shutting down LGFVs may cause new problems, for example, abandoned vehicles bring more potential systematic risks.

The problem is rooted in the management system, so the following measures should be considered. First, we should transform the functions of local governments. We should properly define local governments' investment and financing functions, redefine function scope, establish scientific and democratic decision-making mechanism and prevent improper expansion of functions. We may make strategic readjustments. From the perspective of restructuring, governments should be clear about and abide by the boundary of their functions and narrow down investment sphere. They should liquidize government assets, shift from "out of boundary" sectors to needed sectors, and transform from all government to public service-oriented government.

Second, we should improve the visibility of government debts. We should develop local government balance sheet that includes the investment and financing activities scattered in the vehicles, to effectively control the source, use, structure and total amount of investment. We should establish local government construction budget system to develop uniform investment budgets according to disposable income, including investment sources and use. We should set up local government early warning system to control investment and financing risks through monitoring indicators and risk signals. We should also control local government debt ratio to ensure their financing models conform to the objectives of balanced total amount and harmonized structure.

Third, we should promote marketization of LGFVs through issuance of local bonds and asset securitization. Bond-based projects are highly transparent and self-constrained. We should accelerate the standard development of municipal bond market. It should be noted that, the premise is evaluation of local governments' financial status and solvency. In addition, securitization of existing asset or liability is theoretically a choice to divert and transfer risks. However, we must learn from the financial crisis triggered by the US subprime market. We should at least protect the investors and strictly control the leverage. Otherwise, the risks may be aggravated. We should attract private capital to government investment sector. Private capital may be an important participant in government financing innovation in infrastructure and public services which in turn expand the investment channel of private capital.

Fourth, we should enhance the rigid constraint on local government debt. The central financial authority should establish as soon as possible a national debt management system. Banks should promote more prudent risk management of local government financing, including strict monitoring of project capital, solvency analysis, and strict control over the solvency and loan risks. We should establish local government investment accountability system to prevent excessive debt from being passed down to the next government and prevent debt risk from being handed over to higher authorities.

Fifth, we should change the profit source of the "land finance" through full-amount handover of land transfer fee to national treasury, or establishing national land foundation or "special account of paid use of state-owned land" to check the incomes and expenses of land transfer fee and new construction land fee. We should strictly observe the regulation that land transfer fee is subject to budgetary management, include land transfer fee, which used to be shared by local governments, into local budgets to form a complete local financial budget system, and prevent local governments and officials from unauthorized use of the land transfer fees.

In brief, LGFV is an interaction between the government and the market, and is an active strategic means of market development. The problems arising from LGFVs, especially those generated and worsened by the excessively rapid and substandard growth and poor management ability in coping with financial crisis, are caused by the immature interaction between the government and the market. Therefore, to address those problems, we should start with improving such interaction.

Chapter 9

China's Land Reform

How is the evolution of China's Land Reform impacted by the law of equilibrium? The history of social and economic modernization in rural China is indeed the history of the evolution of the rural land system. Farmer economy of a moderate scale is an important phenomenon and trend that have long been in existence in China. On the macro level, it is advisable to consider "separate land management and national land fund" and cultivate different market entities. On the micro level, it is suggested to optimize rural land management system and establish an overall structure consisted of state-owned farms, collectively owned agricultural production bases in city suburbs, rural construction land, household managed land, ancestral land, land of specialized farming households and land of joint-equity cooperative companies. The role of different entities in food security should be further identified and fulfilled and the market operation mechanism of "land to the tillers" should be perfected to accelerate realization of rural economic and social modernization.

"Land to the tillers" is the basic appeal of human society and also an important law of social and economic development. "Land is the foundation of the world" (quoted from *Hsun Tzu*). There are no other ways for humans to survive and society to develop without land. China is in severe shortage of arable land. Therefore, land reform is a major topic in both history and present. Since the agricultural age, land reform has never ceased and it was the starting points of both the New Democratic Revolution of the previous century, the movement of bringing order out of chaos after the Cultural Revolution, and the reform and opening up. Entering the 21st century,

China witnessed rapid economic growth and urbanization and land served as not only the key to agricultural growth, but also a major tool for rapid urban expansion and fast industrial development. Land has a bearing on local governments' financing capacity, revenue growth and social stability, and frequently triggers social conflicts and interest disputes. All problems concerning "agriculture, rural areas and farmers" in China are manifested by land-related incidents. In the modernization process of China in the 21st century, land will be a key to development and even to survival. Therefore, it is a task of strategic importance that we proactively explore ways to ensure "land to the tillers". The following section summarizes the evolution of the rural land system since the founding of the People's Republic of China and presents the characteristics of Chinese classical land systems. This will help us better understand the current rural land system of China and gain insights into the possible transformation of the system in the future. The historical changes of the Chinese rural land system, successful or not, embodied the profound effect on rural social and economic modernization.

I. Rural Land System Reforms in China

China has gone through three stages of rural land system reform since 1949. The years before 1979 was a time of major changes and after that, there were the period of contracted land till the turn of the century, and the period of evolution in the contracting relationships since the new century.

1. *1949–1979*

From the founding of People's Republic of China in 1949–1979, the rural land system of China roughly went through three periods. The first lasted from 1947 when the *Outline of Land Law of China* was promulgated to 1956 when the private land ownership of equal allocation was adopted. The second was highlighted by the collectivization movement which started in 1956 and gradually changed land ownership from private to collective ownership. The third was featured by the *Regulation on People's Communes* ("60-Article Regulation for People's Communes") issued in

1962, clarifying "ownership at three levels with production team as the basis". The system lasted until the household contract responsibility system was instated after the Cultural Revolution. Accordingly, the models of organization and production, and the ownership of means of production and life were also under constant adjustment.

The first period was a short time of fragile private ownership. In September 1947, the Communist Party of China (CPC) held the national land meeting in Xibaipo Village, Shijiazhuang City, Hebei Province and adopted the *Outline of Land Law of China*. The core message of the law was to set up the framework of a "land to the tillers" system by equally allocating land, confiscating the land owned by landlords and the surplus land of rich farmers. Besides, land was to be allocated at the township or village level and owned by rural households. As a guiding document that reflected the general direction of China's land reform, it stimulated the enthusiasm of farmers for reform and production and played a decisive role in ensuring the CPC's victory in the war. It should be noted that farmers were able to own their land only because the state exerted its power to confiscate land from landlords and rich farmers and distributed it equally among farmers. This served to establish the deeply rooted concept among farmers that wealth and resources came from the state, laying the foundation for the legitimization of the CPC's regime in China. It bears far-reaching influence on China's rural land system afterwards and proves it easy for state power to transform rural land system.

The *Constitution* promulgated in 1954 protected the land ownership of farmers in the form of national supreme law, stipulating that the state shall protect farmers' ownership of land and other means of production according to the law. Meanwhile, China launched the campaign of "Three Great Transformations", put forward the strategic objectives of national industrialization and socialist public ownership, and proposed agricultural cooperatives. The government intentionally guided and helped individual farmers to increase production and encouraged them to voluntarily organize cooperation in production, sales and credit, so as to promote agricultural production and rural economic development. As Chen Yun, who was then in charge of financial and economic work, said: "Previous experience told us that developing cooperatives could increase average yield by 15–30%".[1]

It is worth noting that despite the emergence of mutual help groups and primary agricultural cooperatives, land remained in private ownership of farmers and could be traded and leased. While maintaining the private ownership of land and other means of production by farmers, only the right to use the land was invested into cooperatives for centralized arrangement and use and profit was distributed according to labor and land contribution. Mutual aid groups and primary cooperatives hinted at a trend of separation between land ownership and land use in rural China to some extent. At this stage, farmers in cooperatives lost the right to use their land, but still own the land and may choose to quit cooperatives.

Unlike the *Outline of Land Law of China* which legitimized the regime mainly through "equally allocation of rural land", the *Constitution* of 1954 was more about production development and it touched upon the conversion of agricultural land. The *Constitution* stipulated that for public interests, the state could compulsorily purchase, expropriate or nationalize urban and rural land and other means of production as permitted by law. In other words, the state was entitled to expropriation of land owned by farmers by giving cash compensation in the forms of compulsory purchase, expropriation, or nationalization.

The second stage was highlighted by rural land collectivization. By April 1956, primary cooperatives had basically covered all rural areas of China and the progress towards advanced cooperatives and people's communes had started. As a particular feature of the primary cooperatives, farmers joined the cooperatives by handing over the right to use their land and selling their draught animals and tools to cooperatives at a set price for centralized management. Advanced cooperatives practiced collective ownership and centralized management of land and other means of production and distributed benefits according to the labor contributed by individual members. When it came to the stage of people's communes, the right to trade and lease land attached to private ownership was also given

---

[1] Chen Yun. *Selected Works of Chen Yun (1949–1956)*. Beijing: People's Publishing House, 1982, p. 225.

up as rural collective ownership of land was established. Such institutional arrangements, together with centralized food purchase and sales, enabled the government to make use of price scissors to support industrialization with what was gained in agriculture, but inflicted severely negative impact on agricultural production itself.

First, let us focus on the rural land system at the stage of primary cooperatives. The fact was that during the "Three Great Transformations", the government already started to intentionally guide the farmers toward rural cooperative economy, but did not rush to change land ownership. In 1956, primary cooperatives began to take explicit moves to substantially change the rural land system. First, there were signs of restrictions on land transfer, though land contributed to collectives was still entitlement to dividend, it was also stipulated that "the return on land must be lower than the return on agricultural labor" because "the income of agricultural cooperatives was created by the labor of members, not by land ownership". Second, the dividend of land "should generally be decided by cooperatives as a fixed value, not to change along with its yield". Third, land of cooperatives "shall not be leased" because land leasing was an act of "exploitation" just as "hiring long-term laborers" and "lending money for interest". Fourth, it was made clear that agricultural cooperatives in China would develop from a primary one into advanced ones and in advanced cooperatives, "return on land would gradually be abolished" and "land ownership would be converted to public ownership by a cooperative as a whole, i.e. collective ownership by all members". It could be seen that though land was still owned by farmers, with returns entitled to them, land could no longer be disposed of at their discretion.

Next, let us turn to the rural land system at the stage of advanced cooperatives. In fact, primary cooperatives were converted to advanced cooperative in a very short time span. The *Demonstrative Articles of Advanced Agricultural Production Cooperatives* stipulated in Article 2 that the cooperatives shall, according to socialist principles, transfer main means of production from private ownership by their members to collective ownership, organize collective labor, practice "return according to individual members' ability and work" and give the same return for the same work done regardless of gender or age. In other words, member farmers must hand their privately owned land and main means of production such as draught animals and large tools to collective ownership and water conservation facilities such as private pools and wells attached to land must be transferred as well. However, means of living and a small quantity of trees, poultry, livestock, small tools and tools necessary for household sideline production remained in private ownership, so were some small plots of land set aside for personal needs.

The *Articles* also provided that the cooperatives should set aside a certain quantity of land and allocate it to members to grow vegetables. The quantity of land allocated to each household should be determined based on the number of family members and such land for every individual generally should be no more than 5% of the average per capita land area of the region. Thus, farmers had no choice but to join the cooperatives; those who did join the cooperatives would not get dividend for their land; land of the cooperatives cannot be leased, except for the right to use small pieces of land assigned to households. Public land ownership, which was achieved through cooperatives, had the greatest economic significance in that it eliminated the possibility of earning returns through land transfer. Everyone now earned their living only by laboring. Cooperatives members did not sell labor but worked "for the collective self". Of course, except for contributing their land ownership to advanced cooperatives as share, farmers reserved ownership of means of production and set aside some small plots of land for personal needs.

Next, we should have a look at the rural land system during the period of people's communes. As the "Great Leap Forward" movement was progressed, the government set up people's communes on the basis of advanced agricultural production cooperatives. By the end of October 1958, people's communes basically covered all rural areas nationwide, the scope of collective ownership expanded further, and more land was turned into public ownership. Meanwhile, public administration is unified with the communes and the entire population, including those to be born in the future, was members of people's communes. "Collectives" are the major entities in a system of rural public land ownership, and the public property ceased resting on private properties of farmers and turned to base itself on the elimination of farmers' private properties. Commune members all earned their living through labor and the land of communes was never again allocated. Collectives opened up their boundaries and embraced all newcomers as masters of communes no matter whether they had land or not. Homestead of farmers was previously regarded as a "means of living" and thus was not, and also should not be, included in the classic socialist transformation. However, it was indeed turned into collective ownership at the stage of people's communes. Thus, people's communes radically restructured the Chinese countryside, basically eliminating private ownership and making farmers proletariat that earn income only by delivering agricultural labor.

The third stage was characterized by the withdrawal and reinforcement of collective ownership. People's communes took all means of production and living into public ownership and dampened farmers' motivation for production. When this was combined with the prevalence of the "communist wind", excessive boastfulness and natural disasters, rural areas in China sank into extreme difficulties that lasted three years. In order to correct the adverse trend of the "communist wind", the CPC Central Committee drafted the *Regulations of the People's Communes* to standardize rural relationships and reassure the farmers. The Regulations were eventually passed at the Tenth Plenary Session of the Eighth Central Committee of the CPC in 1962 after several revisions, and was more commonly known as the "60 Articles of People's Communes". The most noticeable change was the shift away from the trend of blindly emphasizing rural public ownership which was prevalent in the early days of people's communes.

First, it established a three-tier ownership system. The "60 Articles" stipulated that the basic accounting unit of people's communes was production teams, the lowest of the three tiers. One advantage was that it generally delegated the ownership and management of rural land to production teams, which were small in size and autonomous in production management and benefit distribution. Team members had a stronger sense of ownership to means of production and living, had clearer production expectations and were more motivated in production.

Second, the owner of the rural land was clearly defined. Though the "60 Articles" forsook the inclination of turning all rural means of

production and living entirely into public ownership, it further consolidated the collective ownership of rural land instead and converted all rural
land into collective ownership. Article 21 provided that all land within the
scope of production teams was owned by the teams. Land owned by the
teams, including small plots of land set aside for personal needs, small
areas of mountains reserved for individual demand and homestead, was
forbidden from leasing and trading. This actually announced farmers'
homestead publicly owned in institutional terms.

Most importantly, rural land transfer was totally banned. The
"60 Articles" extended ban on land transfer to cover all rural land including
small plots set aside for personal needs and homestead. One of the consequences of public ownership of farmers' homestead was the "separation of
housing and land" that has been long in existence in rural China. Housing
of team members was permanently owned by the members and can be
traded or leased but the homestead on which the houses were built is in
control of production teams and cannot be traded or leased. Such a system
persisted until the household contract responsibility system was launched.

2. *Reform of Rural Contracted Land*

The household contract responsibility system was a great reform of the
Chinese rural land system. The reform which took place in the late
1970s and early 1980s profoundly influenced the land system in rural
China and even the development of the Chinese economy in future
years. The household contract responsibility system transformed the
dual collective ownership of the right to own and the right to use rural
land in China and established a new system of "collective property
rights and contract-based household operation". It changed the logic
from establishing and changing ownership through a government-led
political movement to forming effective property rights via trade
between government and farmers and between farmers themselves. By
creating an environment where trade could be made within the space
allowed by relevant policies, the state loosened its grip on the rural
economy in exchange for stable tax revenues, a low-cost control system,
and political support from farmers. In the meantime, farmers paid tax
and assumed management responsibility to obtain long-term rights to

use and transfer land and to own what is left after all due payment to the government was made. They also gained full or partial ownership of some non-agricultural resources.

Generally, the attempt to divide land to households in Xiaogang Village, Anhui Province, was regarded as the beginning of the household contract responsibility system. In November 1978, Xiaogang Village of Fengli Commune, Fengyang County, Anhui, held a meeting to divide land responsibilities among households under the framework of collective land ownership and collective production. Three issues were covered: Land was to be divided among households; they would no longer ask the state for money and food; if any cadre was sent to prison, members of the commune promised to help raise his/her children to the age of 18. Land in the village was then divided and contracts were signed, heralding the advent of the household contract responsibility system. This attempt was later accepted by the government and the household contract responsibility system was rolled out across the country with the No. 1 documents of the CPC Central Committee for five consecutive years from 1982 to 1986.

Document No. 1 of 1982 pointed out that public ownership of land and other basic means of production will remain for a long time, and the various responsibility systems practiced at the time were all production responsibility system of socialist collective economy. It also stipulated that production team members who contracted land had only the right to use the land, not to trade, lease, transfer or abandon it. Otherwise, collectives may take the land back. The document attempted to separate collective land ownership from collective land management right, so as to transform ways of rural land management and stimulate farmers' initiative for production.

Document No. 1 of 1983 continued with the responsibility system that separate the rights to own and to use rural land but launched reform on the integration of public administration and people's communes. People's communes were dismissed. It also pointed out that rural China shall pursue balanced development of agriculture, forestry, animal husbandry, sideline production and fishery and comprehensively promote agriculture, industry and business operation in good coordination, which laid the policy foundation for rural enterprises to emerge and thrive.

Document No. 1 of 1984 moved forward with the rural land system reform in three aspects. First, the term of land contracts was prolonged to generally more than 15 years. The terms should be even longer for projects with a long production cycle and for development-oriented projects, such as orchards, forests and reclamation of barren mountains and deserted land. This allowed farmers to anticipate a longer time and encouraged them to increase investment, do more to enhance soil fertility and practice intensive management. Also, various production factors were allowed and encouraged to flow in rural areas. For instance, the transfer of the right to use land was encouraged to a moderate extent and fund of farmers and collectives was allowed to flow across different regions without restriction. Additionally, farmers were allowed to move to cities as long as they brought their own food. This actually acknowledged that as rural division of labor developed, more and more farmers would no longer be attached to land and a trend of de-agriculturalization would emerge.

Document No. 1 of 1985 focused on two reforms. The first was the reform of the unified purchase of agricultural products. Except for a few varieties, the government no longer required farmers to sell agricultural products to the state in a uniformed manner and at set prices; instead, products would be purchased on the basis of orders or market conditions. Therefore, the agricultural production would change along with the market, instead of the government's intentions. The second was to encourage the development of rural enterprises and encourage farmers to work in the mining industry. This loosened the control over the purpose of rural land use and unleashed the vitality of rural areas by further adjusting the land system.

Document No. 1 of 1986 paid special attention to the development of rural enterprises and the occupation of farmland for non-agricultural purposes as a result of the expansion of agro-industrial system. The government started to intentionally keep the rural land system reform within a certain scope. The State Council began to take the decision making and administration over small town planning, construction and management and water and soil conservation projects back into its own hands so as to contain the occupation of farmland for non-agricultural purposes.

The five documents of the early 1980s ushered in a new and important stage of the development of rural China. By legitimizing the household contract responsibility system pioneered by farmers, the government conferred to farmers the right to use, benefit from, and transfer land on the basis of rural collective land ownership, greatly motivating farmers, releasing their long-inhibited productivity and addressing the problem of feeding the large Chinese population. From 1980 to 1985, the annual growth of farmers' income exceeded 10%, which was quite unconventional.

From the late 1980s to the end of the 1990s, the large-scale industrialization and urbanization absorbed a huge number of surplus rural laborers. As the mobility of rural economic factors increased, the transfer of contractual rights became an urgent need. Such changes in reality coincided with farmers' call for "keeping policies unchanged" and the promise made by the government in response. The government hoped to adjust the contract at an appropriate time point as circumstances changed, while farmers were not opposed to property right transfer but did not prefer to see the government use its political power to push for changes from the top down as it once did. Therefore, in the late 1980s, the institutionalization and legalization of the new rural property right structure was put into the reform agenda. It became imperative to file and confirm farmers' rights through legislation, provide more ways to protect people's property rights and increase the cost of infringement, and on such basis, form a procedure for the revision of property right contracts through consultation between relevant parties.

3. *Evolvement of Rural Land Contractual Relationship in the New Century*

The *Law on Land Contract in Rural Areas* promulgated in 2002 safeguarded the land right of farmers in the form of law. Further, the Third Plenary Session of the 17th CPC Central Committee in 2008 proposed to give farmers more rights for contract-based land use and put in place better safeguards for such rights to keep the current land contractual relationship stable and unchanged for a long time. This was the first time that the government announced to keep land contractual relationships unchanged in a document issued at a top-level conference of the CPC. Document No. 1 of 2014 proposed to stabilize household contract right and revitalize land use on the basis of rural land collective ownership. Since then, "the separation of land ownership, contract rights and operation rights" became the latest principle for the rural land contract system.

First, guarantee was offered by law. The *Law on Land Contract in Rural Areas* conferred farmers lasting and guaranteed right to land use and its transfer in the form of law. The state ensured the long-term stability of rural land contractual relationships through legislation and safeguarded the legal rights of land contractors, allowing no organization or individual to infringe. The state also offered protection for legitimate, voluntary and paid transfer of contractual land rights by law. Contract terms were pre-scribed at 30 years for farmland, 30–50 years for grassland and 30–70 years for woodland; for woodland of special tree species, the term could be prolonged with the approval of national forest authorities. The *Property Law* of 2007 further provided that rural land contract and management rights were usufructuary and farming households, as non-owners of con-tracted land, enjoyed exclusive right to the occupation, use and benefits of land.

Second, contractual relationships would be kept unchanged for long. It was proposed at the Third Plenary Session of the 17th CPC Central Committee that the government would stabilize and improve the system that combined collective and household management of land. This involved two aspects: First, farmers would be given fuller and better secured land contract rights and the contractual relationship would be kept unchanged through land right confirmation, registration and certification. Second, the transfer of contractual rights would be encouraged and a mar-ket should be established and improved for this purpose. Forms of the transfer would include sub-contracting, leasing, swap, transfer and joint-equity cooperation so that operations of a moderate scale could be promoted.

Third, the ownership, contract rights and management rights of rural land were to be kept separate. In 2014, the CPC Central Committee issued its Document No. 1, *Several Opinions Concerning Comprehensively Deepening Rural Reform and Accelerating Agricultural Modernization*, proposing the separation of rural land ownership, management right and contract right for the first time. It also mentioned that on the basis of rural land collective ownership, household contract right shall be kept stable and land management right revitalized. Here, the right of management and the right of contracting are clearly separated. The document made another breakthrough by allowing the management the right to use the rural land for mortgage. Thus, on the basis of stable and long-term rural land contractual relationships, farmers were granted the rights to occupy, use, benefit from and transfer their land and to use it for mortgage and guarantee.

On September 29, 2014, the Fifth Conference of the CPC Central Committee Leading Group for Deepening Reform Comprehensively adopted the *Opinions on Guiding the Orderly Transfer of Rural Land Contract and Management Right and Developing Moderate Scale Agricultural Management* and the *Pilot Plan on the Reform of Actively Developing Joint-equity Cooperation among Farmers and Conferring Collective Assets the Function of Shares*, stipulating that under the precondition persisting with rural land collective ownership, separation of contract right and management right shall be promoted and the pattern of separating ownership, contract right and management right and transferring management right shall be formed. The purpose was to explore joint-equity cooperation among farmers and conferment of more property rights to farmers, clarify ownership of property rights and perfect various functions, so as to release the potential of various rural production elements.

On December 2, 2014, the Seventh Conference of the CPC Central Committee Leading Group for Deepening Reform Comprehensively adopted the *Opinions Concerning the Pilot Work on Rural Land Expropriation, Commercialization of Collective Construction Land for Business Purpose and Homestead System Reform*, pointing out that we must protect the three bottom lines of not changing the nature of public land ownership, not breaking the red lines for farmland protection, and not undermining farmers' interests; and press ahead in good order based on pilot programs. The document also proposed several measures for the reform of rural land expropriation: endeavoring to narrow down the scope of land expropriation, formulating lists of expropriation in a well-regularized manner, improving the mechanism of conflict and dispute settlement, making fully public all information on land expropriation and ensuring a reasonable, standardized and diversified guarantee mechanism for farmers with land expropriated.

On November 2, 2015, the General Office of the CPC Central Committee and the General Office of the State Council issued the *Comprehensive Implementation Plan on Deepening Rural Reform*, confirming the principles to be followed in deepening the rural reform and homestead system reform.

II. Discussions on Deepening Land Reform

Land system reform, especially the measures proposed in the past several years such as separating rural land ownership, contract right and management right, promoting pilot projects of rural land expropriation, commercialization of collectively owned land for commercial purpose and homestead system reform, and conferring farmers long-term and secured land use right and legitimate transfer right for land use and management right in the form of law, is an effort to correct the leftist approach during the time of people's communes and a manifestation of individualized land interests. It is of great significance to the regulation of the rural land market, reasonable transfer of agricultural land and the achievement of our goal of "land to the tillers". However, generally speaking, the series of land system reforms so far have not reached the depth provided in the *Outline of Land Law of China* of 1947 and the *Constitution* of 1954, both of which stipulated a framework of equal land allocation and the protection of land ownership by the supreme law. Though the circumstances, tasks and difficulties we face right now are different from the past, the fundamental question remains unchanged: How to stimulate the initiative of farmers in order to realize "land to the tillers"?

One of the priorities of our current land policies is to facilitate urbanization and relevant examples include "houses with limited property rights"[2] and the *Provisions on Transferring the Use Right of State-owned*

---

[2] In the late 1990s and especially since the beginning of this century, as urban construction developed rapidly and cities expanded in size dramatically, urban land supply for construction got increasingly tight. Also, as housing price surged, real estate developers reached out to rural areas and turned their attention to homestead. Many urban residents also rushed to rural areas to purchase and build houses. Consequently, on the homestead in city suburbs commercial residential buildings mushroomed and such housing was known as "houses with limited property rights". In response, the government issued a series of policies and regulations to contain the development of such housing projects.

Land through Bidding, Auction and Listing promulgated by the Ministry of Land and Resources in July 2002. Major difficulties in land expropriation and management mainly occurred in suburban areas of cities and such difficulties were quite different from what happened in the countryside. In the countryside, what we most urgently need to address is how to realize land operation of a moderate scale, improve labor productivity and increase farmers' income. Problems in pursuing moderate-scale land operation had something in common with those encountered in promoting urbanization, yet differences remained. We may need to devise differentiated land policies for the countryside with a view to achieving land operation of a moderate scale, which involves both property right and the spirit of traditional cooperatives. Therefore, an issue that requires in-depth theoretical and practical discussion is how to maintain good factors in the traditional system while eliminating the leftist and unreasonable factors so as to accelerate the establishment of a modern land system suitable for the realities of rural China.

Urban–rural bifurcation remains the basis of China's land management structure. Different land ownerships in urban and rural areas determine that the land right structure, the way of land use and the scope of capitalization are bound to differ. First, land in urban and rural areas is governed by two different legal systems and thus the corresponding right structures are entirely different. Rural land contractors are entitled to land use, returns and land transfer for agricultural land but once agricultural land is converted into land for other purposes, it no longer belongs to its previous owner and all previous rights attached to it will be gone. Second, institutions and individuals are allowed to enjoy use right of urban state-owned land for construction. By comparison, land that farmers can use as collective land for construction is limited to three types, namely land of rural collective enterprises or joint ventures of farmers and other institutions and individuals with land use right contributed as share, collective construction land for public facilities and public welfare undertakings and homestead of farmers. Rural collective land for construction is not allowed for leasing and homestead is obtained in the capacity of members of collective economic organizations, with one house allowed for each household. As housing and land are separated, holders of homestead use right enjoy rights to the occupation and use of collectively owned land, but not rights to the returns or transfer. Third, the use right to state-owned construction land can be used for guaranteed mortgage. In contrast, rural land and attached can be used for mortgage only if it is "the contractual and management right of wasteland obtained through bidding, auction and public consultation", "use right to construction land occupied by mortgaged workshops and other buildings of rural enterprises", or "rural land contracted through bidding, auction and public consultation". It is explicitly stipulated that the use right to construction land of rural enterprises and the use right to collectively owned land such as farmland, homestead, small plots of land set aside for personal needs and small areas of mountains for similar purposes cannot be mortgaged.

Viewing our existing land policies in a the historical and international perspective, we find three more deeply rooted questions: How to address the issue of nominal owners of state-owned land; how to materialize land use right into clearly defined tangible interests available to farmers while keeping land in state and collective ownership, facilitating moderate-scale land management; how to ensure sustainable use of land resources to support sustainable development of the country.

To answer these questions, we need take a broader perspective, be detail oriented in practice, and create innovative mechanisms. After some thinking and comparative studies, I would like to put forward some suggestions on deepening the land reform. The core message here is structural improvement, financial deepening and market building. My suggestions are: to separate land management on the macro level, separate farmland management on the micro level, a "land to the tillers" market mechanism and land funds which combine land and finance. I hope my views can serve as a spur to induce greater wisdom and lead to solutions to issues related to agriculture, rural areas and farmers.

III. Optimizing Land Management Structure and System

According to *the Bulletin on Land and Resources in China* (excluding Hong Kong and Macao), the country has a total of 2,627,068 km^2 of grazing grass, 1,224,443 km^2 of farmland, 112,878 km^2 of garden land, 235,047 km^2 of woodland, 255,327 km^2 of land for other agricultural purposes, 257,284 km^2 of land for residential, industrial and mining,

22,332 km^2 of land for transportation and 35,895 km^2 of land for water conservancy facilities. The remaining of the territory is not in use.

The *Constitution* of China states that land resources are owned by the people and the representative owner is the State Council. However, in reality, land is in direct control of local governments at various levels while the State Council can only execute its power via local governments. Therefore, it is largely fair to say that state ownership is in effect ownership of local governments and the state's ownership is in a way nominal. Consequently, local governments are not prudent enough in land use because land belongs to the state and the more they can use the better. Also, local governments pursue local interests instead of interest for all, which greatly reduces what the state gains from land use, undermines the macro regulation of the central government and affects the country's overall capacity for sustainable development. Therefore, it is imperative that we address the undermined dominance of the state in land ownership. On the national level, separated land management may be an option worth considering. Measures should be taken to improve the national land management system, establish the ownership structure consisted of state ownership, local government ownership, urban collective ownership and rural collective ownership, creating varied market players.

A desirable design for separated land management is to allocate the long-term use right of land for construction purposes (transferable land) to the five levels of central, provincial, municipal, county and township governments by issuing regulations of the State Council based on an appropriate ratio determined according to factors such as national strategies, population trend and industrialization prospect. Land allocated to these five levels of governments should be parallel and can be transferred on the market to some extent. This design has three advantages. First, the dominant role of the state in land ownership is enhanced for part of the land while multiple parties can all be motivated to assume greater management responsibilities over construction and be more proactive in relevant activities. Second, construction through land resources is clearly assigned to local governments and they can make better plans for mid- and long-term development, make better use of the limited construction land and promote local economic and social development according to their specific conditions. Third, since land use right is owned by multiple entities in this system, one entity may purchase it from others when they need more land to use and this will give rise to a land market on the macro level and empower the central government for macro regulation. Such a land management system has long been in use in the United States. In 2004, state and local governments of the US spent USD 1.5 billion on the purchase of land development right and roughly 2 million acres of arable land across the country was obtained through such purchases.[3]

However, it is difficult for China to adopt this system for the time being. First, the country lacks the ideological and institutional foundation for diversified land right holders and it takes time to forge a basic consensus on this topic. Second, measuring and negotiating for the division of land between for central and lower-level governments will definitely be an arduous process that requires enormous political wisdom. Third, the reform of land management system is closely related to the reforms of political and administrative systems and other in-depth reforms and, therefore, requests design from the top level, effective coordination and good timing. Fourth, functional planning for land use must be improved not only on the national level, but more importantly on the provincial and county levels. Land should be categorized carefully into four types respectively for optimized development, prioritized development, restricted development and no development with clear delineation, clearly defined development period, direction and principles. This is the groundwork for this system and will lay solid technical and policy foundations for the land reform.

Despite all difficulties, we may still continue with our research in this aspect and launch pilot programs in some areas. Meanwhile, it is advisable to actively explore readily available improvement measures that suit our current conditions. A possible option is setting up of national land control fund. Following the principle of being government led, market oriented and managed by experts, it is suggested that we focus on mechanism innovation and sustainable development and create a land regulation fund system by putting part of land-generated income into central and provincial/municipal land regulation funds before calculating it as a fiscal

---

[3] Ding Chengri. *Land Development Right Transfer System in the US and Inspirations for Farmland Protection in China*, available at www.caein.com, April 16, 2007.

revenue. This can be a major regulatory tool that promotes the coordinated development of urban and rural areas. See my book *China Base: County-Level Economy and Society* "Land Finance" for more about this.

IV. Optimizing Rural Land Management Structure and System

In China, rural land is collectively owned. After a series of reforms, rights to occupation, use, returns and disposal of rural collective land have changed profoundly, but the collective ownership remains. According to the Ministry of Land and Resources, rural collective land nationwide, including agricultural land (farmland, woodland, grassland and aquiculture area), rural homestead and some other rural land for construction purposes, covers a total area of 6.69 billion mu, among which agricultural land covers 5.53 billion mu and homestead 160 million mu. Agricultural land and rural land for construction are valuable for production and living and even the survival stability of the country. Besides its fundamental role in agricultural growth, land is a major tool that promotes urbanization and industrialization in China. It also has a direct bearing on the distribution of land-generated income and interests among related parties and is regarded as a trigger of social conflicts and interest-related disputes, showing intensively the "agriculture, rural areas, farmers". China has far less per-capita agricultural land than the world average. To feed the current 1.3 billion population and even more in the future with limited arable land requires a grand strategy and strong efforts from multiple parties. On the basis of existing systems and actual conditions, I propose the new thought of separated farmland management. To be specific, efforts should be made to set up an ownership structure consisted of "land of state-owned farms, suburban collective produce bases, collectively owned rural land and shared village land, and household managed land" to ensure "land to the tillers".

The basic principle of separated agricultural land management is to divide agricultural land and rural land for construction across the country into five types. The first is land for large farms, including reclaimed farms. This part of farmland is state-owned and accounts for around 25% of the total area. With good natural conditions and some positive historical factors, this part of farmland is suitable for intensive management on a large scale and serves as the mainstay in China's food supply.

The second is suburban agricultural bases. This part of farmland is owned by local governments and collectives and is mainly agricultural bases in city suburbs, accounting for about 10% of the total and mainly ensuring daily supplies for urban residents under the leadership of provincial and municipal governments.

The third is rural collective land and shared village land. It is owned by local collectives and accounts for 5–10% of the total. Collective land is mainly managed by corporate farms so as to tap into the advantage of scaled intensive management. It can be used to do what single households cannot do or do well, land left behind by migrant workers can be put into better use in this way. This part of farmland serves as the leverage in maintaining sufficient food supply. Shared village land may include collectively owned land for construction, land for village schools or school vegetable fields, river embankment, idle lake basins, barren mountains, etc. This kind of land has been common throughout Chinese history. Fei Xiaotong wrote down what he saw on shared village land in *Earthbound China: A Study of Rural Economy in Yunnan*. Tang Zhiqing also mentioned the following in *A Study on Rural Society and Economy in Modern Shandong*: "In the 16th century in Shandong, wasteland reclaimed by garrison troops or farmers accounted for 9.2% of all farmland across the province, while official farmland and public farmland accounted for 27.2% and 13.6%, respectively. Official and public farmland combined accounted for 50% of the total and private farmland represented the other half". Here, we can see that shared village land is nothing new in China. During the course of industrialization in China, maintaining a certain proportion of public land may probably be an important way to improve our rural land system. For instance, by offering village primary schools one or two mu of vegetable fields each, we give students an opportunity to deliver labor to grow vegetables and provide fresh food for school lunch, improving school conditions.

The fourth part is household-managed land and homestead, rights of contract, management, use and returns which belong to farmers. As circumstances change, part of household-managed land may be transferred to the extensive clan or specialized farming households, and form three models of household land management (by households, clans and specialized farming households). This will account for about 30% of the total farmland.

The fifth is land managed by the corporate. Farmers can contribute land use right as share to set up modern collective agricultural companies and keep the rights to land occupation and returns in their hands and the result is that they gain returns on land share and even wage income if they also work at the company. By doing so, the government will reach its goal of scaled land management and improved labor productivity. This is a way of land management that we actively promote but farmers may fear that this is a disguised way of government land acquisition, that the company may not perform well or the share of their returns may not be reasonable, thus not willing to join such ventures. This part of farmland will account for about 25% of the total.

Though clans and specialized farming households may manage more land than single households, this is still far from what is required for scaled land management and they are still household management. These joint-venture companies are set up voluntarily by local farmers and have larger areas of land than clans and specialized farming households, but they are still unlikely to reach the level of modern scaled agricultural operation. With land managed by households, family clans, specialized farming households and joint ventures combined, land fully at the disposal of farmers will still account for around 55% of all farmland in the country. Therefore, roughly speaking, household managed land is still in majority in China's rural land structure. As rural population migrates and various policies take effect, typical land distribution of "1.3 mu per person and no more than ten mu per household" in China may be changed to "four to five mu per person and 20–100 mu per household". Even so, we still be no more than small farmer economy, or more accurately, small farmer economy of moderate scale. No matter what the prevailing land contract system is to be maintained or land ownership reform further implemented, moderately sized farmer economy will remain a major feature of Chinese rural economy for a long period of time.

Based on the actual conditions in rural China and the law with agricultural development, it is necessary to further optimize rural land management structure and system and clarity ownership and allocation of property rights in the form of regulations, which will help put into full play the position and effect of the state, collectives and farmer households in food security in China, maintain rural stability and protect farmers' interests. The state should greatly develop state-owned large-sized farms and suburban agricultural and sideline product bases and tap into benefits of large-scale management to provide pillar guarantee for food security. Government needs to moderately guide collective agriculture to develop and forcefully support construction of rural infrastructure, such as water conservancy and roads to take its effect for leverage and adjustment. Chinese people should support perfection of household management system of land to put into play the fundamental role of moderately scaled farmer economy in food security.

V. Establishing a "Land to the Tillers" Market Operation Mechanism

As industrialization and urbanization deepen, on the basis of maintaining land management by household, rural China will inevitably be motivated to require farmland transfer. In either private land ownership or any other ownership systems, corresponding market mechanisms will be needed as a prerequisite for normal land transfer. We attempt to design an appropriative land circulation market here to encourage non-tillers to give their land to others and better realize "land to the tillers". Such a market mechanism should include operation mechanism, cooperative service centers and means of trade.

1. *Operation Mechanism of Rural Land Market*

In such a market system, each county shall set up its own land market according to actual circumstances. One county market may reach out to another county or to a lower- or higher-level market in its operation, but should mainly be responsible for land trading within its own territory. Trading centers can be located at the county seat or a township seat and they should be led by the relevant county government, and hosted by township governments. Cooperative service centers should be created to cover all administrative or natural villages within the territory of the township, keep track of the status of rural land use and channel land toward tillers. This is how they are to perform their duty of rural land right management as delegated by higher-level governments and provide services for land transfer. This market is dedicated to the circulation of rural land, especially rural farmland. Other companies or individuals not engaged in agricultural production are not allowed into the market and whoever receives land via transfer is forbidden from using the land for other purposes.

The market will have its supporting systems. For instance, there should be management methods for the registration and measuring for land transfer to specify the procedures, forms, timeline and certification of land transfer. Different management methods will be adopted for varied forms of transfer such as subcontracting, assignment, swap, leasing and contribution as share; short-term farming on behalf of the original land use shall require no formalities. There should also be management methods for land transfer fees. In the event of land transfer between family members and relatives, there can be transfer fees or no fees as agreed upon by relevant parties. Fees for subcontracting, rental and assignment for transfer of land contract and management right should be determined through consultation between all parties involved and the returns on the transfer shall go to contractors and not intercepted or withheld by any other organization or individual. Methods of rewards and penalties for land transfer and use should also be put in place to encourage farmers to transfer land in the form of diversion and use transferred land reasonably. Related compensation policies should be formulated to promote land transfer to true specialized farming households and land joint-venture companies and punish transfer in violations and use of land for other purposes.

The market is also supported by a series of external interaction mechanisms. For example, in order to encourage migrant workers to give up land rentals in the process of land transfer and thus ease the burden on the recipients of transferred land, it is suggested that we devise a "rental for social security" policy, according to which social security premiums will be paid from the national fiscal fund for migrant workers and in exchange, no rental will be charged for farmland transfer from these people. The policy may promote the transfer and concentration of rural farmland.

2. *Cooperative Service Centers for Rural Land*

In the rural land market, rural land cooperative service centers are public-interest service providers at the basic level and serve as a transfer and trading platform for rural land management right that is created for farmers and other market entities under the guidance of local governments. Following the principle of voluntary, law-based and compensated transfer featuring fairness and justice, they are intended to put into full play the decisive role of the market for rural land circulation, provide services for orderly and standard transfer of rural land management rights, promote moderate-scale land management and reasonable land use, offering support to the sustainable development of rural economy. Their responsibilities include publicizing and implementing national and regional laws, regulations and policies on rural land and formulating provisions and rules on the transfer and trading of rural land management rights; organizing the trading business and give guidance to the signing of transfer contracts; coordinating between farmers, village organizations and enterprises in the process of trading, and safeguarding legitimate interests of trading parties; reviewing entities and conditions of trading. They are also in charge of collection of transfer information, market survey, customer archiving and registration, provision of service guide, release of information and notice to the public.

3. *Ways of Rural Land Transfer*

In the foreseeable future, part of today's land contracted by farmers will remain in the household-based economy and hold an important position in the overall rural land structure, while some of it will be transferred according to actual circumstances in multiple ways.

Land may be transferred to a relative for these people are the most trusted and such a transfer may not really change the nature of the land. For migrant workers, relatives will spare no effort to help once they come back home. Therefore, this form of transfer is the most natural and acceptable. Guided properly, such transfers may give rise to family farms which will constitute an important part in modern farms.

Land may be transferred to specialized farming households. This method of transfer is preferred by the government because it put the skills of specialized farmers into better use and builds up the scale of land management. However, in the process of such a transfer, good judgment and encouraging policies will be required. Some land will be transferred in this way in the future, but not too much, because farmers tend not to feel as safe as it should be when they transfer land to a relative.

Land may be transferred to joint-venture agricultural companies. Some land may be transferred to collectives for reallocation, either used as collective public land or used to set up collective or joint-venture agricultural companies, and become an integral part of the rural land structure. This form is also preferred by the government, but seems less assuring for farmers. They may worry that this is a new way for the government to change the status of property rights.

To sum up, household management, clan management, specialized farming farmers and land joint-equity collectives farms are entities of arable land rights paralleled to previously introduced state-owned farms and cultivation fields and suburban agricultural and sideline product bases. They together constitute the foundation of the Chinese agricultural economy. Such a new pattern takes into full consideration the real conditions of farmers and can help solve realistic problems in rural land transfer, avoid agricultural land lying idle, maintain rural stability and promote agricultural sustainability.

Case Study: What Are the Characteristics of Chinese Classical Land Systems and Inspirations[4]

"Land is a great experiment field of mankind, the warehouse of tools and materials of labor and the shelter and foundation of society".[5] Ancient Chinese people depended on land to make a living and created several important land management systems such as the "nine squares" land system, the system of equal land distribution, the system of stationing garrisons to grow food, and the system of permanent tenancy. The systems still have their influences today, more or less, on modern life. Getting to know some classical land systems is instructive for sound designing and building our of land system today.

---

[4] *Review of Economic Research,* 2015, (42), pp. 5–10.
[5] Karl Marx. *The Grundrisse.* Beijing: People's Publishing House, 1975, Vol. 2, p. 91.

(1) **"Nine squares" land system:** The "nine squares" land system was born in the feudal system as a small-scale agricultural economy within a small area.[6] It came into being on the basis of combined primitive land public and private ownership and was related to the five systems at the beginning of the feudal society.

First, it was related to the primitive commune system, under which tribes settled. Evolvement from nomads to settlement was a big step forward for human development. In ancient times, in order to adapt to the environment and overcome harsh conditions, our ancestors settled on land by tribes and ushered in the age of agricultural economy. The economy started from growing drought-resistant plants such as millet in hilly highlands (easy to defend) and gradually developed to growing rice, wheat, sorghum, beans, hemp and mushrooms in flat areas. The often spotted characters "praying for millet" in inscriptions on bones or tortoise shells could illustrate the point. Settlement for living and production involved the issue of planning from the very beginning for which plot of land to be used as residence, which as food production area and which as producing area of public offering for immolation or war. This formed the primitive land public ownership system, on which basis emerged production management institutions, i.e. primitive communes. The communes shouldered three tasks. The first was to organize members to arrange land into even plots of a certain area, dig trenches and build embankment for watering and drainage, so as to have virgin soil cultivated. The second was to manage plots of land used by commune members. As recorded in the *Rites of the Zhou*, a dedicated position was arranged to especially manage land distribution and rotation. The third was to manage public accumulations, public mountain forests and public land and organize worship of ancestors. As primitive communes developed, the "nine squares" plots of land expanded in area to 50, 70 or 100 mu, which were horizontally and vertically arranged into large areas of "eastern mu" and "western mu". The communes undertook increasingly more functions such as household registration, soil, military affairs, labor services and security. It is fair to say that our ancestors established the relationship between people and

---

[6]Lectured by Qian Mu and recorded by Ye Long. *Chinese Economic History*. Beijing: Beijing United Publishing Co., Ltd., 2014, pp. 12–14.

land first with the identity of tribes and tribe members or communes and commune members.

Second, it was related to the establishment of feudal system. As primitive agricultural economy developed, Chinese ancient society gradually evolved from clan tribal society to feudal society, which referred to "enfeoffment into different states". As stated in the *Rites of the Zhou*, a plot of land was planned and water trenches were built around the land, which would make a city-state. At the beginning of the Western Zhou Dynasty, there were hundreds of such city-states scattered everywhere and they became hundreds of vassal states. The states were small in area and even the largest one only covered an area of 100 square li. People of the vassal states grew farm produce within their respective state area and had different ways of life, including nomadic tribes outside the state area. Within each state area, there were city walls, villages, mountain forests, ponds and lakes and not all land could be used as arable land. Land cultivation required water, which came from wells collectively dug. According to the *Spring and Autumn Annals*, "well" was a unit of irrigation. Water of one well could irrigate one well of land, which covered 900 mu in area[7] and was cultivated by eight households. Therefore, the "nine squares" system could be viewed as small-scale agricultural economy under the feudal system within small city-states.

Third, it was related to private land and public land systems. Toward the end of the primitive society emerged "public land" and "private land". The former referred to land jointly cultivated by tribe members or commune members, while in the latter land was cultivated by individuals themselves. The "nine squares" system in the primitive society generally divided one plot of land of one square li into nine small blocks (like the Chinese character "井"), totaling 900 mu. The block in the center was "public land" jointly cultivated by eight households, while the outer eight were "private land" allocated to eight households for cultivation. Private land was lotment or "fixed property" that agricultural laborers held in order to maintain their reproduction, while public land was producing area that provided public accumulations to support wars and sacrifice offering. However, in a class-free society, laborers had to labor on public land and

---

[7] 100 mu back then is equivalent to 23 mu today.

returns of public land were owned by communes. Private land and public land became the relationship between private labor and public labor.

Fourth, it was related to the formation of private ownership system. As primitive agricultural economy developed, products of surplus labor gradually increased. Heads of some communes refused to give public-owned products and land back to the public and claimed this part their own properties or materials exclusive to their households. Land private ownership came into being accordingly and classes emerged, such as feudal landlords, noble class and "king". Therefore, the relationship between people and land was gradually branded with class and private land and public land became the relationship between necessary labor and surplus labor.

Fifth, it was related to the creation of taxation system. The *Zuo Zhuan* (Zuo's Commentaries on the Spring Annals) (594 BC) recorded "initial tax on land per-mu", which might be the earliest record on collection of agricultural tax among farmers discovered so far. The *Spring and Autumn Annals* explained in response that tax for one mu was levied on each mu of land. It also mentioned that it was officials' fault if private land did not have a sound yield (feudal landlords were not doing a good job in sending agricultural guides to coordinate cultivation by farmers) and farmers' fault if returns of public land were poor. After the initial tax on land per mu was levied, classes were further differentiated, with the landlord class becoming the ruling class and the broad farmers the ruled class. Types of taxes were also segmented. Farmers shall pay land rental to landlords for cultivating private land and land tax to government for cultivating public land. Mencius recorded three ancient ways of levying tax. The *gong* levied a tax of one-tenth of the average returns for several years; *zhu* levied a tax according to the actual returns; *che* levied a one-tenth tax in a fixed way, but surveyed the land every year.[8] The three ways nominally aimed at collecting public accumulative labor products as public "sacrifices". "Zhu" was also named Ji. Zhao Qi of the Eastern Han Dynasty mentioned in the Notes on Mencius that *ji* meant borrowing, i.e. borrowing labor of the public to cultivate public land". The so-called "ancient taxpayers paid

---

[8] Dictated by Qian Mu and written down and edited by Ye Long. *Chinese Economic History*. Beijing: Beijing United Publishing Co., Ltd., 2014, p. 15.

one-tenth tax" referred to the labor ratio between private land and public land at roughly 10 to 1. Therefore, emergence and development of taxation system accelerated the change with concept on land ownership and the reform of classical economy and society.

Karl Marx once proposed the famous theory of Asiatic mode of production according to materials collected from India, pointing out that the economic structure in some oriental classical Asiatic ancient countries were featured by "state-owned land, water conservancy and irrigation, rural communes, despotism, and integrated rental and tax". Though the "nine squares" system in ancient China did not show all of these features, but it was clearly branded with the Asiatic style and proved to be a classical Asiatic ancient land ownership system.

(2) **System of equal land distribution:** The system of equal land distribution was a land distribution system based on the number of household members adopted to ease conflicts in the circumstances of wide wealth gap and tense class relationship in a feudal society. As recorded in the *Book of Wei*, during the first year of Emperor Xiaowen of the Northern Wei Dynasty's reign, given that "farmers were exploited by the rich and powerful and got impoverished", Li Anshi submitted a proposal to the emperor suggesting the system of equal land distribution to aid the poor and contain exploitations. In 485 AD, Emperor Xiaowen adopted the proposal and promulgated the Order on Land Equalization later. The Order had 657 characters in total and consisted of 15 articles in five parts. Part one was about mulberry land, providing that the land should be passed on from generation to generation, could not be returned and could be "reserved in surplus and supplemented in shortage". Apparently, this was intended for the owners of large areas of land. Part two was mainly about idle land with no owner, specifying that "each man aged above 15 were granted 40 mu of idle land and women 20 mu, the same for servants and maids". This could help the state maintain a certain number of small-scale farmers. Part three was about the leeway that could be adopted in the process of implementation and included three "flexible bands", namely the band between "state-owned land" and "land multiplied when granted to people", between "broad area" and "narrow area" and between "selling allowed for surplus" and "buying allowed for shortage". This illustrated

that the compiler of the Order had rich administrative experience and was quite considerate in designing the Order. Part four included some customs handed down from the past, such as "first the poor and then the rich" and "offering to relatives", which put into play the role of clan relationship and kinship. Part five was some additional provisions about fields for mulberry, elm, dates, fruits and hemp, residence land and land allocated to officials according to their titles. The five parts together constituted the sum of various production relationships in feudal northern China in the 5th century.[9] They showed the following important features. Small households were advocated; attention was paid to building of village commune system; household registration was encouraged. In policy, tax was levied on both officials and farmers, rich or poor. This was of major historical and cultural value because it helped crumble the trick of "100 families registered as one household" by the powerful and embodied the concept of officials and farmers being equal before tax. The land equalization system was intended not to realize absolute equal distribution of land, but to restrict the rich and provide a bottom guarantee for the poor. In fact, the system was not quite equal and made many exceptions in favor of the noble and large landlords. It was recorded in the *Book of Wei* that "when land was first granted, the fertile was mostly distributed to the noble and the barren to the poor", a phenomenon that was quite unequal. However, generally speaking, the system was an important reform at the time. Some measures, such as "ownership held by the current owners for long pending problems" (*Book of Wei*) and "owners entitled to cultivation regardless of distance of land" (*Chu Jie Ji Lue* by Ding Yaokang) in the event of disputes over property rights, could help ease the contradictions between farmers and the rich, the powerful and other large clans, alleviate the tension between the new noble and the conservative and objectively stimulate the initiative of small farmers.

Exerting profound influence over the generations to come, the system of land equalization became the basic measure adopted by the ruling class to ease social contradictions, the foundation for farmers to make a living and prosper and also the economic proposition of Su Chuo and Wang Anshi.

---

[9]Zhao Lisheng. *History of Chinese Land Systems*. Hubei: Wuhan University Press, 2013, p. 92.

(3) **System of stationing garrisons to grow food:** The system of stationing garrisons to reclaim wasteland and grow food was a land management system that combined defense and wasteland cultivation in an organized way. Out of military needs, the state deployed troops and criminals to be punished to major border areas in the form of forced labor to defend the areas and meanwhile reclaim land for cultivation. The state offered the people provisions, wage and clothing. The people cultivated land full-time or part-time and submitted returns of the land entirely to the state. The system was partially military in nature and "spared the troops for cultivation" in the interval of wars. It successfully solved the issue of feeding and clothing to survive and addressed land management and defense. Up to the Wei, Jin, Southern and Northern dynasties, the system achieved a lot and was successful in many cases. For instance, in order to expel Hun and expand the state area to the Western Regions, Emperor Wu of Han Dynasty deployed 50,000 soldiers to cultivate land along the way and maintain the troop and eventually forced Hun to surrender. During the time of the Three Kingdoms, the kingdom of Wei unified the central plains by deploying garrisons to cultivate land; this method fed and empowered its troop. The kingdom of Shu adopted the same method to successfully defend itself and finally obtained the dominant position. The kingdom of Eastern Wu fed its people and occupied the area to the south of the Yangtze River by using the same system. Over its nearly two-centuries' history (338–538), the Northern Wei Dynasty attached great importance to the system and land cultivated in this manner was further expanded in the area. As described in the *Book of Wei*, there was a line in the north of the Northern Wei Dynasty and it was built up to prevent harassment of nomads in the north. Its military fortified points were "six towns" at the beginning and later developed to "nine towns", roughly extending to Fengning, Hebei today in the east and Baotou, Inner Mongolia today in the west and covering over 1,000 km from north to south across the Yellow River. Besides towns, the line also covered the Great Wall. As the *Book of Wei* says the Great Wall is over 1,000 km, and parallel to the fortified points and the Great Wall was the cultivation line, on which the system of stationing garrisons to cultivate wasteland was practiced. As recorded in the *Tong Dian* by Du You, in the reign of Emperor Ming of the Northern Wei Dynasty (520 AD), the number of households participating

in the system was twice the number during emperor Wu's reign of the Western Jin Dynasty (280 AD) and reached over 500,000.

The system was distinct in the following five aspects. First, military control was implemented. Besides county governance, the "land cultivation administration" system was also adopted to run in parallel. Second, it was widely adopted and started quite early in history. According to the *Book of Wei*, Tuoba clan, a northern minority group, already distributed land according to the number of family members in each household 150 years before the system of land equalization was adopted by the Northern Wei Dynasty. Third, tax was levied. Quota of forced labor or grain ration was implemented. Exploitations were severe and the state apparently assumed the identity of landlord.[10] Fourth, the system of tenancy was carried out. Officials leased public land or wasteland to lower-level clans and farmers and then collected rental from tenants according to lease contracts. Fifth, compulsory management was carried out. For instance, it was decreed that "no infant could be killed or abandoned" in an effort to increase labor force and "his wife and children would be held liable" if a soldier is missing in order to prevent soldiers from escaping. The rules showed a clear sign of the subordination of people.

(4) **System of permanent tenancy:** The system of permanent tenancy was a tenancy system under which tenant farmers permanently cultivated land of landlords. Regarding land relationship, ownership of land was separated from cultivation right. Farmers cultivated land of landlords and submitted rental, thus becoming tenants. They could be categorized to regular and permanent tenants. Permanent tenants could cultivate the leased land in perpetuity under the condition of submitting rental as agreed in tenancy contract and pass the land on to generations to come,[11] while regular tenants enjoyed land cultivation right for a shorter period of time. Therefore, permanent tenants had greater personal freedom and guaranteed production and life. Both regular and permanent tenants only enjoyed cultivation right to land. Landlords could be classified to primary

---

[10] *Ibid.*, p. 77.

[11] Cao Shuji and Liu Shigu. *Ancient Chinese Land Right Structure and Its Evolvement.* Shanghai: Jiao Tong University Press, 2014, pp. 10–25.

and secondary ones. The state could be viewed as primary landlord. In feudal China, the majority of land was owned by imperial power and especially in the prime time of feudal despotism, all the mountains, forests, rivers, grasslands, land unclaimed and land uncultivated were owned by the state. State-owned land, as the connection between means of agricultural production and agricultural producers, was partially distributed to royal families and meritorious statesman and partially leased to regular farmers. Land leased to farmers resulted in the tenancy system that emerged early in Chinese history. Royal families and meritorious statesman that were granted land did not cultivate land on their own, but leased or sold land to tenants and thus became secondary landlord. Both primary and secondary landlords enjoyed certain land ownership.

With land ownership and use right separated, the tenancy system in which tenants cultivate land belonging to landlords came into being. In this system, land with the attribute of ownership was called *tiandi* (literally meaning "bottom of the field") in history and land with the attribute of use right *tianmian* (literally surface of the field). Under a classical permanent tenancy system, landlords could only collect rental and pay the grain tax (tenants may pay the grain tax or landlord and tenants may share it equally), but were not entitled to increasing rental at will or intervening with the cultivation by tenants; tenants could cancel tenancy and sub-lease or sell the right. Both *tiandi* and *tianmian* were allowed for assignment, with assignment of *tianmian* not affecting rental collection by landlords and that of *tiandi* not affecting cultivation by tenants. Under an incomplete or transient permanent tenancy system, tenants could only cultivate land or cancel tenancy at will, but were not allowed to sub-lease or sell tenancy right. *Tiandi* and *tianmian* were respectively priced, but *tianmian* was mostly priced lower than *tiandi*.[12] Consequently, the far-reaching, lasting and rich traditional land market was formed.

In the traditional land market, *tiandi* and *tianmian* could be freely traded. Though tenants did not own the land, they had the use right which could produce "multiple owners for one same plot of land" and result in a

---

[12] Jiang Taixin. A brief review on the relaxing of patriarchal clan relationship in land trading in the early Qing Dynasty and the social significance. *Researches in Chinese Economic History*, 1990, p. 3.

permanent tenancy. The system assured tenants, enabled them to have a stable life and freed them from the threat of "rental rise" to some extent, but also bundled farmers to tenancy of the state permanently. In the trade of regular land owners, once *tianmian* right was separated from direct producers, tenants had to pay *tianmian* rental besides the previous rental (*tiandi* rental) and the burden would be heavier. Landlords could recover *tianmian* through purchase or withholding due rental and then integrate land ownership and cultivation right.[13] Trading and sub-lease of *tianmian* tended to turn into a land right paralleled to *tiandi* and triggered the phenomenon of "dual owners" or "multiple owners for one same plot of land". This would better mobilize and utilize land resources. Therefore, the permanent tenancy system played a dual role in traditional Chinese economic and social development in history.

5. Inspirations of classical land systems: First, the land system is a fundamental system of the Chinese economy and society. The course of economic and social development in China is fundamentally a process of land system changes. The land system determines the nature of a country's economy and society. Within the framework of Chinese classical land systems, though state-owned land and large landlords had a special position, agricultural economy remained the foundation of the society. Chinese economy could only develop once farmer economy advanced and *vice versa*. The economic prosperity in Chinese history was closely related to land systems and small agricultural economy.[14] As modernization in China deepened, traditional land systems collapsed and the role of small agricultural economy in social and economic development weakened, but traditional land systems, as a part of traditional culture, are a treasure passed on from the past and of great value of reference to modern economy. By comparing classical and modern small agricultural economy, between primitive and modern collective economic ownership or between primitive and modern peoples' communes, we find that though

---

[13] Zelin. *Contract and Property in Early Modern China*. Translated by Li Chao. Zhejiang University Press, 2011, pp. 15–50.

[14] Task Force on Chinese Agricultural Land System, Development Research Centre of the State Council. *Reform of Chinese Rural Land System*. Beijing: Peking University Press, 1993.

time has passed by, the principles turn out to be the same. It is mainly reflected by two relationships. The first is the relationship between benefit and equity. Collective ownership is intended to pool forces to get deals done, but it must take equity as a precondition. The second is the relationship between public and private ownerships. A society cannot survive with pure "public" or "private" ownership in place and a mature society must be one with both ownership systems integrated. In the process of developing socialist market economy with Chinese characteristics, we should take into full account interests of individuals and enterprises because economic development could not be inherently motivated without initiative from individuals and enterprises, and meanwhile look after common benefits of the entire society since without common development of the entire society, development of any individual or enterprises will not sustain.

Second, traditional Chinese land systems embody some features of Asiatic mode of production. First, land is nationalized, which is reflected in intervention with private properties (including land private ownership) by state power. Though since the Qin and Han dynasties, Chinese state power intervened with land in an inconsistent manner, the entity of the state as the actual landlord was not changed at all. Second, awareness and influence of primitive rural communes remain in different periods of time. In classical Chinese societies, the awareness and system of rural communes sustained all the time and were always restricting the deepening of private ownership to different extents. Though the communes in varied periods of time differed in intensity, scope and depth, such a flavor of oriental communes was always smelled. For instance, it was seen repeatedly in traditional literature that except for taking responsibility for their private land or land leased to them, farmers shall conduct labor on public land or supply part of their own products for public accumulation.

Third, the land system is the foundation of social governance. In the time of the "nine squares" land system, private land was cultivated by individual households and public land jointly cultivated by a number of households. It was land that connected people to form basic social organizations. The system of land equalization arranged household registration first and established the "three heads" system to manage land and household registration. Land provided the institutional foundation for social governance system. The system of stationing garrisons to cultivate waste-land raised the troops via land and it was land that embodied the strong political and economic power. The permanent tenancy system had land ownership and use right separated and it was land that created more conditions for distributing and exchanging social wealth. Chinese land systems not only safeguarded stability but also encouraged relocation, advocated both land use and fallow and made land both a permanent property and tradable. The household registration system built on the basis of land system has a history of thousands of years, and is an important part of Chinese culture and history and the embodiment of effective state governance. When promoting the land and household registration system reforms today, we should fully realize that bundling farmers to rural areas with rural household registration is not wise. Rural areas need labor force, but the issue cannot be addressed by household registration. A system is always needed in any country to integrate every member of the society. In history, Chinese household registration system based on land played an important role. Now, however, the system has lagged behind the time, not because land is not important, but because land is too important, serving not only the household registration system, but also national security and development. Land system is the important foundation for us to press ahead with land rectification and maintain national sustainability. In the age of information, while the course of modernization is driven ahead, traditional dual systems will inevitably be substituted by the unified civil system. Both rural and urban residents are citizens of the country and identification card of citizens should thoroughly replace traditional household registration management system.

Fourth, land rights are diversified. China had the tradition and experience in "land right diversification" in history and there existed no classical Chinese society, where land was entirely nationalized or privatized. Especially in the prime time of feudal society, land rights, such as ownership, use right, right to returns and right to disposal, were not concentrated, but scattered. The same plot of land could get multiple benefits out of *tiandi*, *tianmian*, tenancy and mortgage. Owners of *tiandi* paid tax for land; owners of *tianmian* leased land for cultivation; tenants could re-lease the land and even divide the land for selling. In this way, at least three to four people could enjoy benefits out of the plot. *Tianmian* could also be used as bridal dowry.[15] Once sold, land could be redeemed (unless otherwise agreed in the lease contract). After several times of segmentation and trading, land rights became so complicated and diversified that by the 20th century, revolutionary pioneer Sun Yat-sen proposed "equalizing land rights" and the Communist Party of China led the land reform to partially change "land right differentiation". These were of great significance at the time.

In Chinese history, land rights experienced the cycle of "decentralized–centralized–decentralized". Each and every farmer instinctively hoped to own land and more land. In the event of disasters, many farmers had no way out but sell their land and houses, which further concentrated land to a few hands. With arbitrary power intensifying land merger, almost all dynasties ended up with land being excessively concentrated to a small number of landlords and rich farmers, which was followed by uprisings staged by farmers, requesting to redistribute land. Therefore, the cycled change of land rights was in nature the result of balancing effect. Excessive decentralization will result in centralization and *vice versa*.

Looking back to the past, we could learn valuable lessons. It is fair to say that Chinese farmers attach great importance to the right to land returns and care deeply about relationships centered on land. Besides, they tend to keep clear boundaries with regard to rights, with no mistakes made at all about the holder of each item of right. The diversification of land ownership in history complicated and also enriched the relationship between people and land and also provided the historical foundation for us to devise diversified land systems today. When deepening the land system reform, while maintaining state and collective land ownership, we should take into account the actual circumstances and clearly identify land use right to generate clearly defined benefits for farmers with regard to equity, lease, pawn, mortgage, pledge and trust. It is also advisable to turn some important rights into property rights and take them as equals to land ownership providing legal support and basis in trading so as to create conditions for regular changes in holders of land-related rights. This can help build a relatively equal

---

[15] Mao Liping. Discussion on Marriage Dowry Land in the Qing Dynasty. *The Journal of Chinese Social and Economic History*, 2007, p. 2.

contract system with rights and responsibilities specified, establish standardized and advanced modern rural land market with Chinese characteristics, enable multiple different entities to share benefits of the same plot of land and promote reasonable transfer of land. It will also push forward combination of new countryside building and capital, solve the dual headaches of rural "moderate scale management of land" and "farmers attempting for non-agricultural employment but unwilling to give up small plots of land due to unsolved concerns" and accelerate realization of rural economic and social modernization.

Chapter 10

Innovation Balance and Manufacture of China

How to accelerate the enhancement of innovation capabilities according to Chinese conditions? Both the Third Industrial Revolution and the Sixth Scientific and Technological Revolution referred to by some scholars are based on the revolutionary breakthroughs in information technology, reflecting that the development of industrial economy is increasingly related to digitalization, e-science, intelligence and network. China is still a developing country in the development stage of industrialization with the coexistence of backwardness and sophistication, and of tradition and modernity. It is necessary to vigorously explore a path of industrialization with Chinese characteristics, featured by combining the promotion of traditional industries with the cultivation of emerging industries, the First Industrial Revolution with the Second Industrial Revolution, e-science with industrialization and traditional handcraft industry with advanced manufacturing industry.

I. Innovation Balance

What is the purpose of innovation? It is to solve problems that are unreasonable, uncoordinated and unbalanced. To attain this goal, innovation balance is needed.

Innovation balance is balance reached after voluntary innovative activities. Nature is full of variations but lacks innovations. It is great to break balance, but even greater and more difficult to make balance. Innovation balance is the only way that could both break and make balance. Scientific innovation originates from the balanced way of thinking, which is to pursue progress voluntarily or involuntarily to turn dissatisfaction into satisfaction. Meanwhile, scientific innovation is also the result of balanced way of thinking to discern invariableness from variation, and to find simple rules from complexity. Therefore, *innovation is the inherent path for the balancing and rebalancing of things*. In 1687, Sir Isaac Newton published the *Law of Universal Gravitation*, explaining the interrelations and balancing effects among things. In early 20th century, Albert Einstein established the theory of relativity, proposing a concept beyond the experimental capabilities at that time, and anticipating that in the universe there are space–time disturbance waves expanding outwards at the speed of light. Those were all great innovations that both broke and made balance, and were the effective application and profound reflections of the law of balance in scientific development.

The balanced way of thinking has always been the method of thinking of our nation in formulating important strategies. In dealing with the relationship between backwardness and sophistication, and between tradition and modernity, China has always considered innovation — particularly the independent scientific innovation — as the core strategy of our nation. In the most difficult times of economic development, Chairman Mao never spared any effort in developing the "atomic and hydrogen bombs and man-made satellites". In the preliminary phase of reform and opening-up, Deng Xiaoping clearly instructed that "science and technology are the primary productive force". In 1995, the Central Government put forward the strategy to establish a "scientific and technological innovation oriented country". The Third Plenary Session of the 18th Central Committee of the Communist Party of China (CPC) further called for "mass entrepreneurship and innovation". So to speak, our country has always attached importance to innovation in terms of both strategies and slogans.

However, although China has made some achievements in innovation, the level of innovation remains low and there is a lack of self-owned brands and proprietary technologies. Apart from the weak foundations, policies and measures should also take the blame. For instance, the division among industries, universities and research institutions — like a chronic disease — still exists; the "two skin phenomenon" remains outstanding in fiscal fund and commercial financing; and the social environment showing genuine respect for innovation is still far from being sufficient. In comparison, there have not been so many watchwords in developed countries in America and Europe, but many practical and effective measures are in place. For example, it is well worth thinking that there has been strong support from various funds for basic research and scientific and technological industrialization. According to the balanced way of thinking, to encourage scientific and technological innovation, we should not only emphasize strategies, but also adopt proper tactics. We should stress on both applied technologies and basic research. We should have the awareness of innovation as well as the spirit of the craftsman. We should stimulate those who are already leading ahead, and focus on crucial aspects among overall development. It is our valuable experience of accomplishment and effective way of development to stay united when tackling key problems and accomplishing great tasks.

We need to discover unbalanced aspect among balance and explore the path to achieve balance among unbalance. From the viewpoint of coordination between human and nature, the basic trend of China's economic development in the 21st century is to seek low-carbon development based on scientific and technological innovation. The low-carbon urbanization featuring low cost, low emission and low entry-level is the new connotation, new mechanism and new driving force for China's modernization. Integrating industries with cities, and unifying various plans, policies and efforts will contribute to the acceleration of economic restructuring, the shifting of the mode of production and the enhancement of development quality. In the primary stage, both the production capacity of land and technologies are needed. It is not only a way of shouldering social responsibilities for companies, but also a characteristic mode of development to rely on companies that are in good conditions to promote urbanization.

From the viewpoint of balanced development between human and society, the deep technologies such as Internet and intelligent robot could enhance human capabilities in understanding the world and developing the society. Nonetheless, the determinative element is human in that critical time and project could not replace human, and products could replace neither the human concept nor the corporate culture. Innovative power could only be released when the metaphysical spirit and physical products are integrated into one. All sophisticated technologies are double-bladed swords. We should learn to draw on advantages and avoid disadvantages to optimize the effects of innovation. Our innovation should give priority to providing services to the public. In some cases, it is not a very cool or advanced technology but a small innovation that could bring brightness and hope to a group of people. Our innovation should put in prominent status the development of real economy, and encourage the successful companies to repay the society. In accordance with the logic of innovation and balance, we need to pay attention to the problem of alienated technology. At the same time, we can prevent the emergence of technology fetishism and become the victim of technology, and lose the dominance of mankind. We need the development strategy of moderate industrialization, effective control of the industrial economy of moderate growth, and vigorously promote the establishment of "resource conservation, independent innovation" society.

All in all, we should highly emphasize the significant effects of the rule of balance in innovation. There are profound theories in all things, and in making innovations, we should learn from everything in the nature and pursue balance in the process of change.

II. Comparison between Made in China 2025 and Germany's Industry 4.0

In 2015, after analyzing domestic and overseas markets, China formulated a 10-year plan and issued the Made in China 2025 strategy to meet the requirements in industrial upgrading and restructuring. The plan and strategy identified 10 key industries, including strategic emerging industry, advanced manufacturing industry, other traditional industries related to national economy and people's livelihood, and their corresponding supply chains and marketing networks. The focus of idea is the in-depth integration of industrialization and e-science, the focus of effort is on promoting smart manufacturing, and the focus of mode is "Internet Plus".[1] Both Made in China 2025 and Germany's Industry 4.0 are important strategic measures for developing the manufacturing industry under the

---

[1] See *Guiding Opinions of the State Council on Actively Promoting the "Internet Plus" Action* from China State Council, http://english.gov.cn/.

background of a new round of scientific and technological revolution and industrial reform. By comparing the two strategies, each has its own characteristics. In addition to their differences in the technological and industrial foundations, there are outstanding differences in the strategic thinking. Industry 4.0 depicts a detailed blueprint for development, reflecting the carefulness and prudence that are characteristic of the German nation. Besides, many aspects, such as its strategic thinking, basic research, technological education, policy institutions and measures are all worth learning by us.

Differences in strategic thinking: In comparing Germany's Industry 4.0 and Made in China 2025, a significant difference lies in the fact that the former is a revolutionary with fundamentally scientific and technological strategies, of which the standpoint is not on upgrading several industrial manufacturing technologies, but on seeking qualitative change of a great leap from Industry 3.0 to Industry 4.0 by carrying out a revolution from the fundamental aspect of manufacturing method, and then the qualitative leap in the whole industries and eventually the leap in economic and social development. Therefore, the core content of Germany's Industry 4.0 is not limited to the "quantitative change" in the data of value of industrial output, but emphasizes more on the "qualitative change" in the mode of industrial production. Compared with Germany's Industry 4.0, the emphasis of Made in China 2025 is to realize the structural changes and increase of output through applying the tool of "Internet Plus" on the basis of the current industrial manufacturing level and technology. It is to seek in-stage improvement and development under the current industrial level and mode of thinking, and there are many qualitative indexes.

Differences in strategic basis: The strategic basis is the basic conditions for the successful implementation of strategies, including basic research, technological education, and talent cultivation. In careful study of Germany's Industry 4.0, it is not difficult to discover that the most important element in the strategy is the research in basic science, and many detailed task goals are attained by relying on the theoretical knowledge that are advanced, sophisticated and high-end. The German government has always attached great importance to basic research and considered it as the demonstration and pillar of a nation's civilization. They hold the opinion that in today's world, the most competitive country is not the one having rich resources or abundant capital, but the one having the most knowledge in basic science. Thus, Germany has been committed to improving the conditions for basic science research and enhancing the scientific research and innovation capabilities. In comparison, China has a relatively weak foundation for basic science research, resulting in weak scientific research and innovation capabilities and rare significant break-throughs. The fundamental reasons lie in both the historical conditions as well as policies. With regard to policy support, China has provided more support to cross-sectional studies than longitudinal studies in terms of both the amount and the scale of support. The result is that China is rela-tively strong in areas of applied research but weak in basic theory research. In addition, China lacks experience and conditions in establishing interna-tionalized industrial standards. Therefore, it is necessary that we exert great efforts in strengthening basic research. China is such a big country that there will not be any lack of people engaging in basic research and there will be potential for development. So long as we adopt proper poli-cies and create environment and conditions that facilitate basic research, we can achieve great progress in China's basic research. At the same time, we should adopt an open policy of cooperation to carry out international cooperation, to become an important importer of advanced network theo-ries and advanced standards, and to share theories, technologies and mar-ket with Germany and other developed countries.

Differences in strategic training: Germany has always kept its status as a country with strong science and technology and has succeeded in increas-ingly fierce international competition. The reason, apart from making suf-ficient investment into research and development to ensure its profound basic research basis, also includes its emphasis on training of reserve talents and technological talents for scientific research. Firstly, it has a tradition of attaching great importance to technical colleges and schools. Different from universities, the technical colleges and schools are inevitable products to meet the requirements of social development. Their majors are set to meet the practical needs of the society by solving specific technical problems to reap direct gains from industries. During the process of industrial develop-ment, in order to meet the demand for high-level talents in industrialization,

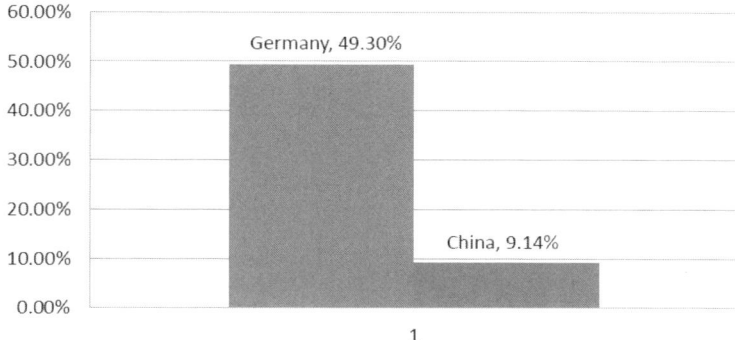

Figure 10.1. Percentage of Population Receiving Technical Education in the National Population (2014)

Note: Basic data originates from the website of the German government and the figure is drawn by the author.

Germany has made great efforts in developing technical colleges and schools, subsidized them considerably and given them the same status as traditional universities. At present, technical colleges and schools are enjoying special status in German education system. For all Germans, those who receive technical education account for as much as 49.3%, but the ratio is only 9.14% for the Chinese (see Figure 10.1).

Secondly, Germany has strengthened efforts in vocational training. The German government formulated the *Law on Technical Trainings for Employees*, making the enhancement of comprehensive quality and strengthening of operational capabilities important elements in the development of enterprises. Thanks to the efforts over the past years, Germany has established the double-track vocational training system with unified standards. The training system makes it mandatory for company owners, company management, start-ups, and all sorts of technical workers and young people to receive training for two to three years so as to engage in certain professional jobs. Half of the training time shall be devoted to theoretical study, and the other half to on-the-job training. Vocational training and technical colleges and schools have provided solid technical foundations for Germany's Industry 4.0.

Differences in strategic measures: Germany pays much attention to the establishment of legal environment and supporting policies for Industry 4.0, and timely makes adjustment to regulations related to responsibilities of enterprises, data protection, trade limitations and cryptosystem. They raise the awareness of competition for all people in the nation, and emphasize on self-reflection and self-adjustment. For example, Germany has made systemic evaluations on the subversive impacts on laws likely to be brought about by the new technologies on the one hand and the frequent upgrading of relevant rules and structures triggered by the shortening of innovation cycle on the other, and made amendments to all existing rules and regulations that might affect development. Besides, Germany has set up a unified governmental coordination organ to be responsible for policy coordination. The secretariat of the organ is jointly established by three specialized associations, namely, BITKOM, VDMA and ZVEI. In China, apart from the leading organ and strategy consultancy committee led by leaders from the State Council, we should also strengthen the development of industry coordination mechanism to give full play to the roles of industry associations.

How to speed up the development of China's manufacturing industry? In my humble opinion, we should:

(a) Actively usher in the new era of smart economy: Smart technology can transfer human intelligence and knowledge into a kind of acting capability, and will have profound influence upon human society. Industry 4.0 can bring brand-new changes to the human–technology and human–environment interaction. With the help of cyber physical system (CPS) system, especially the Internet Plus, human intelligence could be enhanced considerably. The economy based on the organic integration of human intelligence, computers, networks and physical world is more efficient and its efficiency could not be achieved in traditional industry. Hence, it will inevitably replace the traditional economy with a new structure and mode and take the form of smart economy. In the era of smart economy, smart environmental protection, smart architecture, smart transportation and smart medical services constitute the different areas of smart economy; and smart home, smart enterprise, smart city, smart region, smart country and smart world constitute the different levels of smart society. In the era of smart economy, the global economic integration becomes more prominent, the interrelations between market entities become closer, and the social economic systems become more open to the outside world. The entry into the era of smart economy means not only a higher level of productive force, but also more freedom for human during the production process, signifying that human civilization will step into a new smart era. The smart economy featured by smart factories will probably be a new stage in the industrial and economic development. This stage will last for a long term, which could be 50 years or 100 years. We could not anticipate what might happen, but the balanced way of thinking tells us that smart era has pros and cons and the new industrial revolution will probably bring unpredictable influence to the world and create an extremely active environment. We could expect that there will be more outstanding unbalancing phenomena, more varied ways of competition, and enormous changes in global governance. Therefore, we should be fully prepared to take an advantaged position and foster a favorable development trend in development strategy, scientific and technological innovation, and humanity and morality.

(b) Vigorously explore the path of industrialization with Chinese characteristics: The progress of human civilization is not linear, and the development of social productive forces and the change of industrial structure are in spiral escalation. That is to say, there are advanced factors in tradition and meanwhile traditional factors in the process of advancement. The key is to correctly understand the development stage of society, its historical conditions and trend of revolution. Germany is a developed country, which achieved the first industrial modernization in the 1950–1960s, and the second one in the 1970–1980s. The purpose of Germany's Industry 4.0 is, on the basis of powerful technological and industrial foundation, to lead the trend of world's manufacturing industry, raise German industry into a higher level, and to achieve new industrial modernization with smart manufacturing at its core. China is still a developing country in the stage of industrialization with the coexistence of backwardness and sophistication and of tradition and modernity. Traditional industry remains the pillar of industrial production, with products of high technology accounting for only 12% of the whole manufacturing industry and emerging industry contributing to only 8% of GDP.[2] China should actively explore the path of industrialization with Chinese characteristics, including the integration of the first and the second rounds of industrialization, of e-science and industrialization, and of the manufacturing industry and the service industry. If we think of "Internet Plus" as a process for establishing a new industry on the basis of e-science and smart technology, consider "Plus Internet" as a process for e-science and intelligence-based reconstruction on the existing manufacturing industry or other industries, then we have to include the integration of "Internet Plus" and "Plus Internet" into the characteristics of Chinese industrialization. In order to provide a solid and broad foundation for modernization, China has to, probably in a very long period of time, promote Industry 2.0, Industry 3.0 and Industry 4.0 at the same time, achieve the restructuring of traditional industries as well as leaping development into high-end areas, and establish an industrial system that suits the Chinese conditions as well as the development trend of the world. In sticking to the path of industrialization at the national level, it is not necessary for China to pay so much attention to industrialization to some local areas according to actual conditions. Taking the Tibetan region for example, it has more potential to develop tourism than industry. As a large developing country, China is an important player in the global market having various needs and broad leeway in development. It has to consider the enhancement of labor productivity as well as solving the employment issues.

(c) Correctly understand China's manufacturing industry in the reindustrialization process of developed countries: At present, the manufacturing industry of China faces three types of challenges. Firstly, it faces challenges from the high end of the industry. Through "reindustrialization" and the integration of the reindustrialization with the new industrialization, developed countries further strengthened their advantages accumulated for a long time in such areas as science and technology, information and capital, and became the main beneficiaries of the results of scientific and technological reform and industrial revolution, further solidifying the world's labor division system of "center–periphery" to the disadvantage

---

[2] See the remarks by former Minister of Industry and Information Technology Li Yizhong at the i-China Forum on July 23, 2015.

of developing nations, and widening the gap between developed countries and China. Secondly, it faces challenges of being edged out from the low end of the industry. Other developing countries such as India, Vietnam and Indonesia could undertake the shift of labor-intensive industries into their countries with even lower labor cost and seize the low end of the manufacturing industry, resulting in the possible loss of advantages of the Chinese manufacturing industries in the massive low end of the market. The third type of challenge comes from the difficult situations within China. Comprehensively speaking, China does not have strong independent innovation capabilities, and relies highly on foreign countries in terms of core technologies. The manufacturing industry of China is still at the medium- and low-level, lacking world class enterprises and famous brands, and occupying a relatively small portion in the high value-added links along the global industrial chain. The industrial structure of China is not reasonable with a low ratio of technology-intensive industry and production-oriented service industry, low level of industry cluster and industry agglomeration development, and prominent quality problems. There are severe environmental problems with low efficiency of resource utilization. The low management level and poor efficiency result in high management cost and severely affect the competitiveness of products.

The manufacturing industry of China also has three opportunities. Firstly, there are new chances. The new round of scientific and technological revolution and industrial reform has converged historically with the accelerated shifting of China's economic development pattern. The international industrial division of labors has entered a reshaping stage, and there are a lot of unknown aspects at the initiating stage of new theories, new technologies and new patterns. All these, to some extent, have drawn a new starting line for the globe and provided chances for China to catch up and surpass. Secondly, there are new types of supply and demand. The effective integration of the "reindustrialization" of developed countries and the new round of industrial revolution have demonstrated to us brand new mode of production, production factors and organizational patterns that differ from traditional assembly lines and centralized mass production with machines. As a result, new types of supply and demand will be created inevitably and those are areas where China could give full play to its capabilities. Thirdly, the "deindustrialization" and "reindustrialization" of developed countries have left us experience and lessons. The excessive "deindustrialization" and the development of high-risk, high-leverage financial businesses resulted in the detachment of the real economy from the virtual economy. If China, learned their lessons and policy measures in the "reindustrialization" strategy that were foresighted and followed the development trend, put forward structural optimization with focus and in differentiated manner according to characteristics of different fields, and handle such critical links as R&D, designing, marketing network, brand and supply chain management that hampers industrial structure upgrading, there will be great chances for China to speed up the upgrading of the manufacturing industry.

To turn challenges into opportunities, China might have to consider adopting strategies and plans that help the Chinese industries to strive for high-end and low-end product market and ensure their share in the medium-end product market, and establishing a modern industrial system with Chinese characteristics. On the one hand, we should gather talented people to vigorously strengthen the integrated innovation capabilities and cultivate original innovation capabilities, so as to possess a batch of core and key technologies rapidly and to strive for a place in the high-end product market sectors. We managed to achieve that in the past, and should do so in the future to have our say in the international core club. On the other hand, we should continue to fight for a share in the low-end market and create more jobs. It should be our main choice to endeavor to achieve upgrading the long-term bottom-end market to the medium-end market in that it is unrealistic to upgrade all to the high-end product market. We need to overcome the dilemma of "made in China" and become one of the important providers of quality (medium-end) products and services in both domestic and overseas markets.

(d) Highly emphasize the organizational reform of Internet Plus enterprises: "Internet Plus" is the organic integration of science, technology and economy. In implementing the "Internet Plus" strategy, the organizational reform of Internet Plus enterprises has great significance. As the main market entities and economic cells, in addition to promoting coordinated network manufacturing and services by using Internet to strengthen the interrelations with market, enterprises should also make great endeavors to enhance its endogenous power and fully tap its internal vitality. A problem that influences the overall strategy of Made in China 2025 is how to make use of information technology to improve and restructure the production factors, deepen the organizational reform of enterprises, make innovations to the mode of production, and enhance the quality of assets and service functions to meet the market demands and adjust to its changes. To answer this question, first we should correctly understand the relationship between technology and organizations. The relationship between technological structure and organizational structure of enterprises is a mutually promoting and mutually constructing one. In particular, the Internet technology has organized into computer networks all relations including the consumers, suppliers, partners and employees of enterprises, leading to the high speed and convenience in the obtaining, processing, transmission and application of information. This will inevitably require the enterprises to make relevant adjustment and reformation in their mode of production, mode of management and organizational structure. Under this situation, only by deepening the organizational reform of enterprises to integrate the Internet technologies with the production mode of enterprises and establish an efficient information communication and feedback mechanism, can we achieve the favorable interaction between technology and organization, enable the development of Internet technology to work for enterprises, and make enterprises the major force in advancing the technological progress of enterprises.

The industrial revolution with smart manufacturing at its core is a long-term development strategy which could not be achieved in a short time and is still at the starting stage of conception and planning. According to a recent survey by Ubisense,[3] an authoritative European and American organization, even in developed countries, nowadays only 10% of manufacturers are seeking smart equipment and products, 80% of the manufacturing technique improvement still needs to be under the observation of people, and 85% of quality-related problems are caused by people. However, it is beyond questioning that the new industrial revolution is

---

[3]The Promise and the Risks of IIoT and Industry 4.0 [available at] http://www.plant engineering.com, June 2015 Plant Engineering, p. 16.

emerging, the prospect of smart world is unfolding, and the horn of fierce competition has been blown. We should not be in a hurry, nor should we be too slow-paced. The key is to conduct through study, make correct judgment, make plans for long-term benefits, and take proper measures. "Made in China" should be shifted from element-driven to innovation-driven, from competitive advantage of low cost to that of quality and efficiency, from the extensive manufacturing of large resource consumption and high pollution emission to green manufacturing, and from production-orientated manufacturing to service-oriented manufacturing. The core idea in the "Made in China" development strategy should be as follows: taking the acceleration of integrating a new generation of information technology with manufacturing industry as its main theme, the enhancement of quality and efficiency as its focus, the fulfillment of the needs for important technology and equipment in economic, social and national defense development as its goal, and the opening up and cooperation as its means, strengthen the industrial foundation and capabilities, enhance the standard of overall integration, improve the multi-level system of talents development, promote the industrial restructuring and upgrading, cultivate the manufacturing culture with Chinese characteristics, and do a good job in the protection of industrial cultural heritage, so as to accomplish the historical leap from being a large manufacturing industry into a powerful one. It is very possible that China could be the important beneficiary of the new industrial revolution and contribute greatly to it.

All in all, it is not completely the money and technology but the ideas, plans and mode of industrial development that determine the upgrading of China's manufacturing industry and the development of innovation-oriented enterprises. We should include the upgrading of manufacturing industry into the overall planning of national industrial restructuring as well as the chains in global industrial restructuring, and seek a balancing point between structural optimization and resource integration. While forging ahead and drawing back in proper pace and doing the work that is necessary, we should adhere to the new path of industrialization, vigorously cultivate the key industries, and speed up the development of national strategic industries and industries of livelihood projects, so as to increase our competitive advantages and ensure that China play a significant role in the global competition of the 21st century.

III. Economic Mode of the New-type Home-based Manufacturing Industry

1. *Characteristics of the New-type Home-based Manufacturing Industry*

The new-type home-based manufacturing industry, or sometimes called "desktop factories", "micro-manufacturing", "new-type family handcraft" or "new handcraft", means a new-type handcraft manufacturing industry connecting creative ideas, production and services by using home as the workshop, and making use of digital tools, Internet and logistics network. Here, the word "handcraft" means to make tailored products to satisfy customer needs to the maximum extent through the combination of traditional technique with advanced technologies (for example 3D printing technique).

The new-type home-based manufacturing industry has the characteristics of being "digital, home-based, sharing, individualized, networked, and socialized".

Be digital: The manifestations of being digital are that people could set up factories and control the production tools by clicking on the mouse and using the digital tools. They could combine the creative ideas and entrepreneurship by designing on computer screens and manufacture models and products with 3D printers. The transfer between the real world and the virtual world can be realized at any time. Meanwhile, the process of products manufacturing increasingly resembles the process of data making, and the symbol analysts become manufacturing engineers.

Be home-based: The manifestations of being home-based is that people could establish a manufacturing company at bedrooms or even create an empire in the dorms without big investment in factory construction and the employment of a large number of labors, so long as people have good ideas, a credit card, a computer, a 3D printer and a few other devices.

Be sharing: The manifestation of being sharing is that entrepreneurs could easily find reference materials they need through search engines and fully tap the good conditions of knowledge sharing.

Be individualized: The manifestations of being individualized are that it reflects individuality and reality in that they make tailor-made products in small scales, and focuses on niche markets at a small customer base and products those big factories do not produce, such as the manufacturing and sale of parts of classic cars, or making fashionable pinup pictures for MacBook with CD/DVD writer.

Be networked: The manifestations of being networked are as follows. Through network configuration, people could select ideas, raw materials and services in the global market, either produced by desktop factories, or by other companies after sending them design document standards. Besides, rather than waiting for orders or selling products through dealers, entrepreneurs who invent products on their own and possess their own mini brands could become sellers on large-scale websites or sell directly through their own websites.

Be socialized: The manifestations of being socialized are as follows. The manufacturing of new products is no longer a specialized area for a small group of people, but becomes opportunities for a majority of people and everyone could have a try. People connect and communicate through the culture of Internet and cooperation to attract the participation of more people to generate more ideas and creativity, fostering "maker movement". The new-type home-based manufacturing industry could initiate creative activities from bottom to top. Its distribution pattern shows the new characteristics of socialized production since it does not reflect the complex pattern combining the traditional large-scale industrial chain and industrial park, but the natural geographic pattern of entrepreneurs and their creative ideas.

2. *The Mode of Technical Operation*

The mode of technical operation process of the new-type home-based manufacturing industry is a process converting from information into material objects as well from material objects into information. The smallest unit of information metric is bit, and the smallest unit constituting material objects is atom. Therefore, the conversion between information and material objects is also the conversion process between bit and atom. By adopting the designing ideas of digitalized coordination, people change the design process and enhance the manufacturing efficiency. Through "reality capture" — or object scanning to produce a group of data and recover them into point cloud composed of atoms by using computer, people can then connect the point cloud into polygonal net

Table 10.1.   Basic Equipment and Prices

Equipment	Price in 2013 (RMB)	Estimated price in 2010 (RMB)	Estimated price in 2020 (RMB)
3D printer	3,000–30,000	2,000–20,000	1,500–20,000
Numerical controlled robot	22,000–1,000,000	20,000–1,000,000	12,000–1,000,000
Laser cutting machine	12,000–1,580,000	10,000–1,400,000	10,000–1,300,000
3D scanner	59,000–1,000,000	45,000–800,000	30,000–800,000

structure through programming, and design and revise the 3D structure on screens.

At present, there are various types of printers in the market for choice. Other relevant equipment, such as numerical controlled robots, laser cutting machines and 3D scanners all have comparatively proven technologies. With the further improvement of technology, the market prices of the above equipment are gradually accepted by normal home-based entrepreneurs. Table 10.1 shows the list of basic equipment and their prices.

3. *The Mode of Operation and Management*

Spirit of "hybrid creation": The greatest advantage and peculiar characteristic of the new-type home-based manufacturing is digital information. People do not have to start from zero in original design, but could get inspirations by searching the existing documents, or could redesign the existing products. Products of hybrid creation could be manufactured by modifying the existing products through combining people's own and local features, reflecting the spirit of innovation. It is the key to the development of the new-type home-based manufacturing industry that in "hybrid creation", people explore all possible changes of one product, constantly improve the product, and then promote the product in the most rapid speed through the website. In the world, there are various changing elements and unlimited forms of changing, hence the boundless space for the development of "hybrid creation".

The principle of "choosing, revising, manufacturing and marketing on one's own": That means to conduct 3D scanning on the object, revise it by using computer aided design (CAD) software, print it with 3D printer, and sell it on the Internet. In this process, product designing does not mean graph making, but means the generation of mathematical formula for this product. Each product should have its corresponding algorithm. The more varied the patterns are, the better quality the product will have; and the bigger the change of speed is, the more complex its algorithm will be.

Basic qualities of people: People should have entrepreneurship, ideas, creativity and perseverance, master some knowledge about computers, network operation and software, have some preference for local handcraft and some experience in real economy, and understand traditional culture and network culture to enhance the cultural taste of the new-type home-based manufacturing industry.

4. *The Mode of Market Operation*

The new-type home-based manufacturing industry is forming a unique mode of market operation. That is, "taking creativity as the premise and niche market as the target; based locally and allocating resources globally; interacting between supply and demand, and pricing reasonably".

"Taking creativity as the premise and niche market as the target" means that people should first have creative ideas, which do not target normal mass market, but a special market of small scale to fulfill individualized demand in the market.

"Based locally and allocating resources globally" means that people should fully consider the local content in the creative ideas to reflect its specialty. Meanwhile, people should procure proper components at reasonable prices through large-scale free classification websites such as craigslist.org. If necessary, some parts, such as welding, could be contracted to other companies.

"Interacting between supply and demand, and pricing reasonably" means that the new-type home-based manufacturing industry does not attract customers with cheap prices, but negotiate a reasonable price with customers when customers show their preference for the product, so as to keep the sustainable development of companies. In small market that has a bias towards tailor-making, the product itself is the basis for higher prices. Customers choosing small market usually have peculiar likings and are shrewd and capable. Thus, they know well about their choice and are prepared to spend a bit more.

Three factors should be taken into consideration in reasonable pricing for products of the new-type home-based manufacturing industry. Firstly, the cost, including labor cost, cost of materials and postages. Secondly, the profit margin, which usually could be 50% to cover extra cost that was not considered in starting-up stage, such as extra insurance cost. Thirdly, the profit of retailers, and it could be 50% for third-party retailer. Therefore, the price of new handcraft products is roughly 2.5 times its cost, much cheaper than the price of traditional products, which is about four times its cost. The reason is that the links of centralized procurement and wholesale are deducted in the marketing of products from the new-type home-based manufacturing industry.

5. *Organizational Structure*

The new-type home-based manufacturing industry needs to have contacting and supporting system, and needs community cooperatives that suit urban development to provide services such as personnel training, equipment maintenance and repair service, establishing demand and service record, encouraging innovation and giving awards to model workers.

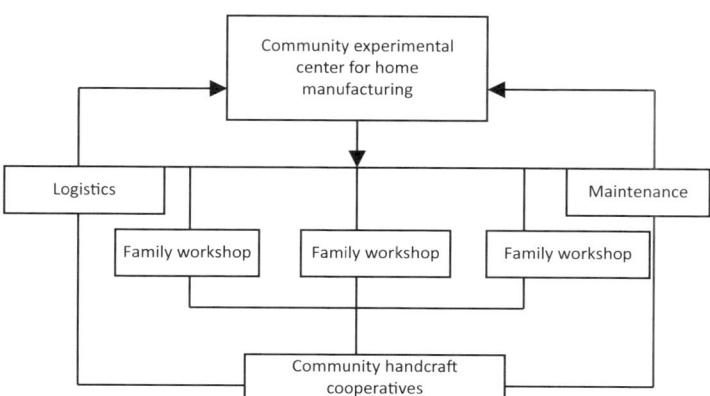

Figure 10.2. Organizational Structure of the New-type Home-based Manufacturing Industry

Figure 10.2 shows organizational structure of the new-type home-based manufacturing industry.

Experimental centers for the new-type home-based manufacturing industry: In cities and towns, we could set up centralized community manufacturing and experimental centers according to conditions and needs, so as to centralize the use of some large-scale manufacturing tools increase the service efficiency of tools and enhance the community innovation capabilities. Some complicated and large-scale processing work could also be handled in those centers.

New-type home-based manufacturing workshops: Family workshops or desktop factories could be set up according to their own characteristics. Those family workshops are full of freedom, vitality and happiness for entrepreneurs who could focus on starting up business, and fully tap their technologies and techniques. The location of company could be the place where entrepreneurs settle and is covered by courier service networks. Those new types of family workshops are in small scale, non-official, and nearly virtual. Most participants do not have to be regular employees bound by affiliation and obligations, but are driven by capabilities and needs. Therefore, it is a kind of "small and loose combination", a kind of free recombination of social construction forces, and a new type of industrial development mode. This mode differs greatly from traditional companies assumed by Ronald Harry Coase. In this mode, different groups of people form teams through Internet and beyond the corporate environment regardless of their geographical location, educational background and employment status, and could make contributions to the company anywhere around the globe so long as they have talents and ideas. This type of company may not have minimal transaction cost, but could enable the best people with the same goal to work together. And seeking homogeneity through website constitutes the new driving force in economic operation.

Logistics support system: The community manufacturing and experimental centers and a large number of family workshops conduct online sales and are globalized from the very beginning, rendering services to transborder customers unlimited by traditional distribution and geographical locations. Thus, certain logistics support system should be established to serve the import and export of the new-type home-based manufacturing industry.

Equipment maintenance and repairing system: It will be one of the responsibilities of the new-type urbanization management to organize some specialized personnel to provide equipment maintenance and repairing services in urban communities.

Cooperatives for the new-type home-based manufacturing industry: They are responsible for providing public services such as home-based manufacturing industry management, skill training, course upgrading, updating web pages and evaluation and awards.

The ecosystem for the new-type home-based manufacturing industry: This system does not only include special customer base, but also other families engaging in handcraft and relevant entrepreneurs. All of them constitute different links along the industrial chain. Those mutually-depending and mutually supportive links enhance the quality of products, and become the systematic basis for the success of the new-type home-based manufacturing industry and their genuine defending advantage.

6. *Financing Mode of "Mass Funding"*

"Mass funding", or "crowd funding" means that entrepreneurs describe their start-up projects by posting in websites, and then interested individuals or customers would collectively fund certain project for product manufacturing. This mode enables start-up companies to obtain financing from various investors in the form of equity return, and promote the projects by designing creative returns for the projects.

The characteristics of "crowd funding": Firstly, investors do not only include professional investors, but also individuals and potential customers who are interested in certain ideas. Secondly, the fundamental purpose of investment is not purely for innovation capital, but more importantly for the creation of new products to people's likings. Thirdly, there is no limit on the amount of investment, fitting for fragmented capital investment in the society. There are various forms of crowd funding, such as donation, leasing and loans. The investors can provide money to one project, and are likely to order the products of the project when the investment surpasses a certain amount.

Rules of "crowd funding": Firstly, set the lowest limit and deadline for project funding. If, by deadline, the lowest limit for funding is not achieved, then the project funding fails and the project cannot be launched. Secondly, adopt third-party mandatory enforcement. The promotion and fundraising advertisement of projects shall be conducted on websites or by a third party recognized by regulators. If the fundraising of project could not be completed, we have to ensure that the invested capital is not embezzled before the stipulated time limit, and that money will be returned to investors at a proper time. Thirdly, establish an incentive system to award investors of different scale to enjoy the ownership and fruit of projects.

Effects and functions of the mode of "crowd funding": Firstly, it can raise funds to ensure that money is available when it is in need. Secondly, it serves the function of market survey. Through the fundraising activities within established time and at established limit, people could know the popularity of the products before they are developed and manufactured, minimizing risks for entrepreneurs. Thirdly, it can foster a special group of customers. Fourthly, it is a way of advertising and promotion, in that the interaction between entrepreneurs and investors could stimulate people to participate in product development.

All in all, the new-type home-based manufacturing industry is the extensive application of the results of new scientific and technological revolution in economic and social development. Compared with micro-enterprises in normal sense, the new-type home-based manufacturing industry embodies the revolutionary changes in such aspects as the mode of technical operation, the mode of operation management, the mode of fundraising and organizational structure. The new-type home-based manufacturing industry is a mode of Internet business that could create markets, and is likely to become a new force creating new room for development and achieving robust growth in the manufacturing industry. This new force will break industrial monopoly, change the pattern of manufacturing and push the industrial economy into a new era. At present, this new force is still in full swing with no sign of slowing down in the international market. In view of the international trend and the strong power of new science and technology, we should vigorously support the development of the new-type home-based manufacturing industry by seizing opportunities to take full advantages of the latest development of science and technology, and taking into consideration of the actual conditions especially the requirements in the new-type urbanization of China. By doing this, we could not only provide strong new-type industrial support to the new urbanization process, but also play a significant role in creating jobs, narrowing the gap between urban and rural areas, giving play to non-governmental resources, and promoting economic growth and social stability. Thus, it is of strategic significance to the national economy and social development.

The following three decades will witness robust growth of the new-type home-based manufacturing industry and will be a critical time for China's urbanization development. China should seize opportunities to facilitate the mutually complementary development of the new-type home-based manufacturing industry and urbanization. When the new-type home-based handicraft industry flourishes, the new-type urbanization will definitely be in sound development.

Part III

The Balance Diplomacy for the Opening Up

The world where we live is filled with intertwined conflicts: the globe is becoming more imbalanced, unstable and uncoordinated; along with the wide application of information network technology, the occurrence of all kinds of unexpected events has increased; conflicts, such as the one between global economic integration and cultural diversity and governance pluralism, are becoming more and more acute; a global geopolitical landscape where multiple systems coexist, one hegemonic country stands along with several powers and the world is an integral whole with great diversity taking shape; and the competition between traditional powers and emerging powers is increasingly fierce. The US as a representative of traditional big countries is active, while China as the representative of emerging countries is passive. The competition between traditional powers and emerging countries is inevitable, long-lived and intense. The US pursues hegemonism and pragmatism and the core of its strategy is increasingly self-centered. Its basic goal is to prevent the emergence of any power that can challenge its global control and values within a certain historical period and continue to maintain its status and interests as the world's only superpower. After comparing various strategic means, the US currently chooses to control the energy supplies of China and other emerging powers, hijack their energy security channels, disrupt their surrounding environment and increase the cost of China's development as a deterrent. It may also deploy local wars and other strategic means. China as a big emerging and developing country, is now shouldering two important historical tasks: national modernization and rejuvenation. China and other emerging countries are an important force in promoting the balanced development of the world and maintaining world peace; however, they may have some impact on the existing international situation, which will not change as people wish. Competition without the basis of a unified perception has huge potential risk. Therefore, it requires insights into international politics and in-depth discussions.

The change of world landscape is a long-term process: The change of the international geopolitical landscape is a result of the interconnection and interaction of a variety of forces; however, it is a relatively long-term process. The existing international governance structures and mechanisms while being constantly improved will continue to play an important role in and exert great influence on the stable and healthy development of the international community in the 21st century. The international community needs the US, the superpower, to continue to play the role of a global system leader; meanwhile it also needs other important countries to share the common responsibilities. The BRICS nations will lead the development of their respective regions, become an emerging force of world sustainable development and promote the balanced development of the world. In the next few decades, the US and Europe will remain the center of modern civilization. The US in particular will remain the world's number one power with the strongest innovation ability, the strongest consumption power and the greatest international influence. The US dollar will remain the major currency in the world. Still, the US cannot solve any of the world's problems alone. Cooperation is the dominant form of social development in the 21st century as well as the common aspiration of all mankind. There is an urgent need for the world to work together to address global challenges and manage a series of crises.

Mutual trust and cooperation are the ideological basis for global peace and development: During the development course of human society, mutual trust and cooperation are an instinct, a need, an awareness of the bigger picture and the spirit of putting overall interests first. Every step of human survival and development are inseparable from the spirit of cooperation. In the 21st century, it will be very difficult for those who are not good at cooperation to survive, for the companies that are not good at cooperation to survive and for countries and regions that are not good at cooperation to develop. The jungle law and winner-takes-all are considered as the golden rules in Western international relations theories, while traditional moral doctrines of the oriental countries emphasize selfless dedication: even though you need only my hands, I will offer both my hands and feet to you. Both ideas are extreme. In the new era, we should seek and expand a path of win–win cooperation between these extremes. We should hold high the banner of mutual trust and cooperation, and establish a long-term relationship of honesty and trust under the principle of righteousness before benefits; we should establish a sincerity-based relationship of honesty and trust under the principle of giving more and taking less; we should establish a mutual assistance relationship of honesty and trust by keeping up the practice of exchanging resources; we should make use of the philosophies and methods of international cooperation with Chinese characteristics to promote international cooperation and mutual trust because honesty can get us everywhere.

We should make active efforts to improve the international strategic layout: The international strategic layout for China in the 21st century is that neighbors are the priority, multilateral relationships are the stage, bilateral relationships are the channel, big powers are the key and Sino-US relationship is the axis. China will remain a developing country for a long period of time. It is pursuing industrialization, urbanization and modernization and thus it is faced with the common problems like other developing countries. Therefore, we need to inherit and promote the international strategic thought of Chairman Mao, Premier Zhou Enlai and other revolutionaries of the older generation to consolidate and develop the international united front based on our relationships with other developing countries, especially the ones with developing countries in Asia, Africa and Latin America. As an old Chinese saying goes, a distant relative is not as helpful as a near neighbor. The surrounding countries are as important as the barrier of our national strategic security. Therefore, they are the number one priority in our foreign relations. We should take full advantage of river basin economy to deepen the cooperation in the surrounding area. When it comes to the exchanges with the neighboring countries, Chinese have always honored the spirit of "giving more and taking less". In the future, China will continue to uphold the spirit and pay more attention to cooperation as well as making friends with the whole world. International multilateral cooperation and multilateral forums are important platforms for us to increase our voice and exert influence. At the same time, we should rely on bilateral channels to solve the deep-seated problems with each other. In international affairs, the role and responsibility of big powers are growing; therefore, to properly handle the relationship with major powers, especially the Sino-US relationship, is of paramount importance.

We should advocate the ideas of international relations with ecological awareness: Generally speaking, we should observe the overall situation, be honest and trustworthy, pursue balanced diplomacy and pragmatism, devote ourselves to planning, give more and take less, and focus on domestic development while continuing the international exploration. We should actively promote balanced international trade and participate in the building of global health network, the global relief network, the global gene pool and the global environment network. Faced with complex and capricious political situation, we should judge the situation calmly, identify the root of problems, catch up with the trends and actively build ecological relationships of politics, economy and culture featuring symbiosis, coexistence and sharing. We should join hands to protect the earth, the home of all mankind, cope with various crises faced by the international community and enjoy the various achievements of civilization. We should hold high the banner of mutual trust, cooperation and peaceful development and adhere to the pursuit of balanced diplomacy. We will not compete for the leading position nor take the path of extreme. We will take the interests of developing countries into full consideration, use political and diplomatic means to forge cooperation with a number of countries, enhance our relations with both developed and developing countries, consolidate our ties with countries both nearby and far away, defend China's core interests and make more contributions to world peace and development.

We should make active efforts to improve the international economic and trade strategy in the new era: Since the founding of the People's Republic of China, our foreign trade and economic cooperation has gone through two stages. The first stage was from the founding of the People's Republic of China to the implementation of the reform and opening-up policy. This government-guided phase featured government-led foreign aids regardless their economic costs. Such assistance was unsustainable. The second phase was from reform and opening up to the first decade of the 21st century. This market-led stage featured market-oriented "bringing in" and "going global" activities. These activities focused on economic benefits but lacked international political awareness. In the new era, we should stick to the philosophy of pursuing balanced economic and trade relations and properly handle the relationship between politics and the market as well as the relationship between the development of China and global development. We should improve international economic and trade strategic thinking by anchoring it to the axis of the Belt and Road Initiative, interconnectivity and regional economic integration and thus usher in a new era of international economic and trade cooperation which incorporates the pioneering faith with the pragmatic spirit.

We should make active efforts to build the new international economic and trade cooperation platform of Belt and Road: The Belt and Road Initiative, the strategy of opening up for the new era which is based on the construction of infrastructure, is introduced to give full play to the role of enterprises, the main players in the new era, respect economic laws, focus on connectivity, broaden all-round economic and trade cooperation, seek cooperation in planning, building, trade, and development for win–win outcomes, and promote the balanced development of international trade as well as the balanced development of the North and the South.

Diplomatic relations are a complex and systematic project. We hold that the world is a global community of common destiny. We are active in building "a new type of partnership featuring political equality and mutual trust, economic win–win cooperation and cultural exchanges and mutual learning". We should stick to the thinking of balanced diplomacy and avoid overconfidence or resting on our laurels. In doing so, we will contribute to national security and development, national well-being as well as the world peace and development.

China needs multifaceted diplomacy and multi-angled trading philosophy. To consider economic and trade relations from the perspective of balance is not only an important part of China's economic pillar, but also an important way to create a favorable environment for the Chinese economy. In this Part, which includes four chapters, I will explain my understanding of the international economic and trade relations from the concept of mutual trust, economic and trade history and the world landscape. I will then introduce the practices of balanced diplomacy with the cooperation in infrastructure planning and construction as the main part. A few years ago, I was responsible for organizing the formulation and drafting of some plans, which gave me an opportunity to reflect on these problems. Here I will comb through these main points, which I hope can provide reference for the readers.

Chapter 11

Mutual Trust and Cooperation
in Globalization

What is the essence of balanced diplomacy? What is most needed in modern-day international community? The answer for both questions is mutual trust and cooperation. Mutual trust and cooperation is the dominant theme of the 21st century and the mainstream spirit in the development of human society. In this chapter, the author will analyze the significance and characteristics of the spirit of mutual trust and cooperation and propose that the relationship between cooperation and competition is symmetrical. The author will also point out that mutual trust and cooperation is a special scarce resource, the basis of peaceful development, the chorus of human destiny and a universal spirit. In the special period and special environment, to advocate this universal spirit will help us gain better moral, policy and technical high ground in international relations.

In today's world, globalization has entered a new phase. On the one hand, all countries and regions have become more interdependent and interconnected, and the trend toward integration and a flat world has become more evident; on the other hand, geopolitical struggle and market competition have become more intense and factors of instability and uncertainty have increased significantly in number. Under the new circumstances, there is a pressing need to recognize the great significance of mutual trust and cooperation, hold high the great banner of mutual trust and cooperation and promote the healthy development of human society.

I. The Significance of Mutual Trust and Cooperation

Mutual trust and cooperation is the foundation of the peaceful development of human society: In the development course of human civilization, mutual trust and cooperation reflects humans' instinctive need, a pattern of behavior that humans reply on for survival and development, and human society's development pattern: symbiosis, coexistence and sharing. Along with the extensive application of network technology and the emergence of innovations, members of human society have become closer to one another and the need for mutual trust and cooperation is getting stronger. And mutual trust and cooperation has become an important way that we depend on to solve worldwide problems.

Mutual trust and cooperation is a special kind of scarce resource: This kind of scarce resource exists independently, connects with one another through a network and has no mandatory obligations. However, for it to have significance it has to be mutual, reciprocal and a long-term thing; unlike other scarce resources which will reduce if used, it becomes more abundant and richer as a result of use. As an old Chinese saying goes, more visits bring relatives closer and fewer visits turn them into strangers. This kind of scarce resource is closely related to charisma, political wisdom, friendship and experience accumulated over time. It is also the basis of the healthy development of society. Without the basis of trust, every society will face a substantial increase in the cost of development.

Mutual trust and cooperation and competition are symmetrical: Competition is essentially an interactive process. Interaction requires mutual understanding, mutual reliance, mutual trust, equality and mutual benefit. Competition among masters is often another form of interaction: your value is equal to that of your opponent and self-improvement can be achieved through understanding and assisting each other. Therefore, real and effective competition is inseparable from mutual trust and cooperation.

Mutual trust and cooperation is also at the heart of everything and the central solution to problems. For example, in the commodity market, mutual trust and cooperation is not only the lubricant that facilitates the production and exchanges undertaken by members of the society, but also at the heart of all transactions. No matter how many layers of structure exist among market players, mutual trust and cooperation are at the heart of all relationships, including both longitudinal and horizontal economic relations. It is especially the case in the political sphere. Mutual trust and cooperation is not only a political expression and a behavioral strategy, but also the key to understanding political behavior.

Therefore, in the era of healthy human development, the spirit of mutual trust and cooperation should be ubiquitous like the air.

II. The Spirit of Mutual Trust and Cooperation

Everyone is an owner of the society, every enterprise is a participant of the market and every country is a member of the international community. In the symphony of human civilization, it is necessary to establish the common awareness of sharing the same lot and the common destiny, pursue mutual respect, cooperation and assistance, advance along with the pace of the world and develop riding the wave of world development.

For a person, the spirit of mutual trust and cooperation is the stepping stone to personal improvement. Mutual trust and cooperation can help one get out of their comfort zone and make up their lack of knowledge by being in a community. It can also help them make up their deficiency with others' strengths, communicate with one another, encourage one another, motivate one another and develop creative and innovative thinking. As one plus one is greater than two, so is mutual trust plus cooperation. There are no two leaves that are exactly the same. One person is completely different from the other. It is the difference that brings about the need for mutual trust and cooperation. It is because of mutual trust and cooperation that human society can thrive in harmony.

For an enterprise, the spirit of mutual trust and cooperation is the soul of corporate culture. Mutual trust and cooperation between enterprises is the requirement of economic laws. As George Bernard Shaw once said, "If you have an idea and I have an idea and we exchange these ideas, then each of us will have two ideas". Competition among enterprises is the norm in market economy, but fundamentally speaking, competition cannot be separated from mutual trust and cooperation, and mutual trust and cooperation always accompanies competition. In a sense, competition, as an interactive behavior, is also a form of mutual trust and collaboration. This form of mutual trust and cooperation brings about chain reaction from the opposite direction, inspiring the generation of collective wisdom and strength and leading to further increase in quality and quantity.

For a country or region, the spirit of mutual trust and cooperation is a prerequisite for the participation in international affairs. The spirit of mutual trust and cooperation between countries is the requirement and basic condition for creating a democratic, equal and healthy international environment. Every country or region has their own strength, such as the innovation awareness of the US, the rigorous attitude of Europe, the indomitable will of Russia, the simple style of the developing countries in Asia, Africa and Latin America, the pragmatic style of Latin America, and the industrious spirit of Asia. They are all worth learning. Only by promoting tolerance and integration of different civilizations, exchanges and mutual learning, cultivating the awareness of mutual trust and cooperation, promoting the spirit of mutual trust and cooperation and establishing mutual trust and a cooperative attitude can the symphony of world civilization become more beautiful and powerful.

Human society is an open society. Open society is closely related with the environment and is part of everything. It is especially the case with modern society. With the constant revolutions of science and technology, humans have acquired a deeper and broader understanding of the world, and the openness of the society has become greater. In an open society, people are free to hold different views and make different choices; in an open society, human cognition still has limitations and there is no absolute empirical truth and no absolute perfect theory or existence for everything is relative; in an open society, the objective existence is of even greater importance and human beings cannot dominate the core of objective reality, and therefore, to gain the correct understanding of the laws that determine the development of things is the only way to make the circumstances in favor of human development; in an open society, everything is subject to the law of balance, and the force and counterforce in the physical world and the balance in economics are demonstration in different areas of how both sides of conflicts are affected by the law of balance.

With the development of modern market economy and politics, traditional social relations, especially the relationship between people, are becoming the political and economic foundation of the public. The characteristics of the new era have major impact on people's behavior. But no matter how the world evolves, the innate social qualities of trust, integrity and cooperation are always the most valuable human qualities that are worth our continuous efforts to preserve in modern society and in the future.

In summary, faced with a complex situation, we should vigorously promote the spirit of mutual trust and cooperation, and hold high the banner of mutual trust and cooperation. We should establish a long-term relationship of honesty and trust under the principle of justice before interests; we should establish a sincerity-based relationship of honesty and trust under the principle of giving more and taking less; we should establish a mutual assistance relationship of honesty and trust by keeping up the practice of exchanging resources; we should make use of the philosophies and methods of international cooperation with Chinese characteristics to promote international cooperation and mutual trust.

Chapter 12

The History and Prospects
of China's Foreign Trade*

How does the law of balance affect China's economic and trade relations with other countries? How to speed up China's transformation from a major to a competitive player in international economic and trade activities? Next, I will apply balanced way of thinking in analyzing three major longitudinal issues. Firstly, I will summarize the history of China's trade and economic relations since 1949; secondly, I will analyze the economic and trade situation and challenges faced by China in the new era; thirdly, I will discuss the ideological system and significance of economic and trade relations envisaged in the Belt and Road Initiative. The Belt and Road Initiative is a new and pragmatic framework of international relations for the new era with cooperation in infrastructure construction as an important part. This initiative builds on the Five Principles of Peaceful Coexistence, embodies the innovative and pragmatic spirit and is conducive to international trade and economic development.

Since the founding of the People's Republic of China, foreign trade and economic cooperation has gone through two stages. The first stage was from the founding of the People's Republic of China to the implementation of reform and opening-up policy. This phrase was government guided. The second phase was from the reform and opening up to the first decade of the 21st century. It was market oriented. In the new era, China

---

*This chapter was introduced on the first issue of *International Financing* in 2016.

needs to hold high the banner of mutual trust and cooperation, stick to the thinking of pursuing balanced economic and trade relations, and properly deal with the challenges that cannot be ignored. In addition, it should usher in a new era of international economic and trade cooperation which incorporates the pioneering faith with the pragmatic spirit by directing the main effort to promote the Belt and Road Initiative, connectivity and regional economic integration.

I. Government-led Economic and Trade Relations

From the founding of the People's Republic of China to the implementation of the reform and opening-up policy, China's economic and trade relations with other countries featured government-led foreign aid regardless its economic costs. China tightened its belt and provided substantial economic assistance to Asia, Africa, Latin America and some socialist countries. Such assistance bears not a slight resemblance to loans and is pure help. It not only lent help to the poor countries in Asia, Africa and Latin America when they needed it but also gave them new options, new hope and the strength to develop. China established a united front in the international arena, proposed the theory of three worlds, and reassumed the position as a permanent member of the UN Security Council, the significance of which is beyond the word "important".

After the founding of the People's Republic of China, Mao Zedong, with China's national conditions in mind and based on the Marxist theory of world market, made an unremitting exploration in how to develop economic relations with foreign countries, how to develop China, and who we should choose as our development partners. As for the relationship between China and the world, Mao had two basic views: firstly, he always stressed the importance of independence; secondly, he always emphasized that China's revolution cannot be carried out alone. "In the era of capitalism, and especially in the era of imperialism and proletarian revolution, the interaction and mutual impact of different countries in the political, economic and cultural spheres are extremely great". With all things linked together in the world, it is impossible to separate any part from it. Therefore, "China should take the initiative to do business with foreign countries" and develop economic relations with them. These views reflect

Mao Zedong's basic position and policy on opening up. At that time, in addition to the business ties with the Soviet Union and the socialist countries in Eastern Europe, China also used Hong Kong as a base and conducted economic cooperation and trade with the US and countries in Western Europe. Making use of the conflicts among Western countries, China also established economic and trade relations with some Nordic countries (such as Sweden, Denmark and Finland). China also followed closely the cause of national liberation in Asia, Africa and Latin America.

However, due to the special nature of the special era, Western countries adopted a policy of political containment and isolation toward China and imposed economic blockade and embargo on us following the outbreak of the Korean War in June 1950. For instance, in December 1950, the US announced that all Chinese public and private properties in the US would be placed under control, and all vessels registered in the US were banned from sailing for China. China was left with nothing but the foreign policy of "leaning to one side", namely doing business mainly with the Soviet Union and socialist countries in Eastern Europe. At the same time, due to the influence of Stalin's "two parallel markets" theory, China did not think opening up from the perspective that the world economy is a systematic whole, which limited the breadth and depth of opening up. And the development of economic relations with foreign countries could be based only on natural and planned economy. Mao Zedong's emphasis on strengthening economic relations with foreign countries was not well implemented, leading our country to consider issues from the political perspective of the united front and devote our major efforts to class struggle. Therefore, China's economic relations theories back then were tinted with strong political overtones and thus economic openness did not receive enough attention. During the Cold War, China's economic relations were mainly limited to the socialist countries and Third World countries. Strictly speaking, confined by domestic and international environments, China did not really "go global" at that time.

However, under the guidance of Mao Zedong's thought of opening up and fulfilling international proletarian obligations, China did reach out beyond its border and engaged in international economic activities, which were mainly in the form of aids to some countries in Asia, Africa, Latin America and the socialist countries. By doing so, China accumulated valuable experience for the formation and development of the subsequent "going global" strategy.

The Asian-African Conference took place from April 18 to 24, 1955 in Bandung, Indonesia. This was the first historic event held in Asia, Africa and Latin America without participants from the colonist countries. At the conference, Premier Zhou Enlai, the head of China's delegation, clarified the Chinese government's position and policy and proposed the Five Principles of Peaceful Coexistence, namely mutual respect for each other's sovereignty and territorial integrity, mutual non-aggression, non-interference in each other's internal affairs, equality and mutual benefit, and peaceful coexistence. These principles enriched the UN Charter, strengthened the legal basis of the new modern international relations, and to a large extent, reshaped world politics and international relations.

During Premier Zhou's visit to the 10 developing countries in Asia, Africa and Latin America from 1963 to 1964, he further elaborated on China's eight principles of foreign aid[1] to the world. The announcement and implementation of these principles indicated that from the start the Chinese government has tried to establish a partnership with the developing countries in Asia, Africa and Latin America based on the characteristics of their respective cultures and realistic strategic needs. Such relationship was built on mutual respect, equality and mutual benefit. In particular, the principle of non-interference in each other's internal affairs has become the cornerstone of international cooperation.

After entering the 1960s, China was on unfriendly terms with two superpowers, the US and the Soviet Union, at the same time. In order to reverse this unfavorable diplomatic situation, Mao Zedong put forward the theory of "two intermediate zones" and included Asia, Africa and Latin America in the first intermediate zone. They were an important strategic support for China's opposition to the two superpowers. The aid to Asia, Africa and Latin America was the most important means for China to win the diplomatic support of these countries.

By the end of 1970s, China's aid to Asia, Africa and Latin America had reached a considerable scale and their cooperative relationship had

---

[1] *Selected Works of Zhou Enlai on Diplomacy*. Beijing: Central Party Literature Press, 1990, pp. 388–389.

taken on its own characteristics and started to exert influence. In the first half of the 1970s, China's foreign aid accounted for 5.88% of the government expenditure over the same period and reached a record-high of 6.92% in 1973.[2] Only in Asia, Africa and Latin America, China had provided economic assistance of USD 2.4 billion for a total of 36 developing countries by 1978, and completed the Tanzam Railway and other 200 plus projects which covered a wide range of areas including agriculture, transportation, meteorology, health, sports and education. It is fair to say China spared no effort to help these countries. The Tanzam Railway signified the beginning of Chinese-led South–South cooperation. Known as "the Railway of Freedom and Friendship" among developing countries in Asia, Africa and Latin America, it carries "epic symbolic meaning" in the modern history of international relations. During the Cold War, such sincere and equal partnership between China and the developing countries in Asia, Africa and Latin America undoubtedly had great appeal for both sides that had for long suffered from the humiliation of Western colonial rule. Within the 10 years of the 1970s, 24 developing countries in Asia, Africa and Latin America established diplomatic ties with China, which brought about a significant improvement in China's diplomatic environment back then.

On February 22, 1974, Mao Zedong proposed the Three Worlds Theory based on the ever changing international situation. This theory was not improvised, but gave a new insight into the division of the complex international political forces after a long period of observation and reflection. It had tremendous impact on China's diplomatic and international cooperative relations. In April 1974, then Chinese Vice-Premier Deng Xiaoping explained Mao's new strategic vision to the world in a speech to the United Nations at the sixth special session of the General Assembly on the problems of raw materials and development. "Judging from the changes in international relations, the world today actually consists of three parts, or three worlds, that are both interconnected and in contradiction to one another. The US and the Soviet Union make up the First World. The developing countries in Asia, Africa, Latin America and other regions make up

---

[2] Shi Lin (ed.) *Foreign Economic Cooperation of Contemporary China*. Beijing: China Social Sciences Press, 1989, p. 69.

the Third World. The developed countries between the two make up the Second World".[3]

Mao's Three Worlds Theory was proposed mainly for the diplomatic need to oppose the hegemony of the Soviet Union. And one of the main ways to achieve this diplomatic goal was to seek extensive contact and cooperation with the Third World countries. The Third World here refers to the numerous developing countries of Asia, Africa and Latin America regions. Most of them used to be the colonies or dependencies of imperialist countries and had thus long suffered from colonialist and imperialist oppression and exploitation. A number of Asian, African and Latin American countries won independence in the sweeping movement of national liberation after World War II, and became an important force in the international arena. Like China, these countries also faced the common historical task of national rejuvenation after their independence, which put China and them in the same historical stage of development. Mao also included China in the Third World, "because China cannot compete with rich countries and big countries in terms of political system and economy, it belongs with these poorer countries". According to this theory, the Third World is not just a group of countries that are relatively independent and in need of China's sympathy and support, but also an indivisible whole where China is part of and shares the common interests. Mao's development and enrichment of the Third World concept not only identified the foothold of China's foreign policy and foreign relations, but also enhanced China's prestige in the Third World.

From the founding of the People's Republic of China to 1978, China made great headway in its political and economic independence. But when compared with the world development, many problems emerged during China's development. One of them is that it was difficult for China, which needed to undertake large-scale construction back home, to continue the provision of a significant amount of aid to the Third World countries. It had been done with an all-out effort for proletarian internationalism had been elated to a supreme position, though it was beyond its own ability. Another problem is that after the war economic globalization picked

---

[3] Speech by Deng Xiaoping at the sixth special session of the UN General Assembly, *People's Daily*, April 11, 1974.

up speed for the world economy went through great changes and experienced a new industrial revolution. The capitalist world represented by Western countries made great headways for they were willing to ride this trend. In contrast, socialist countries as a whole lagged farther and farther behind in their development for their departure from the world's economic system. And the situation faced by China seemed grimmer. There are two main reasons. Firstly, for a long time, we failed to shift the focus of our domestic work to economic development. The second reason is the closeness of our development path. All of this was about to change at the turning point of the Third Plenary Session of the 11th Central Committee of the Communist Party of China (CPC). After nearly 30 years of tortuous economic and social development and deep reflection on the socialist path, China finally confirmed reform and opening up and prioritizing economic development as its basic national policies. These national policies promoted the development of foreign relations featuring foreign trade and foreign direct investment absorption. This turning point was the result of the shift in our understanding of the international environment.

II. Market-oriented Economic and Trade Relations

After the launch of reform and opening up, China's foreign trade and economic relations featured "bringing in and going global" activities dictated by the market. China connected with the world and let the market decide its economic and trading activities. As a result, Chinese enterprises attracted a huge amount of foreign investment, meanwhile foreign trade and outbound investment also grew by leaps and bounds. At the beginning of the 21st century, China joined the World Trade Organization (WTO), which was also of great significance.

With the deepening of reform and opening up and facing the challenges and opportunities entrusted by history, how should we set the course for future development? How do we seize the high ground in international competition? Can we discover new sources of economic growth? The central government has always attached great importance to these questions. Deng Xiaoping, the chief architect of reform and opening up, put forward that we should actively expand foreign trade, explore international labor service cooperation, foreign project contracting and international tourism so as to incorporate the Chinese economy into the world economy, and proposed the strategic policies of promoting import, facilitating export and attaching equal emphasis to these two. In his report delivered at the 15th CPC National Congress, Jiang Zemin proposed that we should "encourage Chinese investors to invest abroad in areas that can bring China's comparative advantages into play so as to make better use of both domestic and international markets and resources". In 2000, the central government confirmed "going global" as the opening-up strategy of the new era. In 2001, in the *Outline of the 10th Five-Year Plan*, "going global" strategy was listed along with foreign trade and the utilization of foreign investment as the three pillars of the open economic development during the 10th Five Year Plan period. At the Third Plenary Session of the 16th CPC Central Committee in 2003, Hu Jintao proposed that we should "implement the strategy of 'going global', improve the service system of foreign investment, give enterprises greater autonomy in oversea operation and management, and improve regulatory mechanisms of oversea Chinese-invested enterprises to promote the development of China's multinational companies". Since the 18th CPC National Congress, the central government has pushed forward the implementation of the "going global" strategy and included foreign economic cooperation as an important part of the economic and diplomatic strategy by putting forward a series of strategic visions including the Belt and Road Initiative, regional integration strategy, China–Pakistan Economic Corridor (CPEC), the Bangladesh–China–India–Myanmar (BCIM) economic corridor, and the "461" cooperative framework for the China–Africa strategic partnership and facilitating financial arrangements, such as Asian Infrastructure Investment Bank (AIIB), New Development Bank (NDB), and Silk Road Fund. It also accelerated the formation of a new pattern of opening up in all directions and drew up a new blueprint for the economic cooperation between China and other countries. These strategic ideas, which are coming together as a relatively complete theoretical and practice system of "going global" with Chinese characteristics, can provide strong policy support and momentum for China's opening up to the outside world.

Based on national conditions and the actual needs, China focused on foreign trade and foreign investment absorption for a very long period of time in the early stage of reform and opening up. Although some companies sought overseas expansion on their own during this period of time, most of such endeavors were small in scale. By the end of December 2002, a total of 424,200 foreign-invested enterprises had been approved, contractual foreign investment reached USD 828.06 billion, the actual use of foreign capital reached USD 447.966 billion, and the actual use of foreign direct investment registered USD 52.7 billion. China overtook the US for the first time and became the biggest destination of foreign direct investment (FDI) in the world. In comparison, outbound investment took up only an insignificant share in China's economy. With the changes in domestic and international situation, China had gained a better understanding of the importance of vigorously exploring the overseas market. At the beginning of the 21st century, China restored its position as a signatory of the General Agreement on Tariffs and Trade and officially became the 143rd member of the WTO. While China's foreign trade grew by leaps and bounds, the flow of its outbound investment started to accelerate. For example, the total trading volume between China and Latin America stood at USD 14.9 billion in 2001, surpassed USD 100 billion in 2007, and reached USD 241.5 billion in 2011.[4] In 2014, Chinese investors directly invested in a total of 6,128 foreign enterprises in 156 countries and regions. In the same year, its annual accumulated non-financial outbound direct investment totaled USD 102.89 billion, up 14.1% year on year.[5] With its accumulated outbound direct investment reaching USD 646.3 billion, China overtook Japan and Britain, traditional outbound investors, became the largest outbound investor in developing countries, made it to the top of the world's biggest capital exporters, and became a new force in the field of the world direct investment.

At the UN Sustainable Development Summit on September 26, 2015, Chinese President Xi Jinping announced a series of new assistance programs for countries of the South: China would set up a fund, with an initial contribution of USD 2 billion, to support South–South cooperation; it would also do its best to raise its investment in the least developed countries

---

[4] *Source*: CEIC Database (updated on January 30, 2012); ECLAC, FDI in Latin America and the Caribbean 2010, 2011, p. 107.

[5] *Source*: http://ccn.mofcom.gov.cn, "China's Outbound Investment and Economic Cooperation in 2014".

(LDCs) to USD 12 billion by 2030; it would exempt the debt of the outstanding intergovernmental interest-free loans due by the end of 2015 owed by the relevant LDCs, landlocked developing countries and Small Island Developing States (SIDS). Xi also announced that in the next five years to come, China would provide 100 projects in six categories for developing countries, including 100 poverty alleviation projects, 100 hospitals and clinics, and 100 schools and vocational training centers. This shows that China's foreign trade and economic relations have entered a new stage featuring both "government assistance and investment by businesses".

III. Challenges that Cannot Be Ignored

In the new era, the situation faced by China's economic cooperation and foreign trade has undergone significant changes. One major change is that the competition among large countries has become more complex. The contradictions or conflicts between the world's major countries, especially Western powers, and China in the developing countries of Asia, Africa and Latin America are not only disputes of economic interests, but also, in some cases, competition of values and strategies. In most cases, they are the three rolled into one or the causal result of the interaction among the three. While China and the developing countries in Asia, Africa and Latin America are the parties of their cooperation and the driving force behind it, they cannot fully control the numerous variables that affect the development of bilateral relations due to their own limitations. In addition, China is gradually transitioning from a planned economy to a market economy; the main players of China's economic cooperation and foreign trade are turning from government bodies to market participants; and the principles of market economy are gradually taking over the dominant role in guiding China's economic cooperation and foreign trade. Meanwhile, China fails to pay enough attention to the particularity of its cooperation with Asia, Africa and Latin America. In particular, in the implementation of their project, some enterprises see only the immediate interests while neglecting the long-term interests and the traditional friendship, and seek only economic interests while neglecting social benefits. This has come as an unpleasant surprise to some countries in Asia, Africa and Latin America who have been on friendly terms with China, and even made them question China. As a result, some Chinese enterprises have run into predicament in their exploration of the developing countries in Asia, Africa and Latin America. A large number of the developing countries in Asia, Africa and Latin America are no longer the same countries for both their internal situation and foreign relations have gone through some changes. Therefore, they have become more sensitive of interests. In addition, China and these countries differ in their level of development and thus have different interests. In some cases, they may even have no common interests. As a result, the number of frictions is about to increase. This phenomenon is more prominent in today's Sino-African cooperation. Specifically, there are eight challenges that cannot be ignored in China's economic and trade relations with foreign countries.

First, it is not easy to control overseas risks. Today's international market has become an arena of political, economic, religious, cultural, scientific and technological competition, creating a more complex mix of risks for enterprises, such as trade friction, intellectual property protection, outbreaks of violence and hostage incidents. Some political risks, such as the unrest and civil war in Libya, Syria and other countries, are even more difficult to manage and control, which bring enormous difficulties to companies. Besides, affected by the international financial crisis, the "economic nationalism" that originated from the United States and Europe, has expanded to Australia, Canada and New Zealand, and been copied by India, Latin America and Eastern Europe. Countries have adopted different policies to attract foreign investment. Some governments begin to re-emphasize the necessity of control and set hidden investment barriers in the name of national security, environmental protection, public interest, and employment protection, thereby increasing the risk and costs of transnational investment. In addition, for political reasons, Chinese state-owned enterprises have been treated unfairly by some countries for their assumed government background, which has also exerted negative influence on their international operations. The international community has always kept a vigilant eye upon China's overseas investment. Though it is unlikely to formulate obvious discriminatory provisions, implicit means such as cumbersome censorship system and arbitrary standards or procedures may be used to pose obstacles.

Second, China is still at a disadvantage in the formulation of rules and standards. Like most developing countries, China is in a subordinate position in the formulation of international standards and rules. As a result, many Chinese enterprises are at a disadvantage when going global due to the applicability of rules and standards. The reason why the US and other developed countries are trying to implement new technical regulations and standards is because they want to lead the new round of international competition and curb the development of developing countries. We need to be wary of the moves they are taking for this intention. Turning back to ourselves, the vast majority of Chinese businesses that are going global are more adept at winning product competition through cost advantages, but they do not pay enough attention to fighting for the right to formulate international rules and standards nor have the capacity to do so. Therefore, there is an urgent need for them to participate in the formulation of international rules and standards and increase their relevant capacity.

Third, Chinese enterprises are relatively weak in international operations. Chinese enterprises have a short history of international expedition and thus lack relevant experience. Their ability to control risks in face of different languages, cultures, institutions and legal systems is relatively weak. Chinese enterprises still have a lot to catch up with foreign multinational companies. And the domestic business community also lacks talents who are up to the task of international operations in the global market. In addition, the current international business environment is no longer the one where Western countries made their early exploration of the international market. That is to say, the approach to seek the maximization of one's own economic interests is no longer feasible. This also means that companies need to be competent in fulfilling their social responsibility and blending into local society, which happens to be one of the weaknesses of Chinese enterprises.

Fourth, financial support and supporting service system for Chinese enterprises need to be improved. Domestic financial institutions have had a late start in their going-global endeavor and lack capacity and experience in overseas operations. In addition, only a small number of overseas branches are set up around the world and their distribution is uneven. There is no sound system of financial support to keep up with the international expansion pace of the enterprises or meet their financial needs. Chinese enterprises lack the basis of creditability in foreign countries, resulting in a series of difficulties including low accessibility, high interest rates, small credit amounts and short payment terms when they try to take out a loan from foreign banks. Because of the difficulty in accessing sufficient financial support and supporting services, Chinese enterprises have to bear the risks of going global alone. If we look at the US, we will find that if it were not for support from transnational banking groups it would not have been able to remain the biggest outbound investor after World War II. These banking groups not only provided US multinational companies with a series of services including financing arrangements, global fund transfer, cash management, consulting, foreign exchange transactions, leveraged buyouts and insurance, but also helped them search the targets for merger and acquisition (M&A), design M&A plans, act as M&A consultants and negotiators as well as facilitate the closure of projects. China needs to draw upon international experience and accelerate the pace of financial institutions' global expansion.

Fifth, there is a pressing need for China to enhance its ability to control the expectancy of foreign aid. With the growth of China's gross domestic product (GDP), it is necessary to increase the amount of foreign aid. Only by sharing its success with the world can China go farther. China tightened its belt to help poor developing countries when it was in a very difficult time. In fact, what China did back then helped itself accumulate valuable diplomatic resources. Now, China is more capable of providing foreign aid, and the Third World countries have also raised their expectancy. Therefore, this is a great opportunity for China to consolidate its relationship with other developing countries. However, China changed from a pure recipient of foreign aid to a widely anticipated provider in a very short period of time. It is important to properly handle the relationship between its capacity of providing foreign aid and the international expectancy. A large area of China remains underdeveloped and in great need of financial support so we need to coordinate domestic and international opinions on foreign aid and our priorities. Facing the international community, we need to stick to the truth. We should let the world see that there are still underdeveloped areas in China; therefore, we should adjust the international community's expectancy of foreign aid to match with our true capacity. In China, we need the right strategy and careful organization as well as the guidance and support of the public opinion to avoid undermining the public's confidence for revealing its backward side.

Sixth, the way China conducts its foreign trade is relatively simple. Firstly, there is only one investment model and the early warning and guarantee mechanism is relatively weak. Therefore, Chinese companies face greater investment risk in countries with poor economic investment climate, immature legal system and unstable political situation. Secondly, our cultural exchanges lag behind economic cooperation. There is much China can do to enhance mutual understanding and create a favorable cooperation environment by enhancing cultural exchanges, educational cooperation and themed publicity based on the characteristics of multiethnicity, multireligion and multiculture of the developing countries in Asia, Africa and Latin America. Thirdly, the way China participates in local development efforts is homogeneous. Apart from infrastructure projects, China should help local people improve their well-being and consolidate the social basis for China's cooperation with Asia, Africa and Latin America through a variety of forms including education, training, labor service cooperation, exchanges and medical assistance.

Seventh, the international cooperation mechanism needs to be improved. The cooperation between China and Asia, Africa and Latin America has great potential; however, the cooperation mechanism is lagging behind. Firstly, China relies too much on their ruling party and overlooks the diplomatic work on other levels for it has not conducted enough "development-oriented cooperation"[6] with the out-of-power parties, non-governmental organizations and the general public. Secondly, comprises only mutual visits, China's political cooperation mechanism is relatively rigid. And there is no flexible, diverse and multi-layered system of cooperation. Thirdly, the impact of the existing forums and other

---

[6]Ai Ping. Development-Oriented Cooperation between Chinese and African Political Parties. *Development-Oriented Finance Research*, 2014, Vol. 3, pp. 68–74.

mechanisms is neither far-reaching nor sustainable, and there is a lack of timely coordination from bilateral and multilateral political and economic cooperation institutions. Fourthly, China is not sufficiently involved in regional multilateral cooperation. There are many regional development organizations in the developing countries of Asia, Africa and Latin America and they play an important role in regional economic and social development. However, China has not engaged in sufficient exchanges and cooperation with these organizations.

Eighth, there is a pressing need to improve the tactics of economic and trade strategy. When it comes to the tactics of the strategy, we often confuse strategic approaches with construction plans, domestic theories with international practices, and Party ideologies with the public opinion, which is detrimental to the development of economic and trade cooperation. As for the systematicity of the strategy, we should enhance the relevance of strategies, the complementarity of policies and the response measures for changing circumstances in a timely manner. As regard to the details of the strategy, we should formulate detailed and specific implementation plans after conducting feasibility studies and considering geopolitics, geography, traditional culture, market foundation, risk management and control and other factors of both cooperation parties to ensure mutual benefits. These efforts are essential for the proper implementation of the central strategic thinking.

IV. Improving the Ideological System of Economic Cooperation for the New Era

Ideas decide action; therefore, we need to discuss in depth and constantly improve foreign trade and economic thoughts to guide the better development of economic and trade relations. We should further define and consolidate the core of the thoughts of economic and trade relations in the new era. The ideological core includes three aspects:

Firstly, the world is a community of common destiny. The reason that we advocate the concept of "a community of common destiny" is to emphasize the necessity for human society to uphold the international relations thoughts with ecological awareness and to achieve common, comprehensive, cooperative and sustainable security and development. We should abandon Cold War thinking and uphold mutual trust and cooperation in order to build a path of global safety and development featuring joint development, sharing, and win–win outcomes. We should also promote the integration and exchanges among different civilizations. Regardless of their size, national strength and level of development, all nations are equal members of the international community and are entitled to participate in regional and international affairs as equals. Issues that involve all the countries should be discussed by them together. With greater size comes greater responsibility for regional and world peace and development. To promote mutual respect and equality, we should first respect each country's right to choose their own social system and development path, respect each other's core interests and major concerns, be objective in viewing other countries' growth and development policies and philosophies, and strive to seek and expand common ground.

Secondly, we should stick to the thinking of pursuing balanced economic and trade relations: in developing foreign trade and economic cooperation, we should firmly establish the concept of balance, analyze factors that cause imbalance, and take balancing measures to achieve mutual benefits. We should take China's national security and development, national well-being as well as world peace and development as our purpose, national modernization and rejuvenation as our priority, and facilitating the implementation of the national strategy as our task. In the international arena, we will not go to extremes, become overconfident or rest on our laurels but instead we will make as many friends as possible and promote political pluralism, mixed economy, network society as well as mutual trust and cooperation.

Thirdly, it is important to further clarify the goal of our international trade, which is to advocate and promote the establishment of "a new type of strategic partnership featuring political equality and mutual trust, economic win–win cooperation and cultural exchanges and mutual learning". China should promote the new international economic and trade relations with win–win cooperation as the core and always act as a staunch force in safeguarding world peace and promoting common development by unswervingly pursuing an independent foreign policy of peace, following the path of peaceful development, implementing an opening-up strategy of mutual benefit, upholding justice and pursuing shared interests.

We should enrich the "going global, bringing in" strategy.

For nearly 30 years, China's export trade has been growing at a double-digit rate, providing fuels to sustain China's rapid economic development. However, it is impossible to maintain such a high-speed growth in export. Therefore, China must go beyond its borders, make use of overseas resources and explore overseas markets. To achieve that, we can consider the following ways: with regard to trade and economic relations, we should properly handle the relationship between politics and the market, attach equal importance to politics and benefits, intergovernmental cooperation and non-governmental cooperation, as well as cooperation in large projects and small businesses; on the mode of operation, we should transition from the low-risk mode of physical production and trade to the high-risk mode of capital investment; in overseas exploration, we should start from countries nearby and then make our way to faraway ones, and begin with countries with small economic and cultural differences to countries with large ones; in terms of market development, we should enter one market after learning about it and its local communities of overseas Chinese and make informed exploration of the market; on investment, we should rely on both investment from state-run enterprises and private investment. In particular, we should raise money from the local Chinese communities. The financing of overseas Chinese is an important part of financing arrangement and a way for them to gather strength and deepen development. From a strategic perspective, China should implement the strategy of pursuing slow-growth in export and mass overseas production for the medium and long term.

In the process of exploring overseas markets, finance is an indispensable supporter of Chinese enterprises' "going global" endeavor. With the development of economic globalization, especially the financial globalization, the core status of finance in the world economy is more evident, which requires China's financial diplomacy to play a bigger role in safeguarding national interests. The core idea of China's financial diplomacy should be "building credibility around the world to enable international financing". Given the central role and working principles of finance in the national economy, we should build a trustworthy partnership in line with

China's long-term development needs to promote mutual trust and benefit and achieve national development goals by fully mobilizing various resources related to finance and coordinating the development of domestic and international markets. To enable finance to play a supporting role in the "going global" endeavor, we should adhere to the principle of "letting people be the actor of finance, letting finance serve the people and letting people enjoy the service of finance", hold high the banner of "joint planning, joint development, joint construction and common development" and implement the strategic thinking of "going global and settling down". Firstly, we should put justice before benefits and establish a long-term relationship of credibility; secondly, we should give more and take less in order to build a genuine relationship of credibility; thirdly, we should exchange resources in order to build a relationship of credibility featuring mutual help; fourthly, we should provide comprehensive support in order to establish a comprehensive relationship of credibility; fifthly, we should diversify our services in order to establish a flexible relationship of credibility, create a variety of business models and credit structures, and provide a variety of financial services. Sixthly, we should focus on developing a relationship of credibility with the Third World countries. We should inherit and carry forward the theory of "Three Worlds" and further consolidate and expand the united front.

While Chinese enterprises are dedicated to their "going global" effort, we should also promote the strategy of "bringing in". We should bring in not only capital and technology, but also advanced theories and mechanisms, which we should modify according to China's realities. We should also properly handle the relations between the domestic market and the international market. Domestic development is not only the foundation of our diplomacy but also its purpose. China's market is an important part of the international market. All the major economies rely on their domestic markets. In its development, China should rely mainly on its own strength and resources and take the strengths of other countries as supplementary support; likewise, the development of other countries and regions should rely mainly on their own strengths and China can only play a supporting role. With a population of 1.3 billion, China is an enormous market. Its urbanization, industrial development, resident consumption and environmental protection can create many huge markets. Fundamentally, China must rely on its own market and strength to develop. Therefore, we should make great efforts to build and develop domestic market that runs in a virtuous cycle and implement the strategy of "expanding domestic demand and supply".

V. Correct Understanding of the Strategic Vision of the Belt and Road Initiative

China proposed the building of Silk Road Economic Belt and 21st-Century Maritime Silk Road, known as the Belt and Road Initiative, which reflects that China is actively adapting to the new global geopolitical environment, focusing on the cooperation in infrastructure construction and going all out in the implementation of the strategic thinking of "going global and bringing in". Like the Five Principles of Peaceful Coexistence proposed by China in the 1950s, this is also the overarching design for economic and trade cooperation put forward by China for the new era. It includes a number of specific projects and transport routes; however, it is more than that. It is a new pragmatic framework and initiative of international relations with the cooperation in infrastructure construction as the main part. With the cooperation platform of the Belt and Road Initiative, all the countries involved can propose cooperation projects, cooperation areas, and cooperation models, and establish cooperation mechanisms between governments and peoples to promote more balanced and sustainable regional development. An open and inclusive attitude is adopted towards cooperation ideas, cooperation space, cooperation areas, cooperation approaches, cooperation mechanisms and partners so as to resolve risks and achieve win–win outcomes for all parties involved through mutual trust and cooperation.

The Belt and Road Initiative is a throwback of the history. It reminds us of the merchant caravans that traveled through the Pamirs in Central Asia, which helped link East Asia with China and create the splendid civilization of East Asia through the joint force of multiculture and multiethnicity. It reminds us of the fleet led by Lin Luan, a navigator in the Tang Dynasty, the one led by navigator Wang Dayuan in Yuan Dynasty and the one led by navigator Zheng He in the Ming Dynasty, which sailed along the Maritime Silk Road in mighty formation, brought forth ancient marine economy and greatly strengthened China's links with the world. However, today's Belt and Road Initiative is not only a reminder of the glorious past but also an ambitious vision for the modern time. This means to build a modern economic corridor in the Eurasian continent to ensure the smooth flow of trade; to build the 21st-Century Marine Silk Road to connect the ports and straits around the world, spread the information and technology that China uses to seek peaceful development and prosperity to all the countries and regions, and share the fruits of development with the rest of the world. The Belt and Road Initiative can be divided into three parts. The first is the all-round economic and trade exchanges which cover all the countries and regions in the world. As long as there is any promise of business, China will strive to establish channels of cooperation. The second part is to focus on the economic and trade relations with Asia, Africa and Latin America and promote exchanges, mutual help and common development. The third part is the connectivity and friendly exchanges with the neighbors, which is the core of the initiative.

The Belt and Road Initiative reflects the diplomatic thought of "prioritizing neighboring countries". There are three subcontinents around China: Central Asia, Southeast Asia and Northeast Asia. Geographically speaking, Central Asia is the center of the Eurasian continent, its hub of land transport and an intersection of cultures. Its unique geographical advantage makes it much coveted by major powers. American political scientist and geostrategist Zbigniew Brzezinski and British geographer Halford Mackinder both believe that Central Asia is the center of the world; therefore, whoever controls the Eurasian hub area, including Central Asia, will control the world.[7] However, in my opinion, it is an overstatement. In modern world with great technology advances and so many ways of interaction among countries, the fate of a nation and the world cannot be determined by a single place nor an intersection. Moreover, South Asia and Northeast Asia surrounding China are also areas with great strategic importance. Therefore, the cooperation between China and its neighboring countries is comprehensive. For example, the

---

[7] Zbigniew Brzezinski. *The Grand Chessboard: American Primacy and Its Geostrategic Imperatives*. Translated by China Institute of International Studies, 1st edition. Shanghai: Shanghai People's Press, 2015, pp. 26–40. American and British strategists emphasized the strategic position of Central Asia. It may be a strategic vision, or an arrangement of distraction.

New Eurasian Continental Bridge, China–Russia and China–Myanmar pipelines, the Kunming–Bangkok Highway, the Trans–Asian Railway, China–Pakistan Railway, China–Myanmar–India Railway, and other economic corridors that are yet to be put on the agenda. These are all efforts made by China to seek progress in the joint infrastructure projects of various areas in different countries and regions in order to build a connectivity network with its neighboring countries. At the same time, China has promoted the cooperation with Southeast Asia, Northeast Asia and Central Asia to facilitate regional economic integration.[8]

It can be said that the Belt and Road Initiative is also a vision and initiative of globalization. It can draw attention from all the countries and regions and motivate them to explore more convenient navigation courses, such as the Northwest Passage, Red Sea–Mediterranean high-speed railway project and the major thoroughfare starting from the UK running through Europe and Asia and reaching North America and South America through the Bering Strait. "China–Russia–Canada–America line" would run for 13,000 km across Siberia and pass under Bering Strait through 200 km tunnel. People can come up with bolder plans and designs. For instance, we can construct a railway/highway from the Cape of Good Hope at the southern tip of Africa which will run through Africa from the south to the north, extend into Europe and Asia through West Asia, stretch northward to the Americas through the Bering Strait and run through Alaska in North America and then Canada until it reaches Cape Horn, the southern tip of Chile in South America. This route is not impossible, because Africa is where human ancestors originated before they moved to other parts of the world. They achieved the globalization of the primeval ages of human society through their footprints. Modern humans are more capable of employing advanced technology to build this major thoroughfare that runs across the world. Such major thoroughfare can bring the world closer.

In April 16, 2017, Beijing hosted Joint Communiqué of the leaders roundtable of the Belt and Road Forum for International Cooperation, which affirm the need to prioritize policy consultation, trade promotion, infrastructure connectivity, financial cooperation and people-to-people

---

[8] Lixing Zou. *Lessons from Development of US: Mechanism of Financial Imperialism,* 1st edition. Hunan University Press, 2009, pp. 397–398.

exchanges, and highlight the concrete actions in accordance with our national laws and regulations and international obligations.

To sum up, there are two stages of trade and economic development since the founding of the People's Republic of China. One is government-led and the other, market-oriented. With the economic globalization, economic cooperation and foreign trade are taking up an increasing share of the core national interests. The government has become more like a market player which takes economic cooperation and trade as its priority task while enterprises now have political considerations and have become increasingly aware of national interests. In addition, the competition among great powers is getting more and more complex. In the new era's trade and economic relations, we should hold high the banner of mutual trust and cooperation, promote the interaction between the government and companies as well as political and economic integration. We should enrich the "going global and bringing in" strategy and consolidate and expand our united front. The Belt and Road Initiative is the overarching design for trade and economic relations in the new era. It should mean more than specific projects and transport routes: it is a new and pragmatic framework and initiative of international relations with cooperation in infrastructure construction as the essential component. This framework and initiative, which inherits and carries forward the Five Principles of Peaceful Coexistence, is an embodiment of the innovative and pragmatic spirit. In pushing forward the Belt and Road Initiative, we should focus on the cooperation in infrastructure construction and make achieving mutual benefit as our goal. We should also propose cooperation projects and areas to provide different modes of cooperation for different economies. In addition, we should increase China's voice in the new landscape of international trade and the formulation of international rules. We should promote balanced and sustainable regional development and China's transition from a major to a competitive player in foreign trade to accelerate its modernization drive.

Case Study: How to Establish a Cross-border E-commence Mode for Industrial Goods

Recently, I worked closely with Raymond (Beijing) Technology Co., Ltd. (hereinafter referred to as Raymond) on how to establish a cross-border circulation mode for industrial products to serve the Belt and Road Initiative. Through thorough investigation and research, we want to help Chinese enterprises think broader, deeper and more comprehensively, and provide some reference opinions for them so that they can solve the problems they have and will encounter during their development; meanwhile, by analyzing the typical case of Raymond like dissecting a "sparrow", we actively explore at a macro level how to improve the relevant policies to give full play to the role of Chinese enterprises so that they can become integrated into the global market; we will further promote the cross-border circulation of Chinese industrial products and better implement the national strategy, and based on these efforts, we can further participate in international governance and enhance China's right to speak in coordination with international macroeconomic policies.

With the in-depth development of economic globalization and regional economic integration, the international economic and trade pattern is undergoing great changes. The Internet-based cross-border logistics and production are becoming an important new form of international trade, having a profound impact on international governance. Based on a global interoperability, China's Belt and Road Initiative is a new framework of international economic and trade relations and a new strategy for China to actively participate in international governance. Chinese enterprises are important implementers, pioneers and promoters of the Belt and Road Initiative. Through cooperating with system experts and exploiting the traditional advantages of research and development, production and marketing, Raymond actively innovates in the cross-border circulation mode for industrial products and builds a new core competitiveness to serve the Belt and Road Initiative, which is of great significance.

1. *About Raymond Co.*

Raymond Co. was founded in 2009 with a registered capital of RMB 116.8 million. Engaged in research and development (R&D), design, production, sales, service, import and export, it is a new and high-tech enterprise identified by Beijing Municipal Science & Technology Commission. Currently, the company has 400 employees, and six technical R&D departments responsible for the R&D and production of high-performance pipeline ball valves, metal hard seal ball valves, special valves, regulating valves, instrumentation valves, and regulating and metering pries. Its products are widely used in petroleum and natural gas transmission pipelines, oil refining, the coal chemical industry, municipal construction and other fields; it is a typical enterprise engaging in R&D, production and marketing of traditional industrial parts. In 2015, the company achieved sales revenue of RMB 500 million and a net profit of RMB 67.99 million, with an annual growth rate of more than 50% both in profit and revenue. Raymond was listed on National Equities Exchange and Quotations (NEEQ) with ultra-high indexes, and it is now a fast-developing enterprise.

During the process of development, Raymond keenly monitored the opportunities and challenges brought by the Belt and Road strategy, as well as the deep contradiction between the huge overseas market demands for Chinese industrial products and the difficulties for Chinese industrial products to go overseas, so it thought that only by innovating in the cross-border circulation mode for industrial products could we effectively remove the choke points of the industry and open up a channel for Chinese industrial products to enter internationals markets, especially the markets of the developing countries in Asia, Africa and Latin America. In October 2015, based on its original overseas sales team, Raymond built a professional cross-border e-commerce team, and accelerated its strategic transformation by utilizing its experience and advantages in overseas sales. It founded Beijing Raymond International Electronic Commerce Co., Ltd. to build a cross-border e-commerce and logistics platform for industrial products, which would cover many countries and regions around the world. Raymond International E-commerce Co., Ltd. has set up 12 subsidiaries and 12 logistics centers in countries such as Tanzania, Kazakhstan, Vietnam, Kenya, Cote d'Ivoire, Zambia, Mozambique, Malawi and Uganda. Currently, the company has accelerated the construction of its online platform and offline overseas layout, and plans to start the one-stop service of its e-commerce platform for industrial products by the end of 2016, to realize the platform's normal functions, such as industrial products' online management, ordering, payment and delivery, as well as integrated management of the background and warehousing logistics management. The company plans to set up 30 subsidiaries and 30 logistics centers overseas in 2017, with an expected total investment of about RMB

7.5 billion. The company's new strategic developmental goal is to build a cross-border e-commerce platform for global industrial products in line with China's overall strategy for foreign cooperation, in order to lead made-in-China products to global markets.

In general, Raymond, who is building a new cross-border e-commerce mode for industrial products, needs urgently to answer some questions related to the "safety, efficient and sustainable development". These questions include: at the strategic level, how to design its overseas layout in line with the Belt and Road strategy to confirm its positioning, deepen development and go into maturity; at tactical level, how to draw on the mature experience cases for reference at home and abroad to devise its business layout and roadmap; at the countermeasure level, how to properly address the issues such as credit and structure financing, business risk control, business network construction and localized operation; how to utilize available resources and favorable conditions under the new international trade rules to avoid risks. We have held an in-depth discussion with Raymond's top managers about these issues.

2. The Features of Raymond's Cross-border E-Commerce Platform for Industrial Products

Building a mode of "traditional industrial products + Internet". Based on its strengths and relying on its production and processing services, Raymond utilizes Internet technology to combine Online to Offline (O2O) and Factory to Business (F2B); by giving full play to the role of its business teams in the developing countries, it builds an extensive network of overseas distributors so that local distributors can pick up goods nearby; after local distributors place an order online, goods will immediately be shipped to them from Chinese ports to achieve timely and convenient delivery and finally realize centralized distribution, which can help enhance the deliverability of Chinese industrial products.

Establishing an innovative mechanism of domestic and overseas interaction: Raymond's cross-border e-commerce platform for industrial products connects two groups — one is Chinese manufacturers and suppliers, and the other is big customers and domestic and foreign distributors who have extensive and details demands for industrial products — to form a new logistics channel based on "domestic bonded warehouses and overseas transit warehouses". Through online payment and offline commodity delivery, Raymond integrates domestic delivery warehouses and overseas "five-in-one", which includes warehousing, displaying, purchase and sale, delivery and after-sales service, to build a super cross-border e-commerce platform for "made-in-China" industrial products. Relying on domestic and overseas interaction, simplified procedure and convenient service, the platform can help enhance the level of the e-commerce service for Chinese industrial products in the international trade.

Building a one-stop purchasing platform for Chinese industrial products: Because many Chinese e-commerce platforms for industrial products are small and provide limited products, and on the platforms, there are many problems such as randomly set prices, varied quality, inconsistent information and imperfect after-sales service, Raymond makes efforts to integrate product resources, build a large-scale, well-chosen cross-border product library and train competent supply chain teams in order to help Chinese industrial products break through time and space limits. Raymond's e-commerce platform also continuously increases new products, manages to lower purchase costs, and timely publishes after-sales policies to achieve integrated supply chain management, which can enhance the efficiency of supplying Chinese industrial products.

Expanding the markets for Chinese traditional industrial products in the emerging countries in Asia, Africa and Latin America: In terms of product positioning, in accordance with the trend of expanding international markets for China's strategic industries such as petroleum, petrochemical, municipal administration, natural gas, water, coal chemical industry, heating and electricity, Raymond pays attention to exporting high-end valves and regards the industrial products such as valves, bathroom accessories, building materials, electromechanical equipment, labor protection products and complete sets of machinery as its basic products; in terms of market positioning, Raymond regards the developing countries in Asia, Africa and Latin America as its targeted markets, reflecting its strategic vision. At present, in Africa, the Middle East, South America, Central Asia and some parts of Southeast Asia, where industry is underdeveloped, a large number of local industrial products and articles of daily use come from China; local distributors generally purchase these products from China and retail them locally, but the groups of local distributors are scattered and have a weak ability to bargain, resulting in high purchase costs, long periods for customs clearance and shipping and imperfect follow-up services, so this is still a low-level international trade, affecting the bilateral trade development between China and these countries. Raymond takes the initiative to carry out e-commerce service for industrial products in the developing countries in Asia, Africa and Latin America, which is not only adapted to the current changes in international trade patterns, but also in line with China's new strategic deployment.

3. *To Create a New Platform Embodying the "Spirit of Sharing"*

Raymond's cross-border e-commerce platform for industrial products is a new type of operation that appears during China's economic transformation and upgrading and implementation of the globalization strategy. In order to promote the safe, effective and sustainable development of this new type of operation, and help it become a world leading cross-border e-commerce platform for industrial products, Raymond should occupy a high ground in ideology, strategy and technology.

In ideology: Raymond should adhere to the principle of "co-building, co-management, sharing and win–win", promote and shape "shared culture", and give full play to the advantages of Internet technology + traditional manufacturing to vigorously develop the sharing economy. In the process of building the platform, some space should be left for future co-building, co-management, sharing and win–win; welcome large enterprises such as PetroChina and Sinopec to take full advantage of this cooperation platform to guarantee the supply of production materials, save costs and improve efficiency; welcome small- and medium-sized distributors to rely on this cooperation platform to expand industrial trade channels and increase additional value; welcome local and international capital, technology and management teams to join the platform construction to achieve a powerful combination and enhance the platform's influence and operational sustainability. Through platform services, Raymond should build a cross-border e-commerce ecosystem based on coexistence and common prosperity.

In strategy: Raymond should adhere to the principles of "good faith and unlimited logistics, online and offline combination & internal and external interaction, highlighting key points and seek intensive development, based on local conditions to achieve win–win cooperation". By regarding customers as God, Raymond should strengthen the construction of credibility and provide all-weather pre-sale, sale- and after-sale responses and services for customers; the online display of made-in-China products should be combined with offline logistics and warehousing centers and sales centers; Raymond should ensure an organic connection of domestic suppliers and international distributors to truly realize the transnational e-commerce service; focusing on the high-end valve products that it is relatively familiar with, and based on the markets of southern Africa, where the economic and social development is relatively stable, Raymond should develop local strategic partners and distributors, try its best to pay attention to and help local social and economic development, and share its development experience and achievements to build an ecological and permanent development relationship with local countries.

In technology: Raymond should adhere to the principle of "normalization, standardization, refinement and integration". The cross-border e-commerce platform for industrial products should achieve "unified management, procurement, pricing and service", fundamentally different from the store-renting mode commonly adopted by current e-commerce platforms. Raymond should adhere to the principle of first seeking common points and then reserving differences. First, to seek common points internationally, it should follow the existing standards and norms to enter the market, and reduce its characteristics to further gain all aspects of recognition; based on in-depth analysis of the rules and *status quo* of the new type of operation, it should try to integrate the existing rules into the platform services, and actively explore and participate in the formulation of new rules; based on the "personalized needs" of different markets and the increasing "flexible specialization" of enterprises, it should build a "professional, precise, special and new" service line; it should design and implement a set of advanced plans for the cross-border e-commerce service of industrial products, to form a flexible vertical integration that can be recombined within enterprises, a horizontal integration between enterprises and a digital integration of the end-to-end projects for social value chains, so that the platform can lead the market development and become an engine of the times.

To sum up, by using technologies of the Internet and the Internet of Things as well as modular technology, Raymond's cross-border e-commerce mode fundamentally changes the dissociation of production, R&D and trade in traditional industry due to geographical locations; as an exploration of transformation and upgrading of traditional industrial product trade, it can adapt to and lead the market and occupy a high ground in the market. First, it extends the foreign trade industry chain to the areas that cannot be reached formerly. Raymond's e-commerce mode changes the limitations of traditional foreign trade and extends foreign trade service to the international trade front — retail terminals, or even directly to the end users. Second, it enhances additional value for foreign trade and can gain profits that could not be gained formerly. Relying on the cross-border e-commerce platform, Raymond integrates suppliers, warehousing logistics, customs clearance, clearing payments, foreign distributors and users with the most efficient way and the lowest cost; the platform can reduce intermediate costs, and increase additional values and profit growth points of the products. Third, it improves the level of foreign trade services by providing services that could not be provided formerly. Raymond's e-commerce mode removes the asymmetry of domestic and international market information faced by foreign trade enterprises, enables them to directly face foreign distributors and end users, and relies on large data algorithms, advanced stocking and offshore warehouses to improve supply chains' responsiveness to demands, so as to greatly shorten the delivery cycle, improve service efficiency and lead foreign trade into an optimized development track.

Chapter 13

Changes in International Situation
and the Countermeasures

What is the main balancing force in today's international arena? Is the "one hegemon and multiple great powers" phenomenon the result of all the wrestling of balancing forces? How to improve our international strategy? In this chapter, I will apply balanced way of thinking to analyze these three important horizontal issues based on the previous analysis of the longitudinal issues. The world is in an age of intertwined conflicts. Under the influence of all kinds of interacting forces, the international geopolitical landscape is changing, and it will be a relatively long process. World peace and development remains the theme of our time; however, the world is becoming more imbalanced, unstable and uncoordinated. In particular, non-traditional security (new security threats) and changes in some hotspot areas are likely to bring changes to the international situation. There is a pressing need for the world to work together to address the urgent challenges and manage crisis. The changes in the strength of big powers present both challenges and opportunities for China to further implement the "going global" strategy. We should improve the development strategies and mechanisms of trade and economic relations and prioritize the interests of developing countries; we should "promote public–private interaction, interaction between multilateral relations and bilateral relations, as well as interaction between stability and reform"; we should "observe the overall situation, be honest and trustworthy, pursue balanced diplomacy and pragmatism, devote ourselves to planning, give more and take less, and focus on domestic development while continuing the international exploration"; we should serve China's modernization drive and make more contributions to world peace and development.

Under the influence of all kinds of interrelated and interacting forces, the international geopolitical landscape is changing. I will take the perspective of balancing forces to analyze the geopolitical strengths of some major powers as well as some new moves in international competition, including the Trans-Pacific Partnership Agreement (TPP), the Transatlantic Trade and Investment Partnership (TTIP) and the principle of competitive neutrality. I will also share my personal views on how China should actively plan its international layout. In response to changes in the international situation, we need the thinking of balancing the interaction.

I. Important Players in the International Geopolitical Landscape

The current world landscape and order was basically formed under the guidance of the US after World War II. This landscape and order was conducive to the post-war recovery and development and will continue to exert influence. With the economic and social development and scientific and technological advances, under the influence of all kinds of interrelated and interacting forces, the current landscape which is not free of drawbacks is changing; however, a fundamental change is unlikely to occur during this period of time. In fact, the global geopolitical landscape has become clearer after the Cold War. Especially after the September 11 attacks, its basic characteristics have become more obvious. Judging from its overall strength and global strategic layout, the US has not declined as a global superpower; instead it has seized the historical opportunity and achieved some development. It remains the dominant player in the world's geopolitical landscape. However, the gap between the US and Europe and other traditional powers is narrowing, and the one between traditional powers and emerging markets is shrinking, too. But it has only limited influence on the current global geopolitical situation. Therefore, we can say that the world is at a stage where unipolar and multi-polar trends are intertwined and new conflicts and contradictions coexist with the old ones. In the decades to come, a global geopolitical landscape "where one hegemon stands along with several powers, diversity is the norm and multi-polarization is the main trend" will take shape. In terms of strength, resources, size and potential, the US will remain the world's sole superpower and no other country will be able to challenge its hegemony; the

EU, Japan, Russia, India, Brazil, South Africa and China are fairly powerful economies, and they are not only the pillar in the global economic architecture, but also have an important impact on the regional and the world economies. As pointed out by American historian Leften Stavros Stavrianos in *A Global History*, the trend of the world history is global diversification instead of homogenization.

1. *The Daunting Strength of the United States*

America is rich in natural resources and ranks among the top in the deposits of coal, gas, uranium and per capita arable land. Although the US lacks oil resources, it has for long implemented the strategy of "storing oil and developing new energy sources", which does not only help it meet the needs of its sustainable development, but also constitutes an important tool of its global strategy. The US has great room for population growth. Its current population is 300 million; therefore, its population density is only 33 persons per km², lower than 52 people per km², the world average. Its total fertility rate is larger than 2.1, the rate needed for simple population reproduction. It is the only one among the developed economies to have reached this level, which allows them to have continuous population growth. The US has an influx of about one million new immigrants each year, many of whom are high-end personnel and workforce. Therefore, it has been able to maintain a more reasonable age structure for a long time. The US has well-developed education and science and technology sectors. It is home to two-thirds of the world's top 50 universities, leaving other countries in the dust. The US is now leading the digital revolution. Therefore, beside investment, consumption and exports, it can also rely on population growth and technological progress for gross domestic product (GDP) growth. In particular, its scientific and technological progress is the most powerful, long lasting, and sustainable force behind US economic and social development.

In addition to natural resources, human capital and technological strength, the soft power of American political culture is also an important factor that cannot be ignored. And the US pushed forward its phased global strategic security arrangement by seizing the opportunities provided by the September 11 attacks, the democracy movement in West Asia and North Africa as well as the geopolitical and economic reshuffle. It took a softer stance toward Cuba, Iran and other "enemies" to establish itself as a new diplomatic image of an advocate of multilateralism, cooperation and dialogue.

The US also grasped the opportunity created by policy adjustment in the global financial crisis and turned the crisis into an opportunity. It strengthened the effort to deepen domestic reform and improve the system, mechanism and related policies. On the political front, it promoted the social change and enhanced political inclusiveness. As a result, all the black people, women and other non-traditional political groups now are likely to make their way into the White House. In economic terms, the US is pushing through the bills for health insurance reform and the financial reform, which will have a major impact on the sustainable and healthy development of the US economy like the financial reform and the reform of social security system in the 1930s.

Generally speaking, the US strategic objectives remain unchanged. That is to say, the US is now the number one power in the world and it intends to remain this way and continue to lead the world in the 21st century. Its national interests are still three-fold: the core strategic interests (homeland security and development), important strategic interests (the security and development of its allies) and general strategic interests. There are no significant changes to its overall policy. However, the United States has made some changes to its strategy and focus, including the constant changes in tactics to meet the strategic needs: continuing to promote unilateralism in the international community while stressing the importance of alliances and joint action; proposing a series of measures in infrastructure, education and innovation to reflect the self-centered focus. Its timely adjustments and reforms reflect the strategic role and flexibility of its economic policy and mechanisms, as well as the strong vitality of the US economic system. The US is more competitive than expected. The international affairs which the US is reluctant to deal with prove to be the most difficult nuts to crack. The multifaceted nature of the American strength enables it to remain in the number one position among the world's major countries for the next 30 years.[1]

---

[1] National Intelligence Council, *Global Trends 2030: Alternative Worlds.* Translated by Center of American Research, China Institute of Contemporary International Relations. Current Affairs Press, First Edition of June 2013, p. 159.

For the US, the Pacific is the focus of the 21st century and in this century, the era where Europe is the center will come to an end and Asia will rise as the new center of the age. Based on this, the US strategic focus is gradually shifted to the Asia-Pacific region. The return to Asia is an important strategic choice made by the US in the 21st century. The strategy for its return does not only contain military content but also covers economy, trade and culture. And it is an important part of its rebalancing strategy. These are its missions: firstly, to bundle the Asia-Pacific allies and strategic partners with their own interests so as to strengthen the US-led security measures; secondly, to sow discord between China and its neighboring countries in order to jeopardize the regional integration in East Asia; thirdly, to block China's marine expansion in order to control marine resources and channels; fourthly, to take the upper hand in Asia-Pacific market in order to maintain its dominance; fifthly, to further impart American values so as to reinforce the cultural foundation for its sustainable development. The fundamental goal is to ensure US strategic superiority and hegemony. The US rebalancing strategy, mission and objectives may backfire. The US tries to disrupt Asia and undermine Asia's economic cooperation and development by creating and leading geopolitical disputes. This will in fact affect the export of the US and hamper its development. Therefore, China should point out the part that is detriment to all in the US rebalancing strategy, come up with balancing measures for the rest of it, seek common ground and jointly promote the Asia-Pacific economic cooperation and development. The US strategy of returning to Asia is based on its judgment of the international development trend and its belief that the Eurocentric era has come to a close. In fact, the 21st century could be the century of multiple centers. Europe is still the world's major trade center and has immeasurable potential in technological innovation and creative product revolution.

China–US relations are the core of China's foreign relations. Peaceful coexistence will benefit not only both sides, but also the whole world. China and the US are interdependent and our common strategic interests actually overweigh our divergences. But the differences in value orientation and the basic conditions for modernization are still huge; therefore, the wrestle between the two will be long and sometimes fierce. We should be neither humble nor pushy, seek no dominance, stand our ground, seek no troubles, fear no troubles, seek common ground and continuously enhance strategic mutual trust.

2. *European Integration Will Remain the Main Trend*

The EU is a politically and economically integrated regional organization. In December 1991, the European Community in Maastricht, the Netherlands adopted the Treaty on EU (the Maastricht Treaty), which created the European Central Bank and its operational mechanism. It is the first supranational central banking system in charge of monetary management in the world. The Stability and Growth Pact (SGP) which entered into force in 1997 set out the basic rules of the EU fiscal policies. Through great political wisdom, Europe united the countries that act in their own ways into an organic political–economic union, achieved regional "united in diversity", and realized the vision depicted by the French romantic writer Victor Hugo: "all nations in the European Continent would come together in unity, constituting a European brotherhood, while at the same time preserving their distinct qualities and individuality".

However, within this union, the relationship between Britain and continental Europe has been quite special. Britons have inherited a historical awareness of autonomy. They proposed the concept of rule of law in Magna Carta in 1215 and have pursued the "splendid isolation" policy in modern times. After World War II, Britain was still reluctant to get too involved in European affairs and remained absent from the integration process for a long time. Britain did not join the European Coal and Steel Community in 1951, the Economic Community in 1957 or the European Communities in 1967. It was not until 1973 that it joined the European Community. And the European Community was renamed the EU in 1993.

After its establishment, the EU witnessed the rapid expansion of its geopolitical boundaries. The number of its member states has grown from 15 in 1995 to today's 28. It covers an area of 2.42 million km^2 and boasts a population of 502.5 million. Its GDP reached USD 18.46 trillion in 2014 and surpassed the US in terms of economic strength. In recent years, affected by the international financial crisis, Greece's debt crisis has worsened, the European financial market has been thrown into turmoil, the confidence in euro has been dealt with a heavy blow and the economic recovery of the EU has lagged behind the United States and emerging markets. In June 2012, the EU proposed the idea of a banking union and established its three pillars, namely the Single Resolution Mechanism (SRM), the Single Supervisory Mechanism (SSM) and a European Deposit Insurance Scheme (EDIS). This was an attempt to establish a broader and more integrated European economic and monetary union, prevent the crisis in the banking industry from sprawling into a sovereign debt crisis and further promote the financial integration of the EU. The conflict between this ever more powerful supranational institution and Britain's spirit of autonomy is becoming more acute. On November 10, 2015, Britain announced its demands on EU reform, which included the reduction of the EU's excessive intervention in member states' economy and trade to protect the financial center status of London and prevent British foreign trade from being hijacked by the EU. These demands reflected the conflicts between British national interests and the overall interests of the EU intensified. Brexit is the best example of this contradiction.

After leaving the EU, Britain is now facing a series of upcoming tough negotiations with the European continent and other countries and regions, which will increase Britain's political, economic and social fluctuations in the next five years. Brexit will enhance, to some extent, Britain's independence in law, border control and governance. But Britain cannot be completely isolated from the EU. It can learn from Norway, Switzerland and Canada. It is most likely that Britain will find new ways to work with the EU. It will certainly strengthen its ties with the world, which gives other countries new strategic leeway. Brexit, a little setback in the European integration, will not affect the general direction of European integration or the process of global integration in the long run.

Currently, apart from Brexit, the EU also suffers from the protracted crisis in Ukraine, the Greek debt crisis and the influx of refugees. These problems result from the changing international situation, but they also reveal the drawbacks of European economic social and political system and precipitate the European reforms. With the phasing in of emergency measures and risk prevention measures, the European debt problems have been put under control, and the banking regulatory mechanism of the Treaty on Stability, Coordination and Governance in the Economic and Monetary Union and other initiatives will enhance the governance of

European integration. In the 21st century, as an important pillar of the world economy, the EU will set the trend of and exert important influence on global development.

After Brexit, the EU has to increase its reliance on the US, so does Britain. As a result, the global leading position of the US, to some extent, has been consolidated. However, the United States' Strategy of Eastern European Expansion will be affected to some degree. There have been both cooperation and struggles in EU–US relations. After the 2008 economic crisis, the US passed on the crisis by depreciating the euro, and intensified the EU's internal problems, for instance, the strikes in Spain and Greek. The US two-faceted attitude to Brexit may worsen the deep-seated contradictions among the US, EU and Britain. The contradictions and struggle between Britain and the EU will be around for a long time.

Brexit increases global market instability as well as geopolitical complexity. It reflects that European integration is not a smooth ride and that it is necessary to set the proper pace for regional integration. Facing the increasingly complex situation, it is urgent for the ruling party to enhance the ability to control the current situation and correctly handle the relationship between the mass line and political leadership. People are the maker of history but it is the elite that govern the country. Some decisions cannot be made merely by referendum. With regard to its political significance, Brexit demonstrates the trend of diversified development in the international political arena. As for its economic influence, it may elevate the status of the yuan in the international market and help push down the pound, causing uncalled for losses to investors in Britain. Therefore, we should follow closely the far-reaching impact of Brexit across the globe.

Europe is an important cradle of modern industrial civilization as well as a pioneer of regional political, economic and social integration. The EU is an important independent force in the world and a very special player in China's diplomacy. The EU is an independent sovereignty and at the same time a supranational union. To handle China–EU relations properly is important for China to grasp and utilize strategic opportunities, coordinate domestic development and international exploration, and safeguard its sovereignty, security and development interests. In its diplomatic efforts in Europe, China can continue to focus on Germany, France and

Britain, take the relations with Eastern Europe, Central Europe, Southern Europe and Northwest Europe as our foothold and utilize the platform of the EU institutions so as to forge a partnership with Europe featuring political equality and mutual trust, win–win economic cooperation, cultural exchanges and mutual learning as well as mutual support in the international affairs and jointly promote the peaceful multi-polarization of the world. European integration can provide valuable reference for the regional integration in other parts of the world, global governance reform as well as China's reform.

3. *Japan Is Still Ambitious*

Japan is a fine example of integrating modern Western ideas and their scientific and technological advances with its own national conditions. In more than 100 years, Japan, an island of only 378,000 km², turned from a country with few resources and a population of over 100 million into a prosperous society, and once, even a world power which can rival the US in terms of economic scale. In 1995, Japan at its full glory registered a GDP 69.6% of that of the US. It was not short of a miracle for a country that only takes up 0.25% of the land in the world to generate 17% of the world GDP. Although Japan's economy has remained sluggish for over 20 years and its GDP is only 27% of that of the US, its innovation capacity and fine management capabilities will continue to have a significant impact on the world economy.

Japan is a country that honors dedication, and their *kado* (flower arrangement), *sado* (tea ceremony) and *kodo* (the way of incense) are all examples of their ultimate pursuit of perfection. Through such dedication, they have not only created an economic miracle, but also enhanced the confidence of the Japanese people. American strategist Zbigniew Brzezinski argued that Japan would overtake Russia if it made the strategic choice to assume a more active international role.[2] However, Japan also practices *Shinto* (way of the gods) and advocates *Bushido* (the way of the

---

[2] Zbigniew Brzezinski. *Strategic Vision: America and the Crisis of Global Power*, 1st edition. Translated by Hong Man, Yu Huiqin and He Weining. Beijing: Xinhua Press, 2012, p. 19.

warrior).[3] Unable to tell black from white and inclined to bully the weak and fear the strong, Japan does not have moral high ground in the development of human civilization, and thus is unlikely to make great achievements. Especially during World War II, Japan committed the terrible crime of aggression against China, the DPRK, other parts of Asia and the United States and is still reluctant to acknowledge their mistakes, creating distrust in its neighboring countries and regions. In international multi-polarization process, Japan has to reposition itself by "leaving Europe and turning to Asia". Proper positioning will increase its influence and *vice versa*.

The Sino-Japanese exchanges go back a long way tainted with both humiliation and glory. We shall not forget what happened in history, nor shall we be trapped by history. When developing Sino-Japanese relations, we should promote appropriate interaction between governments and the peoples, separate our political ties from our economic exchanges, and improve our political relations through economic cooperation.

4. *Russia Has the Spirit of a Large Country*

In 2015, Russia's GDP reached 2.109 trillion US dollars. With a population of 140 million, it is the world's largest country and among the ones with the richest bounty of natural resources. Russia is rich in oil, gas and coal reserves, the mining and export of which have brought a considerable amount of revenue and foreign exchange earnings to Russia. The solid foundation of heavy industry, a legacy from the Soviet-era, paved the way for its industrial take-off and the sustainable development of its aerospace industry in the new era Russia. However, Russia's frigid climate limited the growth of crops, resulting in low yield per unit area. Though Russia possesses nearly the equal size of arable land as China does, its total crop yield is much less. In 2013, it only produced 91.3 million tons of grain. Russia's climate also affects its industrial production. In addition to the military, aerospace and other heavy industries, Russia can barely produce anything else, which explains why we rarely see Russian products in industrial markets. Russia also has a low population density. With such a

---

[3]Yao Youzhi and Yan Qiying (eds.). *Souls of Major Powers*. Beijing: Chinese People's Liberation Army Press, First Edition of January 2011, p. 290.

large landmass and extremely scattered population, Russia faces high distribution costs, which is detrimental to its economic development.

In recent years, Russia has relied on natural resources, especially the oil resources, for national revival, and employed effective means to regain the control of oil resources in order to use them as a strategic tool in domestic and foreign policies. Russia has developed and implemented the "energy superpower" strategy, which not only helps it solve the foreign debt problem, but also increases its weight in opposing the US and influencing Europe and neighboring countries. However, the United States has undermined the influence of this strategy by exploiting shale gas and taking other related countermeasures.

Currently, Russia is in a difficult political, economic and diplomatic plight and thus is active in seeking all kinds of strategic support. For example, it intends to break the siege of the US and Europe through the BRIC mechanism. At the seventh BRICS Think Tanks Symposium in May 2015, Russia presented the Outline of the 2025 BRICS Action Plan, which included 16 political, economic and social tasks. Eight out of them were political tasks. This reflects Russia's prominent political intention of turning the BRICS into a political group to rival with Europe and America. Russia's action in Syria, to some extent, demonstrates its geopolitical position.

China and Russia are both important forces in safeguarding world peace and stability. Facing the challenges in international relations, as permanent members of the UN Security Council, China and Russia should keep up the communication and cooperation on international issues and work together to meet the challenges, which is conducive to safeguarding world peace, security and stability.

5. *India Is Bursting with Potential*

India has a large population which might surpass that of China and thus has a huge potential market. Therefore, domestic demand will become a major driving force behind India's future economic growth. India abounds in labor force and may follow China's footsteps and become the new "world factory". India's population density is close to that of Japan, but its better climate allows for increases in the population. But the huge population is also the biggest problem of the country. India's computer and software industries are developing rapidly, winning them numerous outsourcing jobs from developed countries. The information technology-based service industry accounts for half of India's economy. India is undoubtedly a country with great potential, but it has no advantages in resources for it has few natural resources except coal, iron, bauxite and mica. India's GDP was 2.117 trillion USD in 2015, accounting for 2.5% of the world's total. India has huge potential for economic growth, but this potential is yet to be released. India is "a country of multiple societies" and is now promoting national modernization just like China. It has the ambition to grow into a large country, but tends to show the mentality of a small country. Influenced by their traditional way of thinking and their religion and culture, they have not gotten over some historical events, which hampers its international cooperation. Its position in the current international governance structure is the foundation for its development. India is an elephant, but is not fully awake yet.

China and India have more than 2,000-year history of exchanges, mutual learning and integration, which helped them create their own national glory and promote their development. Into the 21st century, China and India established a strategic cooperative partnership for common prosperity, and became the main promoters of regional integration as well as important players in the global governance mechanism. However, for various reasons, in the past 200 years, the bilateral relations, the above-mentioned heritage of cultural traditions has not been inherited, the timeless spirit of Buddhism has not been promoted and the brotherhood has not been consolidated. And effective solutions are needed to address the problem of "political trust deficit". China and India should promote the Buddhist cultural heritage, reestablish the cultural cooperation ties and enrich the new strategic partnership to present the new splendor of Oriental culture and resolve historical issue through the power of culture. When developing our diplomatic ties, we should "formulate strategies and implement specific programs, avoid direct confrontation with each other, give full play to the power of culture and enterprises, complement each other with our own advantages and pursue long-term common interests". China and India are important neighbors to each other; therefore, we need to rely on each other as external support to back each other up and promote common development.

6. *Brazil Has Great Advantages in Resources*

With an area of 8.54 million km^2 and a population more than 200 million, Brazil can rival China and America. In 2015 Brazil's GDP reached USD 2.504 trillion, making it qualified as a moderately developed country. Brazil enjoys favorable weather conditions. Most parts of the country lie in the tropics and are plains. Thus, it has huge potential for population growth. Brazil has well-developed mining and agriculture industries. Known as the world's leading producer of raw materials, it is a major producer and exporter of iron ore, sugar, coffee, corn, soybeans and other products. It boasts well-developed petrochemical, mining, steel, automotive and other industrial sectors and is also a world leader in the manufacturing of civil regional aircraft. Commodity trade promotes and hinders Brazil's development. In recent years, Brazil's economic development has lost momentum due to declining demand of commodities and the drop in their prices. Brazil has the potential to develop into a great power. But whether it can shift from a large country to a great power depends on its infrastructure and capacity for scientific and technological development.

China and Brazil have established a comprehensive strategic partnership. There have been frequent high-level visits between the two countries, and rapid developments in bilateral economic and trade relations. In 2014, the volume of bilateral trade reached USD 86.58 billion. Brazil has become China's 10th largest trading partner and China's largest trading partner in Latin America; China has become Brazil's largest trading partner. There is still ample room for China and Brazil to enhance strategic interdependence.

7. *South Africa Leads the Development of Africa*

South Africa is a country of favorable natural environment and climatic conditions. It ranks first in the world in both the reserves and production of gold and diamond. In 2015, its GDP reached USD 402.152 billion. It has a population of 51.77 million (2012) and a population density of 47.5 persons/ km^2 (2012). South Africa is a middle-income developing country and also the most economically developed country in Africa. South Africa has well-developed finance, legal, energy, transportation, agriculture, telecommunications and information technology industries and the most comprehensive infrastructure system in Africa which is crucial to the economy of the country and its neighbors. South Africa has a mature stock market, and 90% of the investment by southern Africa is conducted in the Johannesburg Stock Exchange of South Africa. South Africa has a great potential in developing blue economy. It is the second largest economy and one of the most influential countries in Africa. Its GDP occupies one-third of the economic aggregate of sub-Saharan Africa and plays an important role in leading regional economic development.

In 1948, South Africa began to implement apartheid. It lasted for more than 40 years and created the darkest period of time in its history. From 1990 to 1991, the government repealed all legal provisions of apartheid, successfully resolved the leftover problems of ethnic conflicts and achieved national reconciliation, providing other countries with import lesson in solving ethnic conflicts.

After joining the BRICS in 2010, South Africa strengthened it cooperation trade exchanges with Brazil, Russia, India and China and enhanced its influence on Africa. Compared with other BRICS countries, South Africa's GDP is relatively small, but South Africa will be an important engine of development in Africa. By employing a variety of political and diplomatic means, South Africa will exert growing influence on the world.

China and South Africa have a special relationship: they are both comrades and brothers. They can complement each other and learn from each other in political, economic, social, market, scientific and technological areas. South Africa's racial integration and reconciliation is also worth studying.

8. *China Has Achieved Leapfrogging*

China is rich in resources and ranks first in terms of mineral reserves. But it has a relatively small deposit of important resources such as oil, natural gas, uranium, iron, copper and aluminum. On the whole, China's total amount of resources is less than that of the US. Given its large population and the great demand for resources, China's resources are quite limited. China is very populous with the population reaching 1.364 billion in 2014. It means not only great demographic dividends and market potential but also huge resource consumption.

In recent decades, China has maintained the momentum of high-speed growth. China's contribution to the world GDP jumped from 1.8% in 1990 to 13% in 2014, achieving leapfrogging. In 2014, China's output of crude steel reached 822.7 million tons, accounting for half of the world's total output; its cement output reached 2.476 billion tons, accounting for about 60% of the world's total output; and 8 out of the 10 ports with the strongest handling capacity are located in China. China's GDP reached USD 10.385 trillion in 2015 and surpassed that of many developed countries. But there is a great gap in the per capita GDP. This gap can be narrowed by developing science and technology, but it will be a very long process. It is difficult for China to reach the level of Europe and America where the per capita GDP is USD 45,000 to 60,000. It is likely that China's per capita GDP can reach USD 28,000 to 32,000 at most.

China has limited potential for population growth. Its fertility rate is declining year by year and dropped from 1.39 in 2001 to 1.04 in 2011. After the loosening of family planning policy in 2016, the fertility rate is bound to increase, but the expectation of a substantial increase is unrealistic. In 2011, China's labor force reached the peak and then began to decrease significantly. The labor force shrunk by 3.45 million in 2012, 2.44 million in 2013 and 3.71 million in 2014. China's annual number of graduates will drop by around 1 million per year from 2012 to 2022. Along with the change in the supply and demand structure of China's labor market, labor costs will increase. However, China's human resources are far from exhaustion. Though China is aging, it will not fundamentally change the growth of China's economy within 10–15 years. There are still some dividends of China's reform. With the deepening of the reform of the judicial system, the administrative system, the science, technology and education system, the health care system and the financial system, China can maintain its cost advantage. The Chinese economy will continue to grow though at a lower rate.

In the 21st century, there may be some significant shifts in the status of the countries in the world. It should not come as a surprise that China can rise through peaceful development, because except for a period of nearly 200 years, China has always been one of the advanced countries. It is natural that the GDP of such a large country as China will surpass that of Britain, Germany, Japan and even the US. However, China must realize that even if tomorrow China overtakes the US in terms of GDP, its economic and social development still lags behind the developed countries in quality and soft power, its per capita level is still low and it is still in the primary stage of socialist market economy. China has much to learn from developed countries in legal compliance, inventions, environmental protection and the quality of the population. Moreover, although the GDP of developed countries may be surpassed gradually by China, it does not necessarily mean that they will decline and fall. They have numerous strengths and may stay ahead of the world in many areas for the next 100 years. China needs to take full advantage of the opportunities presented by the changes in the international situation and integrate ideology, culture, strategic planning, institutional change, policy development and implementation into an organic whole to enhance China's development capacity and competitiveness.

The above-mentioned eight countries and regions are critical to the world stability and economic development and are important pillars of the global landscape. However, these pillars are not isolated from each other or from other parts of the world. They are interrelated among themselves and with the outside world.

After the US won the Cold War, in order to build itself into the "world empire", it first established North American Free Trade Zone and united North American countries into an organic union. Today, the US is vigorously promoting the TPP and TTIP, which will have new influences on the world landscape.

European integration is based on a solid historical foundation. Since the founding of the Holy Roman Empire of the German Nation, Europe has always considered themselves as a whole region in spite of frequent wars. The main objectives of the EU after its establishment fell into two categories: political and economic objectives, i.e. to eliminate historical hostility in this region and form an entity against the outside world through political and economic integration.

Brazil, India and South Africa all took their BRICS country status as a stepping stone to the permanent membership in the United Nations Security Council and are active in enhancing their regional influence through the New Development Bank.

India, Japan, China and other Asian countries have many cultural similarities. For instance, Buddhism, Confucianism, Taoism and Islam share a number of doctrines and they do not exclude one other, unlike Christianity which seeks primacy. In the long run, these similarities constitute the cultural foundation for economic integration. Therefore, it is also possible for Asian countries to form a pan-Asian union.

The disintegration of the Soviet Union provided an opportunity for political change in Russia and its allies. After the political change, affected by the influence of global economic integration, they realized that regional cooperation is more important than ever before. It is natural for them to revive the alliance led by Russia.

Therefore, at an appropriate time in the future, the eight pillars of the global economy are likely to change into six growth poles and exert more profound influence on their respective regions: one of the six poles is North American Free Trade Area; one is the EU; one is ASEAN Free Trade Area or pan-Asian union which is taking shape; one is the alliance that Russia and its allies are trying to restore; one is South America led by Brazil; and the last one is Africa led by South Africa. Each pole may develop into a relatively integrated regional economic cycle and connect with the external as a common force. But in a world of globalization and intense turmoil, none of the relations will remain the same forever.

II. Effect of TPP and TTIP on the International Situation

Since 2012, the US has been advocating the US-led TPP and the TTIP. Its purpose is to bypass the WTO to launch a new round of rule-making for international trade and investment and build a new trade and investment framework. Although President Trump said he would abandon the TPP, but we believe that the basic intentions of the US foreign trade and investment strategy will not change. This move will bring about a direct impact on the current international trade and investment landscape, and exert significant influence on China's "going global and bringing in" strategy.

The fundamental intention of TPP and TTIP: The US has been exploring new international space to maintain its leading position. Since 2001, the negotiations on agricultural subsidies and tariff barriers among the major trading blocs have reached an impasse in the Doha Round negotiation process, which hampered the US and Europe's effort to promote trade in services. Therefore, the US has intensified efforts to push forward TPP and TTIP. Through this attempt to avoid possible resistance from developing countries, bypass the WTO and accelerate the development of a new set of trading rules, push forward these two agreements as trading blocs, and force the passive follow-up out of the WTO, the United States is trying to build a super free trade area which covers the Americas, Europe and Asia and runs according to the rules formulated by developed countries, turn a new leaf for both American and European trade and investment cooperation and the trade cooperation between the US and Asia, and provide relevant legal protection for the economic revitalization of developed countries.

Effects of TPP and TTIP on international trade and investment rules: Firstly, they may lead to the reconstruction of international trade and investment mechanisms. If the TPP and TTIP are signed and implemented, the number of member states is expected to exceed 40, and regional trade and their trade with the rest of the world is expected to exceed 40% of the world trade. In addition, the member states will have to participate in the revision and innovation of the international governance regulations, which will have an impact on their internal affairs and diplomacy.

Secondly, they can widen the North–South gap. New political, economic, social and environmental standards in the field of trade and investment are expected to be reached through the developed countries-led TPP and TTIP. They are bound to overstep the WTO standards and become the new foundation of international trade and investment rules. This kind of "high standard" trade and investment activities will primarily rely on unilateral trade and investment as well as the new rules. They aim to ensure that the developed countries can expand their international market through their advantageous industries and technology. It is very likely that they will weaken the momentum of development in emerging markets, constrain the driving force of developing countries' external markets and further widen the gap in development between the North and the South.

Thirdly, they may create new threshold for non-member countries to access markets. The two large free trade areas formed under TTP and TTIP are bound to change the pattern of global trade and capital flows and cause the restructuring of global supply chain through customs supervision and coordination of industrial policies. They will place non-member countries at a disadvantage by adopting new procedures and mechanisms for the settlement of trade and investment disputes.

Fourthly, they will protect and enhance US strategic economic and trade interests in different regions. Once implemented, TPP is bound to break the ASEAN-led mechanism of regional economic integration, replace the APEC free trade zone initiated by China, as well as restore and consolidate US dominance in the Asia-Pacific economic affairs. After the implementation of TTIP, EU export to the US is expected to increase by 28%, equivalent to trade in goods and services worth €187 billion; while US export to the EU will grow by 36.57%, equivalent to trade in goods and services worth €159 billion. Europe and the United States' trade with other parts of the world will increase by 33 billion euros. The total exports of Europe and the US will increase by 6% and 8% respectively, equivalent to €220 billion and €240 billion increases in exports.[4]

Effects of TPP and TTIP on China's foreign trade and investment. Firstly, TPP and TTIP will constrain China's economic growth. Their implementation will trigger a new round of cross-regional economic and trade integration, reverse the direction of trade and capital flow, and seriously limit China's economic growth potential. The global standards on state-owned enterprises, government subsidies and government procurement in TPP and TTIP, which are tailored for China, will become an important tool for Europe, America, Japan and other developed countries to develop a new set of anti-monopoly, anti-subsidy and anti-merger terms. In recent years, some countries have frequently adopted anti-subsidy measures in their trade with China, which are, in fact, the preface of TPP and TTIP's new rules.[5]

Secondly, they will limit China's economic influence. TPP and TTIP are both efforts to formulate new trade and investment rules so as to contain China's status as a "world factory" and block the driving force behind China's economic development. Specifically, in the name of maximizing market liberalization, these agreements promote the establishment of an

---

[4] Joseph Francois. "Reducing Transatlantic Barriers to Trade and Investment: An Economic Assessment", p. vvi, p. 2, March, 2013, Centre for Economic Policy Research, London.
[5] Cui Hongjian. "The Europe and US-led TTIP: Origin, Objectives and Influence", *International Studies*, September 30, 2013.

exclusive protection mechanism of regional trade and investment and propose trade and investment rules in favor of developed countries in technical standards, pharmaceuticals, medical services, electronic product specifications, environmental indicators and other fields. They are devised to build new market barriers, greatly raise the costs for China's participation in international division of labor, cooperation and competition, curb the demand of China's market, restrict the momentum of China's foreign investment development and constrain China's international economic influence.

Thirdly, they will curb the growth of China's foreign trade. After the implementation of TPP, China's export to the US may decline by more than 30%. And TTIP will cause a staggering 35% drop in China's trade with the EU. The implementation of TPP and TTIP will affect China's market share in all of its major traditional trading partners and effectively constrain the growth in the trade between China and developed countries. What is worse, all the commitments China made in order to join the WTO after more than 10 years of arduous negotiations will soon become unilateral concessions. The developed countries can impose sanctions on and take anti-dumping measures against China at any time according to the new rules of TPP and TTIP. In particular, the two core elements of TPP, namely the full realization of zero tariff and the new rules of origin will have very negative impact on our general exports. Take the textile trade for example. According to the new rules of origin, the TPP nation must use a TPP member-produced yarns and fabrics for the garment that will be sold to these two free trade areas and the garment has to be cut and sewed within the territory of TPP member countries. Otherwise, such apparel exports will not receive duty-free access. As a result, Chinese textile products will be excluded from the agreed reciprocity conditions and TPP member countries can reduce or even stop sourcing raw materials from China under the terms of the agreement. According to the estimates presented by US Peterson Institute for International Economics, China will suffer an annual loss of more than USD 20 billion in foreign trade until 2025.

Fourthly, they will set up roadblocks in China's peaceful development. Needless to say, the US does not want to see any competitors successfully implement their national development strategies, nor wants to see the rise of a socialist China. On the economic front, the US regards China as a major competitor and its strategic goal is to strangle China before it grows into a strong power. On the military front, the US has established military and paramilitary alliance with China's neighbors. It is an established principle of the US strategy to keep China in a siege. As for the specific political arrangements, the US has stepped up the deployment of their allies around China, built a "value alliance" and refused to recognize the democratic system with Chinese characteristics. As for the specific economic measures, the US uses the TPP and TTIP as a medium to construct an "invisible barrier" to restrict China's rapid development. As for strategy, the US actively promotes the Asia-Pacific Rebalance Strategy, forcing China to increase its input in national defense and use the limited resources for development. On the technology front, the US has for a long time imposed a technological blockade on China, and has even forbidden the Chinese subsidiaries of American multinationals to transfer their technology to China. With regard to regulations, the US tried to bypass the WTO through TPP and TTIP and force China to participate in the division of labor, cooperation and competition with the US according to the "new rules" which are too advanced for the development stage of China.

III. Actively Planning International Layout

The guiding principle is that we should observe the overall situation, be honest and trustworthy, pursue balanced diplomacy and pragmatism, devote ourselves to planning, give more and take less, and focus on domestic development while continuing the international exploration. Facing the complicated political situation, we should judge the situation calmly, identify the underlying causes, grasp the trend, vigorously promote mutual trust and cooperation in the international arena, and demonstrate Chinese characteristics. We should apply the balanced way of thinking. We should not compete for the leading position nor take the path of extreme. We should take the interests of developing countries into full consideration, use political and diplomatic means to forge cooperation with a number of countries, enhance our relations with countries of both the North and the South, consolidate our ties with countries both nearby and far away, establish an extensive international unified front, defend

China's core interests and make more contributions to world peace and development. Domestic development is not only the starting point of our diplomacy but also its fundamental purpose. We should make use of the Belt and Road Initiative to further develop our economic cooperation and foreign trade relations. In particular, we should take full advantage of river basin economy to deepen the cooperation with neighboring countries. The above conclusions are based on the following basic judgment of the international situation and highlight the interaction between domestic work and international relations. Next, I will elaborate my observations based on the current situation:

1. *The Basic Judgment of the International Situation*

I have reached four conclusions based on my analysis of the current situation and the planning for the future. Firstly, the main characteristics and the trends of today's world are that multiple systems coexist in the same globe, one hegemon stands along with several powers and the world is an integral whole with great diversity. Since the beginning of the new century, there have been some power shifts among the countries. Emerging powers are on the rise. The international situation is evolving into one with no superpower. Global power has started to shift from the west to the east and from the north to the south. Following the US and European countries, some regional big countries with international influence are emerging from Asia, Africa and Latin America. The world power is scattered and the characteristics of multi-polarization have become apparent. Today, no one will say there is only one way in the world and that is the road of capitalism. The interdependence, adjustment, transformation and competition among all kinds of social systems and development models have become the basic trend of international political development. The service economy of developed countries, the urbanization of developing countries and the new global manufacturing industry have become an important driving force behind globalization.

Secondly, the 21st century is still the century of the US. While the world is at an intersection of conflicts, turmoil, adjustment and transformation and the hard and soft power of Western countries are relatively weakened, the Western world continues to occupy a dominant position.

The real integration of global forces will be a long-lived process. The US will remain the world's only superpower in terms of strength, resources, size and potential.

Thirdly, China will remain a developing country for a long period of time. Over the past 60 years, China has blazed a path that is suitable for its own conditions and in line with the world trend after experiencing the ups and downs in the world history. It has also changed the landscape of international political forces through its own "personality" and unique style. China's influence is on the rise. The emergence of China on the world stage is one of the most influential geopolitical events in recent years. "China model" has become a popular and sensitive topic in international politics. There are many different comments about China in the world. Different positions are taken in viewing China and different attitudes are taken in dealing with China. Countries of different types all focus their attention on China. They expressed their support, anticipation, speculation, flattery, questioning, doubts, hostility and intention to contain or siege China. However, we should bear in mind that China's domestic development is still far from being "balanced, coordinated and sustainable". The Belt and Road Initiative and other initiatives proposed by and involving China are a necessary complement to the existing global governance system. China will remain a developing country in a long time to come and can only reach the level of a moderately developed country by the end of this century. The energy security, food security, environmental security and non-traditional security issues confronted by China are becoming increasingly prominent. Therefore, China needs a peaceful international environment, and it needs to deepen economic and trade relations with all the countries and regions in the world and support one another in their own efforts to meet challenges and accelerate development.

Fourthly, competition is the norm and cooperation is the mainstream. On the one hand, China and Western countries are both competitors and partners. In recent years, the rapid development of China's economy has brought a great challenge for Western countries to maintain their original market share and has also affected the formation of a market rules system. China shows great potential to rival the Western countries. However, China and Western countries share some demands on the rules. As the

Chinese economy becomes more global and market oriented, the trade between China and Western countries will become more integrated. On the other hand, China and the majority of the countries in Asia, Africa and Latin America are both competitors and partners. The cooperation between China and developing countries is the mainstream. In the economic arena, they enjoy equal status, pursue mutual benefit, complement each other and rely on each other. They share the aspiration for development. However, there may be competition and conflicts in China's cooperation with the countries in Asia, Africa and Latin America, especially in the cooperation with emerging markets. Both sides need to continue to coordinate their interests, and build and improve a system of trading rules for mutual benefit and win–win outcomes.

2. *Vigorously Expanding International Operations*

With the deepening of global economic integration, international competition boils down to the competition of markets. The boundaries of one country's market are an important indicator of its ability to effectively participate in the global division of labor and cooperation. The expansion of market boundaries is also an important prerequisite for China's becoming a modernized power on schedule. To cope with fierce international competition, we should take the following four important types of strategic interaction into consideration:

First, intergovernmental interaction and the interaction between private sectors. For the governments, we should make full use of China's diplomatic relations, coordinate our policy positions through high-level visits, high-level meetings and diplomatic consultations and build a community of shared interests, cooperation and development. As for the interaction between private sectors, we should take the opportunity of the Belt and Road Initiative to groom China's multinational groups, and cultivate the ability of enterprises to lead or participate in regional industrial division and cooperation. We should establish a regional or sub-regional coordination mechanism of macroeconomic policies led by enterprises or industries, prepare enterprise groups, products and technology for the developed markets, and explore China's economic boundaries according to its stage of development. We should hold high the banner of mutual trust and cooperation, and reach consensus on cooperation with the international community; we should promote the new concept of the convergence of interests and plan cooperative efforts for development; we should use connectivity as our bond to promote mutually beneficial trade and cooperation; we should use the circulation of factors as new medium to promote reciprocal investment; we should pursue industrial cooperation as our new foothold and accelerate our economic growth; we should rely on cultural exchanges as a new means to build connections for regional cooperation; we should use the new platform of cooperative development to achieve regional prosperity; we should use the new means of common development to promote regional security; we should advocate fairness and reason and participate together in global governance.

Second, multilateral and bilateral interaction. On the one hand, we should make use of China's vast market and its growing ability to absorb goods to increase our voice in the multilateral arena. We should also accelerate the development of China's free trade zones, develop trade and investment rules to cater the characteristic of developing countries, as well as plan for and participate in formulation of the "next-generation trade and investment rules". On the other hand, we should adhere to the strategy of "developing both bilateral and multilateral relations" and take advantage of the sound bilateral relations between China and relevant countries to facilitate the signing of free trade agreement. We should establish and improve the bilateral industrial cooperation and development mechanism through strategic dialogue and consultation. We should promote business mergers initiated by either side, and complement each other with respective manufacturing capabilities and technical advantages. We should establish joint research and development centers, form strategic alliances of enterprises for joint production, joint marketing and joint exploration in the third-party markets, and expand the strategic space for China to compete with its major rivals.

Third, interaction between the efforts to maintain the authority of the WTO and to promote reform of economic and trade system. On the one hand, we should maintain the authority of the WTO's governance. We should uphold fairness, impartiality and the sharing of development interests, win support from the developing countries, pursue progress in the Doha Round negotiations, and actively promote the reform and innovation of the WTO multilateral trade system. We should also promote the development of new global trade and investment rules that can balance the interests of developed and developing countries. On the other hand, we should take advantage of China's position in the current global governance mechanisms and work with relevant countries to promote the process of economic governance. We should actively participate in the formulation of new international rules and establish cooperation mechanisms for developing relevant industrial commodity standards through dialogue and negotiation so as to safeguard common interests and plan for common development.

Fourth, the interaction between the efforts to maintain the international political order and to promote the reform of global governance. After World War II, the developed countries led by the US built today's international political order and the world governance model and set up a number of international organizations such as the UN, World Bank, WTO and IMF to solve international trade disputes in the international community through participation, negotiation and coordination. We should not easily deny the merit of this system, but it does need further improvement. With the deepening of globalization, global conflicts, environment, ecology, resources, climate and market issues are gaining growing attention. The requirements for global public goods have become more demanding. In order to jointly deal with irresistible natural disasters and safeguard world stability and peace, human society needs to strengthen global governance and work together to provide global public goods. Under the new circumstances, out of the consideration of the well-being of human society, we should provide global public goods and services through global cooperation mechanism, adjust all the actions and maintain the normal international order. Therefore, the core task of global governance should be to effectively provide global public goods and services. The task can be broken down into the following nine parts. First is the early warning system. We should issue early warning and forecast of irresistible natural disasters. The second is about the market rules. We should regulate market transactions. The third is to set up a rewarding system to stimulate human creativity. The fourth is to set up an emergency response mechanism to respond effectively to economic and social emergencies. The fifth is about global transport. We should ensure the connectivity of land, sea and air transport. The sixth is to combat crimes and let criminals who are hazards to human society have no place to hide. The seventh is to maintain peace and avoid war. The eighth is to uphold justice. We should protect the universal values and human culture. The ninth is about supervision and punishment. We should maintain the normal order of the international community to promote balanced and sustainable global development. There is no world government to perform global governance and there will not be any. Therefore, it is necessary for the world's political forces to establish and improve the working system and mechanism for us to jointly address major global issues on the basis of consultations.

3. *Promoting Domestic Development*

The stage for China is the stage for the world. China itself is an important part of the international community. For China, the international market and the domestic market are inseparable. Certain provisions of TPP and TTIP bear similarities to the thoughts on China's new round of reform and opening up, such as the strengthening of intellectual property protection, the gradual lowering of tariff barriers, the raising of labor standards and the necessity of making government procurement transparent. We should make full use of the special advantages of China's status as a big country, apply international perspective to solve domestic problems, tap the domestic potential and enhance domestic capabilities.

Firstly, we should improve the development theories and mechanisms with Chinese characteristics. To enhance international cooperation, we should further improve the foreign trade and economic development strategies with Chinese characteristics. We should also hold high the banner of mutual trust and cooperation, pursue the thinking of balanced development, encourage the interaction between the government and the market, mobilize all the forces that can be mobilized and unleash all the enthusiasm to build a modern economic system which can create a virtuous economic cycle inside China. We should focus on domestic demand, and tap the full potential and explore the leeway of the domestic market. It may be an important strategy to deal with the economic challenges of globalization.

Secondly, we should speed up our economic restructuring. We should capitalize on the trend and ride the trend. We should use the "next-generation trade" rules of TPP and TTIP as a reference to push us to speed up our effort to devise innovative development mechanisms, update our development concepts and adjust our economic structure. We should improve the socialist market system, promote reform through innovation, pursue development through reform and improve our status through development.

Thirdly, we should consolidate China's position as a manufacturing and consumption power. In the process of developing a market economy, the absorbing ability of the market is a kind of scarce resources, which is much needed in China's drive towards sustainable development and also a strategic tool that enables China to participate effectively in international competition. China needs to accelerate economic restructuring, industrial upgrading, as well as the transition from the "Made in China" stage to the "Created by China" stage. At the same time, we should fully tap the potential of China's consumption and the driving force of China's foreign trade. We should promote innovation and expand consumption at the same time. We should formulate plans to increase government procurement in target countries so as to consolidate our economic and trade relations with some important countries and regions and give China more leeway to respond effectively to changes in the international economic order and market changes.

Fourthly, we should carry out the strategy of "bringing in". During global economic and industrial restructuring, multinational companies sped up the industrial transfer to other countries. In addition, they gradually expanded the scale and improved the level of the transfer. Multinational companies may place China's market at the center of their development strategies. Multinational companies are gradually moving their production, research and development, regional procurement and management centers to China. The world industrial chain is also likely to extend to China. Taking full advantage of the international economic and industrial restructuring and giving full play to China's geographical and resource advantages, we should prepare ourselves for the transfer of multinational industrial chain, which will help China become more involved in the international economy.

To sum up, the world is in an age of intertwined conflicts. It is becoming more imbalanced, unstable and uncoordinated. World peace and development remains the theme of the time, but the world is not a peaceful place. In particular, non-traditional security (new security threats) and changes in some hotspot areas are likely to bring changes to the international situation. There is a pressing need for the world to work together to address the urgent challenges and manage crisis. We should hold a positive and cooperative attitude, follow the international regulations adopted by the UN's, and build and improve appropriate mechanisms for global economic integration together with other countries so that all the countries can benefit from the international division of labor, cooperation and competition and back up each other's effort to promote global balance, stability and development. The power shift among the countries has created both challenges and opportunities for China to further implement its "going global" strategy. We should be prepared for a deeper and higher level of participation in international affairs and be fully prepared to step onto the international stage.

Chapter 14

Strengthening Asian–African–Latin American Cooperation

How will the new era help to further strengthen the cooperation among Asia, Africa and Latin America? Majority of the countries in Asia, Africa and Latin America are developing countries and share similar foundation for the development of traditional friendship and good neighborliness. This chapter analyzes a number of important projects to explore strategic ways to strengthen the cooperation between Asian, African and Latin American countries.

I. Tentative Idea on Establishing a Northeast Asian Development Bank

The current violent fluctuations in the international scenario will not affect the general trend of regional economic integration. Accelerating the preparation for the establishment of Northeast Asian Development Bank (NEADB) to provide investment and financing services for regional cooperative projects on economic and social development may probably become the breakthrough for solving political headaches. Strengthening financial cooperation in Northeast Asia and preparing to establish NEADB enjoy research foundations and are supported by sound political consensus.

Apart from the complicated geopolitical situation and composition of stakeholders due to conflicts between big powers, the Northeast Asia has seen even more complex and uncontrollable situation in recent years with the US launching its strategy of returning to the Asia-Pacific Region. On the other hand, the region boasts the most active economy in the world as well as comparatively advantageous foundation for all-around economic and financial cooperation. Since some of Northeast Asian countries are not covered by the Asian Infrastructure Investment Bank (AIIB), it is of special strategic significance to establish a sub-regional international financial institution (Northeast Asian New financial institution) in this region.

The new financial institution will help to bring about a new situation of peace and development in Northeast Asia. Since the Northeast Asian New financial institution aims to provide financing services to cooperation projects of regional economic and social development, it is of high possibility that it will become a breakthrough to solve political issues, as it demonstrates the sense of Northeast Asian community, advocates equality, mutual trust and inspection, inclusiveness, cooperation and win–win spirit, turns challenges into opportunities and conflicts into cooperation, enhances political mutual trust, improves friendship, and calls for joint efforts to develop a peaceful, stable and prosperous Northeast Asia.

The new institution facilitates infrastructure construction for "two circles and one corridor". The capital economic circle consisting of Beijing, Seoul and Tokyo, multilateral free trade circle with the Tumen River as the core and China–Mongolia–Russia high-speed corridor closely related to the Silk Road strategy demonstrates important features of global economy such as "regionalization, networking and coordination", which may give rise to a new situation of cluster development of the "two circles and one corridor". As a supportive financial institution, the Northeast Asian New financial institution can perform its special financial functions and gradually develop the core and new-type financial support mechanism for regional economic integration in Northeast Asia, to promote sustainable and balanced economic development in this region.

The new fund can contribute to the transformation and upgradation of economic structures in Northeast Asian countries. It can propel the development of the modern service industry and emerging industries of strategic importance, and enhance industrial cooperation and complementation of industrial chains of Northeast Asian countries to provide a lasting drive for long-term development.

The new institution can improve the region's capability of countering the financial crisis. It can establish a communication and information sharing mechanism and an effective risk supervision mechanism, channel cash flow in the region, support cross-border infrastructure projects, resist international economic risks and safeguard financial security.

The Northeast Asian New financial institution has special significance for China. First, it can help China to revitalize its old northeast industrial bases, especially to build Changchun–Jilin–Tumen region into a driving force for sub-regional development. Second, it can promote RMB internationalization, to serve as a window for currency circulation in the China–Russia–South Korea–Japan free trade zone. Third, it can expand international space for China's future development, and help China to play a role in development of the Arctic Ocean and even the North Pacific economic circles and open new Eurasia–North America passage; besides, it can strengthen China's position in the TPP negotiation.

Then how to comprehend the Relationship between Northeastern New financial institution and AIIB? The current international financial system is far from meeting the demand of global economic development. Joining hands with many other emerging economies, China is committed to building a new pattern of the international financial system and has become an important driving force for the international financial reform. China has proposed or been involved in the establishment of AIIB, BRICS Development Bank (or New Development Bank BRICS) and Development Bank of Shanghai Cooperation Organization. Meanwhile, it has made multi-layered and all-around arrangement for regional and sub-regional financial cooperation by setting up the Silk Road Fund and China–ASEAN Investment Cooperation Fund. The Northeast Asian New financial institution is a new mechanism to deepen sub-regional financial cooperation. Cooperation and complementation between the new fund and current international financial institutions, especially AIIB, are mainly manifested in three aspects:

(a) In terms of financial services, the Northeast Asian New financial institution is a strategic complementation for the AIIB. The *Articles of Agreement of the Asian Infrastructure Investment Bank* designs a business mode of "special fund" through which the Northeast Asian New financial institution can start its business. According to preliminary estimates (see the column), between 2014 and 2024, the cross-border financing demand for infrastructure construction in Northeast Asia will total USD 3,300 billion, or USD 330 billion annually, leaving a funding gap as high as 90%, which can be filled neither by the Asian Development Bank (ADB) nor the AIIB. The new institution can give a hand to resolve the fund bottleneck.

(b) Regarding deepening cooperation, the new fund can serve as a strategic tool to promote sub-regional cooperation. During in-depth regional economic integration, especially for energy saving and environment protection, Northeast Asian countries, joined by mountains and rivers, share the same destiny and strong will of cooperation, but lack effective approaches. The Great Tumen Initiative, a project most possibly to promote the transformation of export-oriented economy in the Northeast Asia, has not played its due role yet due to lack of in-depth financial cooperation. While preparing for the AIIB, Japan and South Korea, especially the former, have been looking for cooperation opportunities that fit their benefits. The Northeast Asian new financial institution may become an important way to promote regional cooperation as it complies with the common will of relevant countries and reflects their shared interests.

(c) The Northeast Asian new financial institution is a strategic platform to strengthen mutual trust and cooperation across the region, so as to integrate financial and intelligent support. Considering the complex geopolitical situation and interests, better communication and enhanced mutual trust and cooperation are critical for promoting peace, stability and development in the region. In June 2015, President Xi Jinping pointed out in his congratulatory message to the first annual conference of the Conference on Interaction and Confidence-building Measures in Asia (CICA) Non-governmental Forum that: "One should never underestimate the role of non-governmental sectors when exploring the region's path towards security and promoting mutual respect, mutual trust and harmonious co-existence among different civilizations, regions, races and religions in Asia". The Northeast Asian new financial institution will work closely with forums and academic groups in this region to bridge the gap between the government and the market and connect thoughts and actions. Such cooperation may form the "Track 1.5" communication mechanism and a negotiation channel in the region, and become a new measure to generate new content, new progress and new form of mutual trust and coordination in Northeast Asia and to improve the "North–South" relationship and "South–South" cooperation, so as to make new contribution to world peace and development.

With 57 founding members, the AIIB was established in June 2015, receiving initial capital of USD 100 billion, including 26% from China, 7.5% from India, 5.9% from Russia and 60.6% from remaining 54 members. Its establishment has brought about positive effects to the international community but has given rise to some doubts too. First, its integrity is affected since Japan and the US were not its members. Second, many thought China's real intention is to make more money through this bank, market its excessive capacity and products and expand its influence across the region to make more countries and areas succumb to its influence and to challenge the current international order and governance structure. Third, many concerns have come up, such as the low credit rating of the bank due to dominance of the emerging countries may directly affect the investment cost and term; with poor capability of risk control, the bank may fail to control investment risks of large projects; and lacking experience, the management staff can not adopt an efficient, transparent and equal approach to maintain sustainable development. Such concerns are understandable and can be resolved; as a matter of fact, they are proved unnecessary. It is quite normal to interpret new things from different perspectives in international competition. However, we should explore in deeper strategic research why our goodwill to share fruits of China's reform and opening up to more countries and regions cannot be understood and how such concerns can be resolved. For this purpose, we should take proper tactics. With the help of joint research with Japan and the US for a long time, the establishment of a Northeast Asia fund could act as an important measure, as it may satisfy Japan and the US, and may complement with the AIIB to promote cooperation in Northeast Asia.

What is the path to establish the Northeast Asian New financial institution? The regional and sub-regional development banks have seen increasing relative importance since the 1990s all round the world, rather than undergone a trade-off process. For example, the European Bank for Reconstruction and Development and European Investment Bank, Inter-American Development Bank and North American Development Bank, and African Development Bank and Banque ouest-africaine de développement (BOAD) have all made whopping progress. It is undoubtedly logical to establish the ADB, AIIB and Northeast Asian new financial institution. There are solid research foundations and sound political consensus to strengthen financial cooperation in Northeast Asia and establish the Northeast Asian new financial institution.

International activities to promote the establishment of a development financial institution in Northeast Asia. At the inauguration conference of the Northeast Asia Economic Forum held in Tianjin in 1991, Mr. Ma Hong from the Development Research Center of the State Council first proposed to establish a regional financial institution in Northeast Asia. In September that year, Dr. Duck-Woo Nam, the former Prime Minister of the Republic of Korea, expressed his agreement with Mr. Ma's proposal. In August 1997, Mr. Stanley Katz, the former Executive Vice President of ADB, led the feasibility study and officially finished the first feasibility report of the Northeast Asia Bank,[1] which received great response at the Seventh Annual Conference of the Northeast Asia Economic Forum (Ulaanbaatar). At the Ninth Annual Conference of the forum (Tianjin) held in October 1999, experts from Northeast Asian countries agreed to establish the Northeast Asia Bank.

In the new century, the Institute for North East Asian Research and the Northeast Asia Economic Forum set up in Tianjin-based Nankai University, the Research Center for Financial Cooperation in Northeast Asia, consists of the expert committee of the Northeast Bank and former political leaders of China, Japan, South Korea and US. Through discussion and demonstration of the plan for establishing a Northeast Asia Bank at many themed conferences, experts in the research center reached a consensus and drafted on behalf of the Northeast Asia Economic Forum the *Brief Report on Establishing the Northeast Asia Bank (2011)* and the *Brief Report on Establishing the Northeast Asia Bank (2013)*, which were

---

[1] The new financial institution was first named as the Northeast Asian Development Bank, but was renamed as the Northeast Asia Cooperation and Development Bank (Northeast Asia Bank for short) upon negotiation of experts from relevant countries, in order to mistake it as a branch of the ADB and to highlight its function of promoting economic cooperation and development.

submitted to Chinese, Japanese and South Korean governments, as well as research reports on constructing cross-border infrastructure in the Northeast Asia. All such reports have received high attention from relevant governments.

In recent years, Northeast Asia has suffered from various complicated situations brought about by sharp Japan with China and South Korea conflicts due to the former's rightist outlook of the war and disputes over islands, and recurrence of denuclearization in the Korean Peninsula. However, that does not change the overall tendency of friendly cooperation, and non-governmental organizations in China, Japan and South Korea keep advocating the establishment of a Northeast Asia bank. After three years of suspension, the foreign ministerial meeting between China, Japan and South Korea resumed on March 21, 2015 and ministers reached a consensus at the conference that the cooperation mechanism of the three countries should be further developed as an important framework to promote peace, stability and prosperity of the Northeast Asia.

South Korea held an active attitude. South Korean President Park Geun-hye said in her speech in Dresden University of Technology in Germany on March 28, 2014 that if needed, South Korea would like to work together with the neighboring countries to set up the NEADB to help the Democratic People's Republic of Korea with its economic development. Mr. Park Kwan-yong, the former president of the South Korean National Assembly, interpreted this speech as South Korea's agreement to establish such a bank. The Foreign Minister of South Korea reaffirmed the importance of strengthening cooperation in Northeast Asia at the foreign ministers' meeting of ASEAN and China, Japan and South Korea (10+3).

Though supportive, Japan called for more efforts. Toshiki Kaifu, the former Prime Minister of Japan, admitted the importance of establishing a Northeast Asian new financial institution by taking stock of respective advantages in Japan, China, South Korean, Russia and Mongolia in capital, resources and projects, etc. He agreed to give play to the role of non-government organizations, and wished an early establishment of the bank. Taro Nakayama, former Foreign Minister of Japan, showed support on behalf of himself and many others to Japan's research on establishing a Northeast Asia bank, but called for more joint efforts, given the difficulties encountered and disagreement of other politicians in the Japanese Government.

The US was willing to get involved. Briefed by his economic consultant with issues related to the establishment of a Northeast Asian bank in May 2011, the US President Obama said that the US should get involved, if a new international financial institution is established in East Asia. Mr. Stanley Katz, former Undersecretary of US Treasury Department, held that the AIIB might be regarded as a Chinese bank, since China is undoubtedly the biggest shareholder of this bank. Therefore, the US prefers the framework design of a Northeast Asian bank that can better represent equality and internationalization.

In summary, many political leaders, former political leaders and financial experts from Northeast Asian countries admitted the availability of favorable conditions for establishing a Northeast Asian development financial institution and acknowledged the feasibility and innovativeness of the idea.

1. *China's Preparation for Establishing a Northeast Asian Bank*

In October 2004, former State Councilor Tang Jiaxuan proposed to incorporate establishment of the Northeast Asian Bank into the Northeast Asia financial cooperation framework for coordinated arrangement. A coordination team for establishing a Northeast Asian cooperation and development bank was formed, and the State Council issued the *Minute of Conference on Involving Tianjin into Opening up of Bohai Rim Area and Northeast Asia Regional Cooperation* (Guoyue [2004] No. 129).

In March 2008, the State Council replied to the *Overall Plan for the Pilot Program of Comprehensive Reform in Tianjin Binhai New Area* (Guohan [2008] No. 26) and praised Tianjin for its efforts to further explore the feasibility of establishing a Northeast Asian bank.

Major leaders of the municipal CPC committee and the municipal government of Tianjin have all attached great importance to the establishment of the new bank. With instructions of municipal leaders such as Zhang Gaoli, the former Vice Mayor Wang Shuzu led the Municipal

Development and Reform Commission to cooperate with relevant departments to conduct basic research and help non-government supportive activities. Following instructions from leaders of the State Council and Tianjin Municipal Government, Tianjin Municipal Development and Reform Commission and the Research Center for Financial Cooperation in Northeast Asia sent delegates several times to attend the annual conference of the Northeast Asia Economic Forum and themed conferences on financial cooperation.

Jiang Zhenghua, former Chinaman of the NPC Standing Committee, once reported issues related to the Northeast Asian Bank to State Council leaders. His report was reviewed by Wen Jiabao, Li Keqiang, Wang Qishan, Dai Bingguo and other State Council leaders in April 2009, March and October 2010 and January 2012, who then required relevant departments of the State Council to conduct further research of such issues.

Since 2013, research results have been reported to departments like the Ministry of Foreign Affairs, Ministry of Finance, the Central Bank and China Center for International Economic Exchanges led by Zeng Peiyan. According to the information, research and programs to jointly establish a development financial institution in Northeast Asia serve as an important reference for and play their role in promoting the establishment of the AIIB.

This chapter holds that establishing a development financial institution in Northeast Asia mainly through joint efforts of China, Japan and South Korea is not only a financial breakthrough, but also it might become an important opportunity to implement China's strategy of promoting interconnection and deepening win–win cooperation in Northeast Asia.

2. Thoughts on Accelerating the Preparation for the Establishment of NEADB

The Chinese's 13th *"Five-year" Plan on Economic and Social Development for Beijing, Tianjin and Hebei*, explicitly emphasize to "promote the preparation for and the establishment of Northeast Asian Development Bank". The statement fully recognizes the work that Northeast Asia Forum, Chinese Association of Asia-Pacific Studies and Tianjin municipal government have been doing for driving forward the undertaking and is a major strategic measure of utilizing financial leverage to promote peaceful development of the region. Promoting the preparation for and the establishment of NEADB is a mandatory task that starts off at a high starting point and requires worldwide coordinated arrangement.

Basic guidelines: Dedicated to the tenet of "cooperation, equity, efficiency, innovation and responsibility", the preparation, establishment and operation of NEADB should proceed by following the guidelines of "cooperation in planning, building and development for win–win result" and observing the principles of "strategic mutual trust, policy support, expert management, business mode, risk sharing, social responsibility and common development".

Main measures: The initiative of local government and civil associations should be fully tapped and academic groups such as Chinese Association of Asia-Pacific Studies led by Jiang Zhenghua and Research Center for the Financial Cooperation in Northeast Asia should be supported to grow stronger. Efforts should be made to put into play the important role of civil think tanks, further improve the feasibility plans, provide consultation for related departments and win support from the government of other Northeast Asian countries and the banking sector through people-to-people diplomacy. Meanwhile, with the support of the central government, the effect of local government, such as Jingling province, Tianjin municipal, should be fully exerted in establishing the fund. Tianjin government has been paying close attention to and participating in the study on plans on Northeast Asian development financial institutions. Binhai New Area can serve as the base and have its location advantages fully explored to accelerate the development of NEADB.

China Society of Social and Economic Analysis will continually work hard to accelerate the preparation for the establishment of NEADB by cooperating with all friends in home and abroad.

II. Planning for Power Development Cooperation in Southern Africa

Southern Africa boasts rich resources but suffers from outdated infrastructures, particularly the power industry, which seriously encumbers the economic and social development of the Southern African countries. Adhering to the concept of planning first and combining two markets, China Development Bank has developed the Power Development Cooperation Plan in Southern Africa in collaboration with the domestic and overseas institutions such as China Three Gorges Corporation, and studied the feasibility of developing the hydropower resources of the Congo River.

This plan involves 12 South African countries, including the Democratic Republic of the Congo (DRC), South Africa, Botswana, Mozambique, Malawi, Angola, Lesotho, Namibia, Tanzania, Zambia, Zimbabwe and Swaziland, all of which are member states of the Southern African Development Community (SADC) and the Southern African Power Pool (SAPP).

Research priority: Based on the research on the power development in the Southern African Region, mainly the DRC and South Africa, analyze the construction progress, development mode and supporting power transmission and transformation lines of the Inga Hydropower Project as well as the social, economic and environmental factors influencing the project.

The Congo River harbors ample hydropower resources. The Inga Hydropower Project, planned to reach a total installed capacity of about 43,000 MW, will vigorously drive the economic and social development of the DRC and the entire Southern African Region. However, considering the power market, power transmission and transformation investment, and other factors, it is necessary to reasonably determine the scales of different stages of the Inga Hydropower Project through the technical and economic comparison according to the development and construction of the power market. Besides, the Inga Hydropower Project and its power transmission and transformation project involve large scale and long construction period, and will encounter such risks as the political risk, the environmental risk and the power market risk during the construction and operation of the project. Thus, it is necessary to fully tap China's rich experience in hydropower construction, align such experience to Africa's actual conditions, strengthen the coordination and cooperation with relevant parties, and drive the project development.

For this purpose, we jointly set up a working team in charge of the research and preparation of the Power Development Cooperation Plan in Southern Africa by organizing China Development Bank (CDB), China Three Gorges Corporation and other institutions. We have also conducted research on the three topics of "geopolitics, geo-economy, and cultural & social impact and environmental assessment" to supplement the Power Development Cooperation Plan in Southern Africa. Instructed to help Chinese enterprises actively participate in the bidding for the Greater Inga Hydropower Project in Southern Africa and its power transmission and transformation lines, seize a favorable position in market competition and occupy a good geographical location, so as to promote mutual benefits and win–win results.

1. *Background and Purposes of the Plan*

Development through represents the trend of the world's economic and social development, and economic globalization tends to intensify the inherent consistency of the world economy. Against this background, China and Africa provide with each other opportunities, markets and drive in the course of development.

With stabilizing political situation, sustained economic growth, and accelerating regional economic integration, Southern Africa stands out on the African Continent with a sound investment environment, and represents an emerging growth pole and engine for the global development.

But the shortage of infrastructures in this region has seriously encumbered the economic and social development of countries in this region, and constituted a huge barrier to their modernization. The hydropower resource reserve concentrated in the basin of the Congo River roughly accounts for one-sixth of the known hydropower resources in the world, and the Congo River boasts a theoretical hydropower reserve as high as 390,000 MW, ranking the first among all the big rivers in the world. The developable hydropower resources can reach an installed capacity of some 156,000 MW with an annual generating capacity of 964 billion kWh. Yet, only about 7% of these abundant resources have been developed and utilized. The reasonable development and utilization of the hydropower resources in the basin of the Congo River not only help expedite the development of Southern Africa and implement the "Lighting Africa" program sponsored by the UN, but also contribute to the sustainable global economic development.

The Greater Inga Hydropower Station and its power grid in Southern Africa represent the world's largest hydropower infrastructure project in the 21st century, with the scale twice of that of the Three Gorges Project in China.

China has made eye-catching achievements in infrastructure construction during the implementation of the reform and opening-up policy. China Development Bank (CDB) has developed a set of effective management mechanisms, and gained valuable experience in infrastructure financing. The CDB has applied the development finance's concept of advance planning, and provided financing support for Chinese and African enterprises to jointly develop and construct power infrastructures in Southern Africa. This will be of great strategic significance to back China and Africa to take the commanding point in the global economic reform, promote the construction and interaction between the domestic market and the foreign market, and consolidate the China–Africa strategic partnership.

2. *Planning Concepts and Principles*

Planning concepts: Follow the principle of sustainable social, economic and environmental development advocated by the United Nations, carry forward the spirit of the "South–South Cooperation", and combine China's infrastructure construction experience and Africa's rich resources to reasonably develop and utilize the rich green energy in the basin of the Congo River; establish strategic partnerships with the Democratic Republic of the Congo (DRC) and South Africa, closely ally with the member states of the Southern African Power Pool (SAPP) and the Southern African Development Community (SADC), adhere to the concept of "driving a big power grid with a big power source, and developing and constructing multiple projects jointly", build regional strategic cooperation projects of large power source, large power grid and large market in the positive direction, turn Africa's rich natural resources into the drive for social and economic development, optimize and build the Southern African power network, support the modernization of the region, and promote the fast and sustainable development of China and Africa in the 21st century.

Planning guideline: "coordination, highlighting priorities; relying on experts, cooperating with external resources; resource integration and market operation; step-by-step advance and prioritizing quality programs". The planning must be "scientific, systematic, strategic, international and operable".

Planning principles: Organically combine China's experience and Africa's actual conditions, and observe the principle of cooperation and win–win results; organically combine project construction and social development, and observe the principle of coordinated development; organically combine power service and regional demand, and observe the principle of balanced development; organically combine resource integration and mutual complement of advantages, and observe the principle of market operation; organically combine the construction progress and quality control, and observe the principle of scientific rules; organically combine near-term construction and medium- and long-term development, and observe the principle of sustainable development.

Planning tasks: Analyze the current social and economic status and development prospect of Southern Africa in depth, reasonably forecast its demand to develop the power market, take into account the influence of geopolitics, geo-economy and ecosystem on project construction and *vice versa*, scientifically design the overall objective, concrete tasks and construction steps of the power construction in Southern Africa, work out the measures for power generation, power transformation lines, investment returns, financing mode, market operation and risk control, prepare the Research Report based on profound research, and compile the Plan.

3. *Market Overview and Forecast*

For the purpose of this Plan, Southern African countries refer to 12 member states of both the SADC and the SAPP, and include DRC, South

Table 14.1.   Economic and Social Indicators of 12 SAPP Member States

Country	Population (10,000)	Total area (10,000 km²)	GDP in 2010 (USD 100 million)	GDP per capita in 2010 (USD)	Growth rate over previous year (%)
Congo (DRC)	7,045.9	234.5	131.25	186.3	7.25
South Africa	4,991.2	121.9	3,572.59	7,157.8	2.78
Angola	1,905.3	124. 7	853.12	4,477.7	1.61
Tanzania	4,134.9	94.7	226.71	548.3	6.5
Namibia	209.9	82.4	118.65	5,651.7	4.38
Mozambique	2,158.5	79.9	98.93	458.3	7.01
Zambia	1,325.7	75.3	161.92	1,221.4	7.6
Botswana	183.9	58.2	140.3	7,627.5	8.56
Zimbabwe	1,257.5	39.1	74.74	594.3	9.01
Malawi	1,569.5	11.8	50.53	321.9	6.6
Lesotho	254.2	3.0	21.27	836.7	2.44
Swaziland	116.1	1.7	35.53	3,061.1	1.99
Σ	25,152.6	927.3	5,485.54		

Africa, Botswana, Mozambique, Malawi, Angola, Lesotho, Namibia, Tanzania, Zambia, Zimbabwe and Swaziland. The economic and social indicators of these 12 countries in 2010 are shown in Table 14.1.

Forecast of Regional Economic Development. For the purpose of this Plan, the base year of the current status is 2010 and the target year is 2025 in the near term and 2050 in the far term.

On the whole, Southern Africa is still at the early and mid-term stage of industrialization (namely the early- and mid-term stage), which decides that the fast development in energy, mining, manufacturing and other sectors will still represent the basic direction and feature of the region in the following 10–30 years. At the same time, the development levels vary evidently from one country to another in the region, with the highest GDP per capita being 35 times of the lowest, showing different development characteristics in different areas. As the economic development of the emerging countries proves, the overall economic development of Southern Africa will be faster than the world's average.

The major indicators of the region are forecasted as follows, by using the elastic coefficient method, the Delphi method and other forecast methods in the Plan and taking into account the social and economic development features of Southern Africa.

The natural population growth rate will reach about 2‰, 2.4‰ for the period from 2010 to 2015, and 1.9‰ for the period from 2015 to 2025; the total population of the region will be controlled within 345.69 million by 2025 and reach about 498.48 million by 2050. Table14.2 shows the major forecast indicators.

The GDP of the region will grow by 5.6% annually on average, 6.9% for the period from 2010 to 2025, and 4.8% for the period from 2025 to 2050, and the GDP will reach USD 1,510.4 billion by 2025, and USD 4,907.2 billion by 2050. Table 14.3 shows the major forecast indicators.

Regional Power Industry Development Forecast. The power load is forecasted with the regression analysis method, the growth rate method, the elastic coefficient method and other methods.

Table 14.2.   Population Forecast

Unit: 10,000

Country	2010	2015	2025	2035	2050
Angola	1,905	2,169	2,744	3,340	4,227
Botswana	184	211	234	252	276
DRC	6,650	7,340	8,712	10,866	13,500
Lesotho	254	217	231	240	249
Malawi	1,570	1,800	2,319	2,859	3,658
Mozambique	2,159	2,596	3,119	3,658	4,415
Namibia	210	241	281	316	359
South Africa	4,991	5,168	5,377	5,548	5,680
Swaziland	116	129	146	159	175
Tanzania	4,135	5,211	6,739	8,389	10,945
Zambia	1,326	1,498	1,889	2,291	2,896
Zimbabwe	1,257.5	1,403	1,678	1,900	2,218
Total	25,153	28,384	34,569	40,838	49,848

Table 14.3.   National Economic Development Forecast

Unit: USD 100 million

Country	2010	2015	2025	2050
Angola	853.1	1,312.60	3,687.5	17,368.1
Botswana	140.3	209.4	290.5	435.8
Congo (DRC)	131.3	206.6	324.8	763.7
Lesotho	21.3	33.4	47.2	100.8
Malawi	50.5	74.3	114.2	229.0
Mozambique	98.9	170.3	354.4	661.7
Namibia	118.7	166.6	240.6	485.5
South Africa	3,572.6	4,733.6	8,885.6	26,705.1
Swaziland	35.5	38.2	51.6	83.0
Tanzania	226.7	333.3	510.6	1,174.1
Zambia	161.9	286.9	429.8	720.7
Zimbabwe	74.7	120.8	167.4	345.2
Total	5,485.5	7,853.9	15,104.3	49,072.7

There are rich results from previous researches on the SAPP region, and most of these results forecast the power load level around 2025. For this reason, the power loads of different countries in the near-term target year will be forecasted with the existing results or with the regression analysis method based on the existing results. Considering the uncertainties of the economic and social development and the load growth in the far-term target year, the power loads will be forecasted with the growth rate method and the elastic coefficient method based on the forecast results of the near-term target year (2015–2025) to generate the high load level plan and the low load level plan for the far-term target year of 2050, respectively.

Influenced by the international financial crisis, Southern African countries have rolled out a series of policies to stimulate economic development, which, combined with their unique geographical advantages and resources advantages, will continue to drive local economic and social growth. As a result, the power shortage problem will become more prominent. According to the forecast, the maximum load of the region will reach 72,472 MW, and the power consumption will arrive at 470.7 billion kWh

Table 14.4.   Forecast of Power Loads of the SAPP Region

Item	2010	2015	2020	2025	2050 Low load level plan	2050 High load level plan
Power demand (100 million kWh)	2,837	3,604	4,125	4,707	8,803	11,437
Maximum load (MW)	45,761	55,447	63,335	72,472	133,386	176,055
Number of hours when the maximum load is utilized (h)	6,200	6,500	6,513	6,495	6,600	6,496

in the near-term target year of 2025, and grow to 133,386 MW and 880.3 billion kWh by 2050, respectively, if the elastic coefficient method is adopted. The major indicators are forecasted in Table 14.4.

4. *Construction of Hydropower Stations and Power Transmission and Transformation Projects*

Overview of the Inga Hydropower Station Project. Located in Bas Congo Province of the DRC, the Inga Hydropower Station is about 225 km away from Kinshasa, capital of the DRC, and more than 60 km away from the port city Matadi. It lies in the Nkokolo valley in the lower reaches of the Congo River, and is some 150 km away from the estuary of the Congo River (see Figure 14.1 for the sketch map of the geographical position of the Station). The Project is planned to reach an installed capacity of some 43,000 MW.

After comparing original design plans solicited internationally, as the research moves on, we have found that the installment development plan is more realistic and reasonable, that is, the design plan for the development of the Greater Inga Hydropower Station Project based on the Inga Phase I and Inga Phase II Hydropower Stations that have been built (hereinafter referred to as the "installment development plan").

The development of the Greater Inga Hydropower Station will be divided into seven phases. In Phase I and II, a diversion canal is excavated from the upper reaches of Sikila Island, and to divert part of the flow of Congo River to the Bundi valley. The Bundi Dam with a designed normal

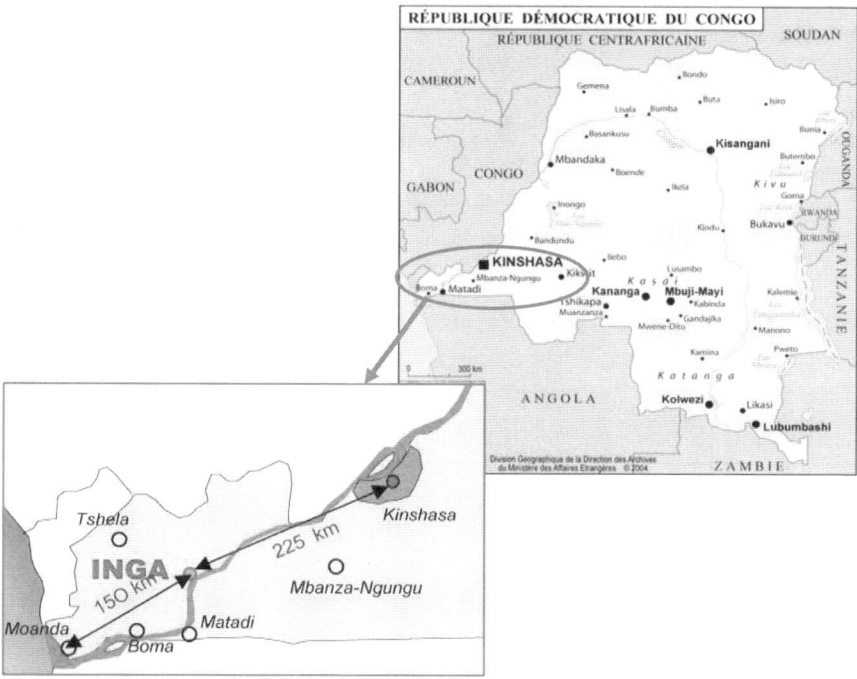

Figure 14.1.   Geographical Position of the Inga Hydropower Project

water level of 170 m will be constructed at the mouth of the Valley. In Phase III, the priority is to raise the normal water level to 205 m, and the main tasks include the dam construction in the upper reaches of Sikila Island and capacity expansion of the units built in Phase I and II. Meanwhile, the construction of the spillway of Bundi reservoir, heightening the Bundi Dam and the construction of the Matamba Dam should also be completed in this phase. In Phase IV till VII, the units should be installed by phase on the basis of the construction completed in Phase III, with eight units per phase. The main characteristic indicators of each phase are listed in Table 14.5.

According to the development-by-phase plan, the total installed capacity of the Greater Inga Hydropower Station will reach 41,100 MW and the annual generating capacity 289 billion kWh. In particular, the Phase 1 hydropower station will have an installed capacity of some

Table 14.5.   Main Characteristic Indicators of Each Phase

Phase	Normal water level (m)	Capacity installed (MW)	Number of units	Average generating capacity in multiple years (100 million kWh)
1	170	4,000	8	280
2	170	4,000	8	251
3	205	11,500		850
4	205	7,400	8	510
5	205	7,400	8	510
6	205	7,400	8	510
7	205	7,400	8	510
Final scale	205	41,100	48	2,890

4,000 MW ("Phase 1 Hydropower Station" for short) to replace the Inga III Hydropower Station.

Judging from the development of the entire Inga Hydropower Project, the construction of the Phase 1 Hydropower Station will mark the beginning of the construction of the Greater Inga Hydropower Station. After its completion and putting into use, the Station will definitely further improve the domestic power grid of the DRC, influence the power grid construction in Southern African countries, and further push forward the construction of the whole Inga Hydropower Project.

Power Transmission and Transformation Project. The domestic power grid of the DRC finds it hard to accommodate the huge generating capacity of the Inga Hydropower Project. Therefore, it is necessary to strengthen the construction of the Southern African power grid in order to fully realize the tremendous power output of the Project.

After Phase 1 Hydropower Station is completed in 2025, the installed capacity of the DRC will increase from 5,500 MW to 6,500 MW. Hence, the ultimate objective of Phase 1 Hydropower Station is to satisfy the domestic power demand by 2025. According to the current status and planning of the regional loads, the generating capacities that will be supplied to three major load centers in the DRC are as follows: (1) Inga–Bas Congo Province, 1,600 MW; (2) Inga–Kinshasa, 600 MW; (3) Inga–Katanga, 1,200 MW.

After 2025, as the development and construction of the Inga Hydropower Project moves on, the domestic installed capacity of the DRC will rise to about 40,000 MW by 2050. By then, the maximum load demand of the DRC is forecasted to reach about only 25,000 MW (low load level plan), so it is necessary to export quite a part of the residual power to other countries. At present, the potential importers mainly include Northwest Nigeria, Northeast Egypt and the southern and south-eastern parts of South Africa. After analysis, the most likely importer is South Africa, and the discussion about the power transmission network will be based on this conclusion.

The major supply capacities considered in the power transmission line plan include: (1) South Africa, 4,000 MW (including: 2,000 MW for the Western Power Corridor and 2,000 MW for the Eastern Power Corridor); (2) Egypt, 3,500 MW; (3) the DRC, 2,000 MW; (4) Nigeria, 1,300 MW; (5) Angola, 700 MW.

It can be seen from the above analysis that the power transmission and transformation project planned at present will not suffice to accommodate all the electric power generated by the Inga Hydropower Project. Hence, besides the aforesaid lines, other lines of far-term planning will be added in due course, according to the construction progress of the following hydropower stations of the Inga Hydropower Project after 2025.

Project Construction Sequence. The electric power generated by the Inga Hydropower Project will be used first to satisfy the domestic power demand of the DRC, then supply to the neighboring countries, and to far-off countries if the project still has a power surplus. The following construction sequence is prepared, according to the aforesaid principle, and by taking into account the large power consumption market of South Africa:

1. Before 2025, build the major power transmission lines from Inga to three load centers, including Bas Congo Province, Kinshasa and Katanga, within the territory of the DRC;
2. Then, build the power transmission line from Inga to Luanda of Angola, which is closer to Inga;
3. Before 2050, build the eastern and western power transmission lines heading to South Africa with investment potentials: first, build the

Inga–Witkop (eastern line) power transmission line, and then the Inga–Gama (western line) power transmission line;

4. If conditions permit, it may consider building the Inga–Calabar power transmission line to export power to Nigeria;
5. If conditions permit, it may consider building the power transmission line from Inga to Cairo Region in Egypt;
6. Finally, plan other power transmission lines according to the progress of the Inga Hydropower Project.

5. *Investment Estimate*

(1) The Inga Hydropower Project. Table 14.6 shows the investment estimates for different phases of the Inga Hydropower Project based on the installment construction plan.

(2) The power transmission and transformation project. In principle, the investment estimate for the power transmission and transformation project only considers new EHV or UHV projects as the first landing points that are required to transmit the electric power of the Inga Hydropower Project and receive the electric power in the destination regions.

At the current stage, the investment in the power transmission and transformation project of the Inga Hydropower Project is estimated by referring to the ratio of the investment in the power transmission and transformation project to the investment in the Chinese hydropower projects like the Three Gorges Project. The price level of the power transmission and transformation project is the same as that of the Greater Inga Hydropower Project. Table 14.7 shows the investment estimates for the power transmission and transformation project of the Inga Hydropower Project.

Table 14.6.   Investment Estimates for the Inga Hydropower Project by Phase

Target year	Item		
	Average generating capacity in multiple years (100 million kWh)	Project investment (USD 100 million)	Unit investment per kWh (USD/kWh)
2025	280	74	0.264
2050	2,890	270	0.09
Total	2,890	344	0.119

Table 14.7.   Investment Estimates for the Power Transmission and Transformation Project of the Greater Inga Hydropower Project

Target year		Item	
	Supply region	Power transmission & transformation investment (USD 100 million)	
2025	Domestic market of the DRC	19.02	
2050	Southern Africa, Nigeria, Egypt, etc.	135.58	
Total		154.60	

Table 14.8.   Investment Estimates for the Inga Hydropower Project and Its Supporting Power Transmission and Transformation Facilities

Target year	Average generating capacity in multiple years (100 million kWh)	Hydropower project investment (USD 100 million)	Power transmission & transformation investment (USD 100 million)	Total investment (USD 100 million)	Unit investment per kWh (USD/kWh)
2025	280	74	19.02	93.02	0.332
2050	2,890	270	135.58	405.58	0.14
Total	2,890	344	154.6	498.6	0.17

Table 14.8 shows the investment estimates for the Inga Hydropower Project and its supporting power transmission and transformation facilities, which are based on the estimates for the above two items.

(3) Financial analysis. According to the preliminary forecast, the total investment in the Greater Inga Hydropower Project and its power transmission and transformation project will be about USD 49.9 billion, including USD 9,302 million in Phase 1 (including the power transmission and transformation project).

In China, capital and bank loans constitute the main funding sources of hydropower stations. Bank loans contribute 80% of the total investment, with the annual interest rate of 6.55% on long-term loans with a term of five years and longer offered by Chinese commercial banks, a repayment period of 25 years, and a repayment of principal and interest in equal amount within the repayment period.

The preliminary financial analysis shows that if the investment in the power transition and transformation project is included, the power price of Phase 1 of the Inga Hydropower Project will reach USD 0.033/kWh when the power is sold to the power grid, and the end power price will be USD 0.050/ kWh after the power transmission and transformation project is included.

(4) Sensitivity analysis. Considering the current progress, the project investment and the power output of the hydropower stations are subject to changes, so the negative influences of these two factors are selected to test the fluctuation of the power price to the power grid and the power price at the terminal of power transition and transformation facilities. Table 14.9 shows the sensitivity analysis of the power price, given the internal rate of return on capital of 8%.

If the internal rate of return on capital is increased to 10%, the power price to the power grid will rise by USD 0.004/kWh and the end power price by USD 0.006/kWh. Table 14.9 shows Sensitivity Analysis of the Power Price of the Inga Hydropower Project.

Based on the sensitivity analysis, in case of negative changes to the project investment and the power output of the hydropower project, the grid power price and the end power price will rise to varying degrees, and when the project investment rises by 20% or the power output drops down, the power price will rise by about 25%. Hence, as the work goes on, it is also necessary to strengthen the research on the power price policy (such as differential price and power options) to ensure the financial benefits of the project.

Table 14.9.   Sensitivity Analysis of the Power Price of the Inga Hydropower Project

Unit: USD/kWh

Variable	Power price	
	Power price to grid	End power price
Basic plan	0.033	0.050
Project investment rises by 10%	0.037	0.056
Project investment rises by 20%	0.042	0.062
Power output falls by 10%	0.037	0.056
Power output falls by 20%	0.041	0.063

6. *Risk Analysis*

Risks facing the investment and development of the Inga Hydropower Project and its power transmission and transformation project mainly include political risk, social risk, environmental risk, project construction risk and investment risk.

Political risk: Currently, the political climate in South African countries is basically stable. But they, except for South Africa, all ail from power struggles, deep-rooted tribal and religious conflicts, which are complex and tend to trigger violence and riots, and cause humanistic disasters, and even power shifts, coups, mutinies, racial conflicts, strikes, rebellions, and even civil wars. In the aforesaid cases, the Inga Hydropower Project and its power transition and transformation project might terminate, the construction might be forced to stop, and the construction period extended, causing losses to all the parties involved. In addition, European countries, the former colonial power in Southern African region continue influencing the political climate and foreign policies of these countries, which, in turn, will affect the construction of the Inga Power Project, and in particular, the power transition and transformation project.

This Plan includes the geopolitics section devoted to the discussion of political security. The preliminary research indicates that political security is the precondition for economic security and sustainable fundamental cooperation. The following six factors must be taken into account in order to achieve political security. First, political object. The cooperation with political parties is very important to protect investment security of the parties. Second, political trend with every possible political event taken into consideration. Third, cooperation mode. For one thing, be honest with African countries about China's actual development level and capacity, and for another, provide them with the cooperation package plan and supporting measures and cleverly integrate economic and social developments. Fourth, attach importance to media influence. Fifth, take the influences from Europe and America into account. Sixth, take possible intervention by some other opposing forces (including the Taiwan separatist force and the Tibetan separatist force) into account. We shall consider all these factors and work out corresponding countermeasures.

Social and environmental risks: First, let us consider the risk posed by historical problems. Though there are only a limited number of historical problems encountered in the construction course of the Greater Inga Hydropower Station, and the migration problem is not that serious, yet the existing problems still deserve our much attention. For example, the government of the DRC has expropriated the land of local residents in Inga for the licensed use of the national power company. But the land in question, according to the local convention, is just leased, and still owned by local residents. After they realize the true nature of the deal, local residents feel hurt for they only obtained a small amount of compensation for the land. Currently, there is no power supply in local villages, and the roads leading there are in poor conditions. Representatives of local residents filed a protest to the government in 2008 and asked for resettlement in the short term. Without proper handling, such conflict will evolve into a public conflict in the near future and exert negative impacts on the construction of the Inga Hydropower Project.

Second, the possible impact of the construction of hydropower facilities upon the environment. Vegetation damage, soil erosion, geological disaster and land occupation and damage caused by the opening of roads; vegetation damage, soil erosion, geological disasters, soil pollution, air pollution and noise pollution brought by dam construction, reserve clearing, civil works and tunnel projects; noise and flying dust caused by damp sites, material yards and material transport; damage to original natural landscapes or historical sites; reduction or extinction of wild specimens caused by human activities and habitat damage; and pollution or introduction of source diseases after the move-in of construction workers. Potential impact on the environment after the hydropower facilities are put into operation, which might trigger or aggravate geological disasters, such as earthquakes, flooding, siltation, scour, collapse, landslide, leakage and water pollution caused by the reservoir; the change of the original land and water ecosystems, and the impact on animals, plants and aquatic organisms in the region; and the change of the microclimate which leads to abnormal weather.

In the section of "cultural and social impact and environmental assessment" of the Plan, the research team borrows experience and methodologies of the World Bank, the Kreditanstalt für Wiederaufbau, French Development Agency and other international organizations, based on the field survey, mainly studies and assesses the influence of the power development in Southern Africa on the local "cultural and social environment" and a series of effects on the local cultural and social environment during the implementation of the project; eliminates the negative factors and leverages positive factors to create a favorable environment for the power development in Southern Africa.

Project construction risks and investment risks: The project construction risks mainly include the power market risk, technical risk and safety risk of on-site institutions and personnel. Apart from South Africa, Southern African countries are relatively backward in economic and social development, along with inadequate experience and capacity in the building of large hydropower projects and power transmission and transportation projects, and suffer from severe shortage of various basic materials and technical standards necessary for the project. This will pose the risk of substantial design change and construction delay during the project implementation.

The investment risks mainly include the inflation risk, exchange control risk, exchange rate risk and taxation risk. At present, the investment policies and relevant laws are not very sound in Southern African countries. Since the Inga Hydropower Project involves a long construction period, the changes in the investment policies of these countries will affect the economic interests of the investors during the construction period and the operation period.

This plan contains the "geo-economy" section as a supplement to the Plan. This research on geo-economy focuses on four aspects. First, the analysis of the social and economic foundation of Southern Africa, with focus on the current status and demand forecast of the power industry. Second, the international technical standard for power planning and design as well as infrastructure planning mode. Third, the research on the "power finance" for Southern Africa, and detailed analysis of the mode of collaboration among the entities of the hydropower project. Fourth, the assessment of the economic, social and environmental impacts of the hydropower facilities, the analysis of the geo-economy and the research on the international standard to help Chinese enterprises cooperate with Southern Africa and the international community, accelerate the power infrastructures in Southern Africa, mainly hydropower, and actively promote the final landing of the Inga Hydropower Project.

7. *Construction of the Power Financial Market in Southern Africa*

Basic framework of the power financial market. Construct the power financial market in the market-based way and based on the demand for electricity and other kinds of energy generated in the medium- and long-term economic and social development in Southern Africa, develop the market-oriented financing and management mechanisms, and make full use of various useful resources to accelerate the development and production of green energy, mainly hydropower, the construction of the power transmission and transformation network and the construction of the power trading market in Southern Africa, and promote local economic and social growth.

Considering that this is the construction of the Greater Inga Hydropower Project, its power transmission and transformation lines and power market in the DRC involves a long period and a large amount of financial input, with more investment in the early stage and less investment in the later stage, but less yields in the early stage and more yields in the later stage, we have proposed the concept of building the power financial market ("power finance" for short) in Southern Africa for the purposes of reasonable use of various resources and stable operation. The basic framework of this concept is illustrated in Figure 14.2.

Design of strategic options: In this section, we believe that the strategic options applicable to the power development in Southern Africa can be divided into three levels; with the ongoing development and utilization of the strategic options, the strategic credit will have broader and clearer space, and the risk exposure will gradually narrow, and finally covering the future risks with the economic development.

Differential pricing: Differential power pricing means to make different price plans based on the different power resources and demands in Southern Africa. The differential pricing charges for the domestic, commercial and industrial power consumption at different rates without

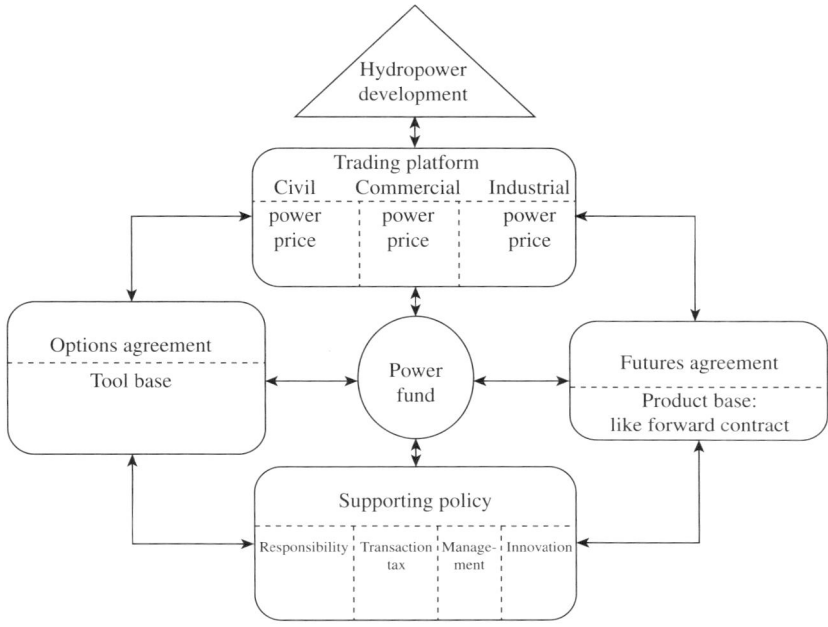

Figure 14.2.   Basic Framework of the Power Financial Market

increasing or by slightly increasing the existing power bill of residents, so as to realize a benign operation of the overall power price level.

This Plan initially designs three different kinds of pricing — the lowest power price for common household users (the price is low but free electricity is not recommended), pricing for commercial power consumption by office buildings and hotels (the price can be adjusted according to the market situation), and the pricing for industrial power consumption. These three kinds of pricing should strike a comprehensive balance so that the overall price level will guarantee sustainable power production. For example, in the DRC, the differential pricing strategy can effectively balance the social welfare and the benign operation of the power market.

8. *Builders of the Power Development Project in Southern Africa*

Cooperate with the Government of the DRC and the Societe Nationale d'Electricite (SNEL), the national power corporation of the DRC, to vigorously carry forward the construction of the DRC-based Greater Inga Hydropower Project, which will rely mostly on local resources, and fully consider the local market demand.

Cooperate with the Government of South Africa and Eskom, the national power company of South Africa, to vigorously promote the construction of the Southern African power grid. South Africa, which is competitive economically and technologically, has a tremendous power demand. Therefore, South Africa will play a positive role in facilitating the construction of the Southern African power grid.

Carry out the planning cooperation with the SADC to accelerate the building of the power market coordination mechanism. The SADC can exert its political and organizational advantages, and leverage its existing ministerial-level energy coordination mechanism to coordinate the power market on the whole.

The CDB shall give play to development finance to vigorously propel the power development in Southern Africa through cooperation. In 2010, China had a total installed capacity of more than 200 million kilowatts, the largest in the world. The hydropower construction technology and quality of Chinese enterprises represented by China Three Gorges Corporation have both reached the world advanced level. In collaboration with enterprises represented by China Three Gorges Corporation in China, SNEL in the DRC, Eskom in South Africa, members states of the SADC member states and other international organizations, the CDB will strive to promote the building of the Greater Inga Hydropower Project and its power grid in Southern Africa, and contribute to the world's largest green energy project in the 21st century.

9. Conclusion

The joint establishment of the hydropower project will drive the economic development. In this Plan, we have studied the feasibility of the cooperative power development in Southern Africa to some extent, with the market demand as the orientation, and by adhering to the principle of sustainable development. The Inga Hydropower Project, with its huge engineering scale and substantial economic and social benefits, will play a key role in promoting the economic and social growth in Southern Africa. By vigorously propelling the construction of the world's largest hydropower green project in the 21st century and facilitating "South–South Cooperation", it is possible to bring Africa from the margin to the center of globalization, cement the foundation of the China–Africa strategic cooperation, promote the modernization process of China and Africa, and contribute to global sustainable development. This Plan reflects the concept of "cooperative planning, development, construction and development", aligns project construction to social development and ecological balance, and seeks positive results out of the combination of Chinese elements and international practices with local actual conditions. Through cooperation in facilitating the construction and development of the Inga Hydropower Station and the power grid in Southern Africa, we will surely create a new bond intensifying the economic and trade ties between China and Africa. Despite all the risks and difficulties lying ahead in the implementation course, we will be able to realize mutual benefits and win–win results through cooperation. This Plan does not represent the final research result, and further research is needed for the revision and refinement of this Plan.

III. Thoughts on Cooperative Development of Caribbean Tourist Infrastructure

Following the third Sino-Caribbean Forum, we conducted in-depth research on the strategic status of the Caribbean region in the international arena and its latest development, and came to the conclusion that it would be a major move to enhance our strategic global presence to increase Sino-Caribbean cooperation on tourist infrastructure. For one thing, we could become more flexible and take more initiative in our relations with the US and Taiwan and also on the issue of internationalizing the renminbi; and for another, we follow the economic development trend of the Caribbean region, which is dominated by its tourist industry.

1. Great tourist market potential lies in the Caribbean region. The Caribbean region enjoys the most unique and high-quality beaches in the world. The Caribbean region is one of the most attractive beach destinations in the world with its fine weather conditions, high-quality beaches, wonderful tropical scenery and a mixture of diversified cultures. Statistics

from the Caribbean Hotel and Tourism Association in 2011 show that it boasts the most developed tourist industry in the world with tourism revenue accounting for more than 60% of the GDP in some countries. Many Caribbean countries consider tourist industry as their strategic economic pillar and offer strong support accordingly.

The tourist industry in the Caribbean region is in earnest need of a diversified development strategy. Although tourism is the most competitive industry of the region, it suffers from high reliability and fragility. First, it relies too much on the US. In 2010, some 36.6% of the total 30.3 million US tourists traveling abroad went to the Caribbean region, accounting for half of the total 22 million tourists it received for the whole year. The economic condition and spending power in the US have decisive influence on the tourist industry of the Caribbean region. Second, economic development is imbalanced. The Caribbean countries generally neglect traditional industries like agriculture, manufacturing and commerce, and there is also a lack of industrial division and cooperation among them. As Mr. Bourne, President of the Caribbean Development Bank, noted in 2009, the Caribbean countries rely too much on foreign trade, their export commodities and services are comparatively simple, their reliability on foreign investment, assistance and remittance is high, and their self-support ability is low in many sectors with most of the food, industrial items and energy being imported. Under the world economic slow-down as a result of the US subprime mortgage crisis and the current European debt crisis, the Caribbean tourist industry has undergone great impact and offers great market potential. The Caribbean countries thus have shown an increasing willingness to cooperate with China.

Infrastructure inadequacy renders great business potential. At the current stage, the Caribbean countries offer great investment potential. First, the development level of tourist resources differs in Caribbean countries and many of them, due to backward infrastructure and lack of capital, suffer from limited service capacities. Second, China has not established diplomatic relations with many of the Caribbean countries, however, with the increase in political contacts and exchanges, these countries are expected to provide more convenience for Chinese investment and entry of Chinese citizens. Third, Cuba, a traditional good friend of China, is underdeveloped in its tourist services and infrastructure due to restriction from the US. However, with the speeding up of its reform process, this largest Caribbean island country releases great development potential and offers prime investment and cooperative opportunities at the current stage.

2. Guidelines and cooperative principles for Chinese enterprises investing in the Caribbean region. The institutional system of the Caribbean countries is greatly influenced by the United States and the European countries, and their tourist services are targeted at the high-end customers of these countries. Chinese enterprises and Chinese investors, as new-comers, will find it difficult to enter the existing market. The difficulty is even bigger for a single enterprise or project. Therefore, before entering into the market, enterprises should have a thorough and systematic consideration and also follow some clear and very important guidelines on business conducts and cooperation abroad.

Follow the principle of "to the people, by the people and for the people". Due to the geopolitical features of the Caribbean countries, there is a remarkably close similarity between their political systems and ideology and those in the US and European countries. Chinese state-owned enterprises, with influence from the Chinese government, thus become sensitive and eye-catching. Therefore, we should follow the principle of "from the people, by the people and for the people", emphasize the full independence of the Chinese enterprise as a corporate entity and adopt cooperative patterns in line with market behavior and business conduct.

Pay attention to environment protection and social responsibility. The Caribbean region, lacking other natural resources, is ecologically vulnerable and rather sensitive to topics like environment protection and human rights, and the public has strong political influence. Chinese enterprises should respect internationally accepted norms, earnestly exercise their social responsibilities there and build up good image and reputation for Chinese enterprises in order to realize sustainable development.

Increase cooperation and create synergies of teamwork. Chinese enterprises should increase coordination, enhance self-discipline, complement with each other, integrate resources and increase self-regulation. The China Development Bank or other leading enterprises could set up a communication platform to unite closely the Chinese enterprises conducting businesses in the Caribbean region, to coordinate their actions and create synergies.

Seek international partners and pursue existence through localization. The primary attraction of the Caribbean tourist industry is its natural scenery and unique culture. Chinese enterprises should stop hugging on to the Chinese elements if they want to seek business opportunities. In the Caribbean region, which features a highly internationalized and westernized business scene, Chinese enterprises should also be internationalized and good at building up their images. They could gradually enter into the Caribbean market by the easiest or most cost-effective means, for instance registering companies abroad and cooperating with local or foreign enterprises by setting up joint ventures, becoming shareholders or through leasing.

Press home the advantages of Chinese enterprises in several industries and adopt a differentiation strategy to gain a competitive edge. The development of different industries is imbalanced in the Caribbean region. China, although lagging behind in hotel services and management, enjoys considerable advantage in industries like road and port construction, architectural building, electricity, energy, petty commodities manufacturing, agriculture, forestry and fishing. Chinese enterprises, as active participants in the transition of the Caribbean region towards economic diversification, will undoubtedly win support and encouragement from the local governments.

3. Ways to invest in the Caribbean region

To systematically design the Caribbean tourist infrastructure construction, highlight the key points in tourist industry development and promote Sino-Caribbean economic cooperation through financing.

The backwardness of tourist infrastructure is the bottleneck restricting the tourist industry development and the overall economic development of Caribbean region. However, besides its shortage of infrastructure, we should give an overall and systematic consideration to the entire value chain of the tourist industry and get a complete picture in order to promote its overall growth. Basically we believe, through a preliminary review, that the entire picture is composed of eight parts: first, the opening of flight routes; second, hotel construction; third, transportation infrastructure construction; fourth, tourist resort construction; fifth, water and electricity facilities construction; sixth, entry and exit service facilities construction; seventh, tourist and cultural development; eighth, building of management platform. We should focus our efforts on these eight points and increase our bank's investment accordingly.

To play well the trump card that China is an enormous market in terms of tourist spending, cooperate with governments and enterprises of the Caribbean countries and engage in infrastructure construction of the region. China, with a population of 1.3 billion, is a huge market in terms of tourist spending and thus of great attraction to the Caribbean countries. In fact, we want to achieve the purpose of "resources-backed projects". On the one hand, we promote our influence in the Caribbean region by engaging in its infrastructure construction; on the other hand, we set up a resources-backed financing mechanism, i.e. having our bank loans guaranteed by resources needed for China's sustainable development. At the same time, we could take into account the expectation of tourists, and design several highly welcomed Caribbean travel routes targeted at the high-end customers in the Chinese and the Asian market as well based on a thorough market survey and scientific planning. With a focus on the travel routes, Chinese tourists could be provided with services through the entire value chain, from exit and entry service, air flight service to accommodation, catering and sightseeing.

To work together with Caribbean countries to develop new tourist areas and to enrich the tourist market. Based on the current development of the tourist industry in the Caribbean countries, we could differentiate them into three categories: the well-developed market, the developing market and the underdeveloped market. Based on the different market patterns, we could choose different investment models and ways of cooperation. However, Chinese enterprises should focus on the cooperative development of new tourist areas. There are major business opportunities. First, to expand tourism further to the hinterland areas. In the Caribbean countries, most of the modern tourist infrastructure is situated along the coast, and the hinterland areas lag far behind, therefore leaving big market potential. Second, to develop tourism in less-developed and developing areas. The tourism industry development levels are different among the Caribbean countries. Some countries, with a low level of economic development, are eager to improve their public service and tourist infrastructure and earnestly yearn for foreign investment. In these countries, it would be quite smooth for Chinese enterprises to enter. Third, to develop tourism in countries without diplomatic relations or with special diplomatic relations. Through their business activities, Chinese enterprises are expected to break through the existing barriers and open up new business opportunities. Countries without diplomatic relations or countries like Cuba, which is relatively isolated diplomatically, could provide even better opportunities for Chinese enterprises with pioneering spirit.

Case Study: How to Establish the BRICS Bank

On July 15, 2014, the Sixth BRICS Summit — Fortaleza Declaration and Action Plan officially announced to establish the New Development Bank (known as the BRICS Bank). The establishment of the BRICS Bank is a big event in today's international community, ushering in a new stage in the cooperation of BRICS countries. It reflects the development of complex contradictions in the globalization process and plays an important role in promoting balanced global economic development.

1. *Infrastructure Development Helps Resolve International Conflicts*

As we enter the 21st century, globalization deepens, mankind interacts with the nature, and economic, political and cultural interactions bloom to the full in the human society. Yet in the meantime, we are caught in a web of conflicts. For example, global warming, the shortage of resources, population growth and other issues are becoming more prominent; some new conflicts, in particular, arose out of the trend of global integration and intensified after the financial crisis in 2007.

To address these complex conflicts, one should approach from multiple angles and infrastructure development is one of them. It helps promote the north–south balance in development and resolve economic, social and political conflicts. Infrastructure development plays such a role mainly in the following two aspects.

The global demand for infrastructure financing is huge. According to estimates released by the World Bank and the ADB, the gap between the demand for infrastructure investment and the funds available is around USD 1.5 trillion annually. In the Asia-Pacific region, for example, an estimated USD 8.22 trillion (in 2008 US dollar), will be needed for infrastructure development in the 32 developing members of ADB during 2010–2020, or USD 800 billion per year, about 68% of which will be used for new construction projects, and 32% for the maintenance or replacement of existing facilities. The demand for infrastructure development in Asia is expected to account for 6.5% of its GDP during 2010–2020, about 49% for energy infrastructure, 35% for transportation infrastructure, 13% for ICT infrastructure; 3% for water and environmental sanitation infrastructure. How to meet the demand of USD 800 billion per year is a great challenge facing the Asia-Pacific region. Asian economies are diverse and each has made tremendous efforts to cope with the global financial crisis. A large quantity of high-quality national and regional infrastructure will be needed to support the continuous growth of productivity and competitiveness, the reduction of trade and logistics costs, the expansion and deepening of product networks and the transition extensive to intensive economies.

Financing for infrastructure development creates conditions for global governance reform. How to attract investment? How to transform infrastructure development plans into projects that can be funded by banks? How to engage investors from different countries and regions in good coordination and cooperation? To answer these questions, we need to make great efforts in the global governance reform or regional cooperation. In other words, we need to establish appropriate mechanisms and frameworks in the Asia-Pacific region to screen, select and arrange projects in different priority levels, and use innovative financial instruments in a flexible way, and establish and improve regional financial markets. We should also make appropriate efforts with regard to policy and regulatory support.

For example, in terms of infrastructure development, the information available to private and public sectors remains seriously asymmetric. To solve this problem, public–private cooperation is an option. Government authorities and the private sectors may cooperate to provide public goods and services or to construct infrastructure projects. They may enter into a concession agreement and sign contracts to define the rights and obligations and ensure the success of the cooperation, so that the parties can eventually reach a more favorable outcome than as expected from acting alone.

Take the issue of bonds denominated in the Asian Infrastructure Currency Unit (AICU) as another example. The implementation of an infrastructure project, from the very beginning to the very end, is often a lengthy process that exposes borrowers and lenders to substantial exchange rate risks. The establishment of the AICU is a way to address this. AICU, consisting of the currencies of major Asian and non-Asian advanced economies, is created for the relative stability of Asian currencies. It is an accounting unit and device for valuing infrastructure investment and repayment obligations.

As our understanding, the AIIB just like the BRICS Bank, is also a new financial institution and a supplement to the World Bank and the ADB, both of which have comparative advantages but neither of which is the solution to infrastructure financing in Asia. Thus, it seems feasible to create a special and new institution to fund infrastructure projects. Asian countries generally see a high savings rate and have ample reserves. However, since there are limited approaches available, most of such funds are loaned to developed countries. The AIIB is expected to direct these funds to regional and other infrastructure development projects. Thus, infrastructure financing will undoubtedly help improve global governance.

2. *The BRICS Bank Lends New Momentum to Balanced Global Economic Development*

All conflicts deeply rooted in the globalization process are fundamentally related to finance and both their causes and solutions can be found in finance. The World Bank, International Monetary Fund and other international financial institutions, established after the World War II, with the US and other developed countries in the dominance, have played an important role in the post-war recovery and development of the global economy and will continue to make their contribution. However, it is proved by the existing deep-rooted conflicts that there are defects in the existing international financial system and that reforms, new forces, and supplements are needed. The BRICS Bank reflects the needs and trend of the times. It is the solution to the deeply rooted conflicts in the globalization process and lends new momentum to balanced global economic growth. It shows some important new features.

It manifests a new paradigm of south–south cooperation. First of all, the Bank is the result of democratic cooperation: Each founding country makes equal initial capital contribution to it; the headquarters is based in Shanghai; the first president will be recommended by India, the inaugural chairman of the board of governors by Russia, and the inaugural chairman of the board of directors by Brazil. Such institutional arrangements render all five BRICS countries equal, allowing no one to seize all benefits alone. Contributions to the contingency reserve fund of the Bank are based on the economic conditions of each country, which also reflects the principle of equality. Secondly, it is also a substantial cooperation. The Bank is a new platform for financial cooperation that propels BRICS countries to consolidate and realize their cooperation plans and expand the scope of cooperation from trade to more economic and financial fields. Thirdly, it is an in-depth cooperation. Such a platform for financial cooperation helps BRICS countries to build closer ties and have better communications.

It is a new supplement to the existing international financial system. Unlike the World Bank, IMF and other international financial institutions, the BRICS New Development Bank places more emphasis on loans and investment in developing countries to support their infrastructure building and its contingency reserve arrangements are mainly aimed to help the member states cope with short-term liquidity pressures and strengthen the global financial safety net. The BRICS New Development Bank shares similar pursuits and ideology, serves similar customers, and holds similar regional goals with sub-regional development banks such as the ADB, the African Development Bank, and the Inter-American Development Bank (IDB), which lays the foundation for cooperation. However, the Bank differs from sub-regional development banks in that, as an intercontinental bank, it allocates resources across different continents for a better resource allocation structure and better connectivity. Therefore, it supplements what the sub-regional development banks do on the macroeconomic level, forming a complementary relationship. Such complementarity is reflected

Table 14.10. Voting Rights in the IMF

	Voting rights (%)		Voting rights (%)
US	16.75	China	3.81
Japan	6.23	Brazil	2.61
Germany	5.81	India	2.81
France	4.29	Russia	2.39
Total	33.08	Total	11.62

not only in the common cooperation principles and spirit, but also in the efforts made by the members of the Bank to promote regional development and cooperation on all fronts. The establishment of the Bank is also a key step made by BRICS countries to mitigate their severe lack of voice in the international financial system (see Table 14.10).

It is a new driving force towards the Millennium Development Goals. The Bank, to some extent, creates a separate investment and financing channel for new economies, offering relatively independent economic supports with a relatively independent organizational structure. Despite the limited initial fund in its capacity, it represents a huge step made by new economies towards a reasonable global economic structure and showcases the positive role of the BRICS countries in global economic governance reforms. In today's world, the potential for economic growth shrinks, investment falls, and trade protectionism is prevalent. BRICS countries are thus faced with new challenges. In particular, as the US winds down its quantitative easing policy, some emerging economies see serious problems such as capital flight, currency devaluation and economic slowdown. The BRICS New Development Bank helps the BRICS countries build their own financial safety net, reduce dependence on developed economies and the impact of the adjustment of international monetary policies, and promote stable, sustained and healthy economic development. The Bank will also serve as an important bridge between the North and the South, and between the developed and developing countries, pushing the world toward a better balance and universal benefits by strengthening dialogue and cooperation.

3. *Promote the Development of the Bank in a Practical Manner*

This section defines the functions of the Bank properly. In the next decade, the newly established Bank will have three main tasks:

Consultancy and planning. It shall study the BRICS countries and the global development strategy and analyze the macroeconomic situations so as to prepare long-term investment and financing plans for the BRICS countries. It shall also organize personnel exchange and training programs to enhance the soft power and promote the healthy, sustainable and balanced development of the BRICS countries.

Loans and investment. It shall establish an effective financing mechanism to support and speed up infrastructure development in the BRICS countries and other developing countries so as to improve their conditions for economic and social development; raise funds for the development of agriculture and small businesses, as well as for environment protection, help the BRICS countries and other developing countries build their capacity to maintain food security and support the development of SMEs and environmental protection; raise funds for the human resource development, increase basic health and education services and improve the basic skills and employability of poor people; raise funds for reform efforts; help the BRICS countries and other developing countries build their governance capacity and establish a political environment and market mechanisms that promote long-term stable development.

Security and risk control. It shall help the BRICS countries and other developing countries turn natural resources into a driving force for development by providing security and guidance; help these countries build their productivity, enhance production levels, and improve working conditions; it shall give priority to more urgent projects via collaboration with other international institutions in terms of lending and guaranteeing so as to promote sustained and balanced international trade growth; it shall cope with financial market instability and maintain financial security and economic security by establishing a risk warning and prevention mechanism.

The basic function of the Bank is to provide fund and pool wisdom to support infrastructure construction and sustainable economic and social development in the developing countries, which should be implemented unswervingly. However, we should also be aware that, in the long run, as concepts, purposes and means of construction and development change, the functions of the Bank may also shift. In this regard, the Bank can learn from the World Bank. In the late 1940s, the World Bank was focused on the reconstruction of Europe. From the 1950s to the early 1960s, the focus of the World Bank shifted to solving economic problems of less developed countries, setting support for the GDP growth in low-income countries as its top priority. Now, the priority of the World Bank has extended to income distribution, poverty mitigation, environment protection, cultural development, etc. The means of development have also changed and the focus has shifted accordingly from capital accumulation, foreign exchange, and large industrial projects such as transportation and power plant construction to small agribusiness, reproduction and the provision of social services in urban and rural areas. This shift reflects economic and social changes and development. It may also be the case for the BRICS New Development Bank and we should have the strategic awareness and practical plans.

Proactively explore the BRICS monetary mechanism. The Bank shall learn from the IMF's experience regarding the Special Drawing Rights and establish its own monetary mechanism through which it can perform its functions. There are options: (a) a basket of currencies composed of the currencies of the five BRICS countries; (b) a basket of currencies composed of the currencies of the five BRICS countries and some other developing countries; (c) create a new basket by adding BRICS currencies (including RMB) to the basket used for IMF's Special Drawing Rights. To implement the reform in a progressive way, a double-track system may be adopted at the beginning. Specifically, the Bank may continue to use the US dollar to perform its functions while establishing a BRICS currency swap mechanism and speeding up the creation of the BRICS currency at the same time. In the early stage, the first option is more appropriate. A basket of the currencies of the five BRICS countries can be created by giving quotas commensurate to the economic scale of each country, and then it can be used by the Bank to perform its functions in various financial activities and daily operations. No matter which option is adopted, the BRICS currency will be a positive supplement to the existing international monetary system.

In addition to a basket of currencies, the Bank shall also establish a unit of account, a clearing system and a reserve system, which can be deemed as a new part of the international clearing system. The Bank can also consider establishing its own clearing system when conditions are ripe.

The BRICS monetary system differs fundamentally from the Special Drawing Rights of the IMF. The Special Drawing Rights are rights distributed by the IMF to its member states to use currencies. The BRICS monetary system is not a right or power, but a system through which the Bank performs its functions. It is a cooperation framework to promote the use of the currencies of the BRICS countries.

Create innovative operational mechanisms for the bank. Mechanisms are more important than institutions. The vitality of the Bank lies in the innovation in mechanisms. (1) Innovative in the fundraising mechanism. China's experience shows that the problem of developing countries is not the lack of funds but the lack of a fundraising mechanism or a mechanism that transforms various resources into funds.[2] We shall find ways to establish an effective fundraising mechanism to transform the advantage of the BRICS countries and other developing countries in resources into an advantage in funds. (2) Innovation in bank governance. The Bank shall combine advanced theories and technologies with the actual situations in the BRICS countries and design proper governance institutions that balance equity and efficiency and act as a bridge between the market and the government. (3) Innovation in operational mechanisms. The Bank shall adhere to the principles of strategic mutual trust, policy support, professional management, business models, risk sharing and common development. The BRICS countries shall develop policies and provide legal support on the basis of strategic mutual trust; pool financial resources to

---

[2] Lixing Zou. *China's Rise: Development-oriented Finance and Sustainable Development.* Singapore: World Scientific Publishing Co., Ltd., 2014, pp. 312–315.

support inter-regional infrastructure construction and the development of fundamental sectors; give full play to professional and technical personnel to build a reasonable business operational model, set up a risk control system and mechanism, so as to ensure smooth operation of the Bank and the healthy and fast development of the BRICS countries and other developing countries, and contribute to global good governance and balanced development.[3]

---

[3] Huang Huaguang and Zhou Yuyuan (eds.). *Adjustment, Innovation and Collaboration — Collection of Papers Prepared for 2012 BRICS Think-Tanks Forum*. Beijing: Party Building Books Publishing House, 2013, pp. 15–20.

Chapter 15

The Underlying Reasons for the Trade Issues of the US and China

Through in-depth analysis of the problems and deep reasons of Sino-US economic and trade relations, this chapter explores the new ways and measurements to improve Sino-US economic and trade relations actively. The problems of Sino-US economic and trade relations should return to economy and trade, which would be more complicated if too politicized. In the market competition, both players should understand each other. Both China and US should concern some problems from the other side, seek common ground, enhance consensus, stand on reality, construct the future and promote the process of economic globalization mutually.

I. Background

Since opening up its economy, China has been actively developing economic and trade ties with the US. China and the US have developed into mutually dependent economies with connected interests that concern the world. The US is the second largest trade partner and the largest export destination of China, while China is the second largest trade partner of the US and the largest foreign holder of US Treasuries. By the end of 2015, total bilateral trade between and the US reached USD 558.4 billion.[1] Economic and trade ties between China and US are the most important bilateral ties in the world today but, despite overall progress, there are still

---

[1] China General Administration of Customs, available at http://www.customs.gov.cn/ publish/portal0/.

disputes. In particular, since the global financial crisis broke out at the end of 2007, and demand in the global market declined, major developed countries experienced an economic downturn, and trade protectionism grew. Since China's GDP ascended to second place in the world, trade frictions between China and the US increased and counteroffensive measures were intensified to protect each other's national interests. This trend of economic and trade politicization has a tendency to aggravate. To some extent, these issues were inevitable and their causes were complicated. Addressing these issues in an appropriate manner and overcoming difficulties are significant for promoting healthy and sustained development of China–US relations.

II. Underlying Reasons for the Politicization of Economic and Trade Issues

Issues like trade imbalance, the exchange rate and intellectual property in the China–US economic and trade relations are merely symptoms, behind which there are underlying reasons:

They reflect the conflict between developed and developing countries amid complex global changes. Amid global changes, the US, as the only super power in the world, and developing countries represented by China are seeing increasingly conspicuous conflicts. First, developed countries remain dominant in the global economy, but with a shrinking share. For example, from 1990 to 2012, the share of goods exported from developed countries dropped from 72% to 50.7% of the world total, while that of goods exported from developing countries rose from 24.3% to 44.7%.[2] Along with such a change in the overall pattern, the driving force for developed countries to dominate the world's economic growth is weakened. Second, developed countries enjoy advantages in science and technology, but it is more difficult to maintain their leading position amid the trend of globalization. This is particularly found in the development of information and network technologies that have facilitated global communication and technology dissemination. Third, developed countries are

---

[2] World Trade Report 2013, WTO.

still vigorously promoting their values and development ideologically, encountering more and more questions and resistance along the way. For instance, developed countries promoted the Washington Consensus to the developing world in the 1990s, but Latin American countries did not get the expected result after accepting it. Fifth, the gap between the south and the north incurs asymmetric market structure and factor flow. Developed countries dominate the concepts, rules and pricing in the market with the US, enjoying advantages in capital, management and technology; while developing economies, such as China, only have edges in primary factors like unskilled labor. The two sets of production factors are asymmetric.

They mirror the complexity of China–US economic and trade relations. The instability and uncertainty of the world's political and economic climate make it more complicated for China and the US to adhere to the principle of opening up in economic and trade relations. Openness is now subordinate while complexity dominates. Different stipulations of various bilateral FTAs and RTAs form the "Spaghetti bowl" phenomenon and further complicate the picture. China and the US differ greatly in their development mode, stage and level; and the gap, while strengthening the mutual complementarity economically, also makes it more difficult for the two parties to reach consensus.

Culturally, China and the US are similarly inclusive and both emphasize national identity, but there are huge gaps in terms of history, culture and law. China, with thousands of years of history, values the Doctrine of the Mean, while the US culture highlights personal freedom and self-value based on the Anglo-Saxon Protestantism. Politically, both countries stress responsibility and hope to contribute to the welfare of mankind but abide by different value standards and logics in thinking. The two have adopted entirely different political systems. China adopts the system of multi-party cooperation and political consultation led by the Communist Party, while the US practices two-party representative democracy, believes that the representative system is the best system of democracy, and is passionate about exporting its values. Therefore, it is easy for the US to politicize economic and trade issues. However, the two countries have entirely different historical and cultural backgrounds and thus different starting points when thinking about issues.

They embody insufficient preparation and the underestimation for changes in world pattern by China and the US. Since its reform and opening up of its markets, China has developed rapidly, to the surprise of many people. In the US, many did not anticipate or were not ready for this, especially that the economic aggregate of China may approach or even exceed that of the US in the near future. With a strong sense of urgency and insecurity, they responded by efforts to hamper China's development and prevent the impact of China's rapid growth on the world pattern and the vested interest of the US and other developed countries. Entanglements of the China–US trade deficit, pressing China to raise the Renminbi exchange rate, pressing the intellectual property issue, controlling high-tech exports to China are all seen by China as acts to politicize China–US economic and trade issues.

In China, some people are carried away by GDP data, believing that with our economic aggregate increased, China will become a developed country or even a super power in the world. In fact, economic and social development of China still lag far behind developed countries in terms of quality, and the per capita figures remain low. Therefore, China will still be a developing country for a long period of time to come. Though the developed world might gradually be overtaken in terms of aggregate, but these powers may not necessarily subside. They enjoy so many advantages that they can still lead the world in many aspects in the 21st century. China will face severe problems if we cannot make correct judgments on the situation and our competitors, set an appropriate position and direction for ourselves, or effectively respond to the complex changes of situation and market competition.

They signal insufficient mutual understanding between China and the US. China, a country that advocates "harmony foremost", does not have the gene and cultural heritage to proactively offend others. In the course of empowering and rejuvenating the country, it has no intention to build a "Central Kingdom", expand its territory or seek global hegemony. Instead, it simply strives to maintain its national unity, territorial integrity, economic prosperity, welfare of the people and independence in the world. China also has the cultural tradition of "regarding authority to be the supreme power", and this culture prevails in all sectors of the society. It is normal situation in official circles that "the winner is the king and the loser is the subordinate", and common civilians usually resort to "personal relationship or acquaintances" in solving problems while likely to ignore the treaties and procedures. This demonstrates that China is still in the preliminary stage of social modernization. However, the development of China will only mean opportunities to both the US and the world. As long as the US adopts a friendly strategy toward China and actively promotes the economic and trade cooperation, China and the US will truly become strategic partners of reciprocity and mutual benefit.

There are two significant characteristics in the strategic culture of the US.[3] Firstly, it has dual characters[3] of being pioneering as well as aggressive, being scientific as well as barbarous, being democratic as well as hegemonic, and being realistic as well as idealistic. They are full of adventurous spirit of exploration, democratic thought of advocating science, legal concept of respecting human rights, and the sense of competition that "the weak serves as a prey to the strong". Secondly, it has the "sense of mission". The US is a quite ideological country particularly in high praise of its own values. It endows people with firm national confidence and has an almost religious "sense of mission" about all affairs in the world. Just as was pointed by Harry S. Truman during the World War II, "Striving for the first place and take the first place" is the soul of American spirit. This national religion is to satisfy people's psychological appeal through eulogizing the glory and honor of the US, and is an important means for maintaining national confidence in the dramatically changing world. This means an appeal is not a bad thing but the driving force for human development and the target rod for world civilization. However, the US shall not be over-confident for there is neither absolute strength nor absolute weakness in the world. China and the US could cooperate with each other in a mutually beneficial way no matter in material wealth accumulation or mental and psychological appeal.

They indicate that economic and trade theories lag behind trade practice. As economic and trade globalization deepens, trade becomes the result of production globalization and consumption globalization to a large extent. With its nature and mode fundamentally changed, trade is no longer taken

---

[3]Yao Youzhi and Yan Qiying (eds.). *Souls of World Major Powers.* Beijing: Chinese People's Liberation Army Publishing House, 2011, p. 35.

simply as the commodity exchange that realizes final product value, but a necessary step to ensure normal global production. The current WTO framework does not eliminate the difference between economic systems of its members. Except for "horizontal trade" and "vertical trade", there is a large amount of "circular trade".

In "horizontal trade", commodities are horizontally exchanged, with each side providing what the other side lacks, and a balance is easy to reach. Under the new international division of labor, vertical trade replaces horizontal trade, and trade occurs more in the circulation process of global production. Thus, the upper and lower-stream sections disconnect easily, and trade can be imbalanced.

Today, a circular mode is taking form in international trade. Guided by the dominant forces in the international market, the starting point and the ending point of trade are connected and develop in a spiral manner. As the dominant force in the international market, the US holds the power for market construction, design and raw material pricing on the one hand, and as a major consumer, controls major consumption channels and product prices on the other. Developing countries such as China are only involved in the production and processing links in the supply chain, merely providing some value-added. Production and processing in developing countries are subject to the control of multinationals of developed countries. Take "designed in California and made in China" for instance. For every assembled cell phone that sells at USD 300, China only gets a profit of USD 6.5.

In circular trade, multinationals of developed countries actually have under their control key links along the chain such as market concept, product design, factor purchase, wholesale pricing, retail pricing, consumption and market feedback, leaving only the processing, export packaging and delivery to the so-called manufacturing bases. This causes the long-term trade conditions for developing countries to deteriorate severely, further complicating the international trade imbalance.

III. Measures to Improve China–US Economic and Trade Relations

During the next period of time, trade imbalance between China and the US will remain. The imbalanced savings–investment pattern in the US will find it difficult to change in the short term. China will still enjoy its comparative advantage as a "world assembly factory", and except for traditional basic consumables such as electronics and mechanical equipment, will take bigger export shares. As competition expands and intensifies, China–US trade frictions tend to become regular, and the two countries have both common grounds and disputes in terms of ideology, political views and economic interest. Therefore, new ideas and measures are needed.

Seeking common ground while reserving differences and constructing win–win China–US economic and trade relations. In the short term, the US is not likely to change its way of viewing economic and trade relations with China from the perspectives of politics and security, and it may even take some new measures to press the Chinese market and hamper the development of China. While trade frictions between China and the US are increasing, common interests in the bilateral economic and trade ties are rising as well. Therefore, the two countries need to explore common grounds while reserving differences, try to coordinate with each other, and seek solutions to their respective problems through development on their own.

Disputes between the two countries include those in ideology, political concepts and economic benefits. Ideology, or values formed under different historical backgrounds, especially beliefs, is something intrinsic that needs to be mutually respected without any judgment or any argument of good or bad. Strategies and policies involving national security and development require careful analysis and planning on a long term basis. The major disputes between China and US probably lie with the contention for market for actual national economic interests. In market competition, we should resort not only to political forces but more importantly, market power to solve economic and trade problems with economic methods and carefully devised strategies. Both global vision and specific benefits are required to realize win-win results.

Seeking correctly positioning, and a complementary manner to promote sustained and healthy economic development of the Chinese and the US. China will remain in the primary stage of socialist market economy in the 21st century, and has a lot to learn from the US and other developed countries. It also has quite a few outstanding achievements and experiences to share with the international community. In the 21st century, the US will remain the most powerful and innovative country in the world, exerting the most important influence on world economy. Close and complementary cooperation between the largest developing country and the largest developed country of the world will not only be conducive to the sustained and healthy development of the two countries but also play an irreplaceable role for the stability and development of the world.

Enhance exchange and mutual respect to jointly driven forward economic globalization. China and the US should respect each other and treat each other with equality, appreciate each other's advantages, and draw on them to promote economic growth and social progress. The US should be inclusive, not only at home but also in the international arena, value the positive factors of China, and learn from them to promote development. Regarding the issue of trade imbalance, it is necessary for the US to decrease the various limitations on its export of advanced technologies to China and create favorable conditions for Chinese enterprises and capital to enter the US market, so as to facilitate balanced economic development in China and the US. China shall make good use of the reasonable aspect in the dual characters of American strategic culture, to integrate itself with international norms and values that are of general sense and are advocated by the US and western countries, and to promote reforms and development during the integration process. China should try its best to follow international criteria recognized by the US and the Western world, and advance its reform and development amid integration. Only by doing so can we truly show the vision of the two countries to contribute to the welfare of mankind.

Strengthen research and increase consensus to jointly improve international trade rules and multilateral mechanisms. At the G20 Summit in Hangzhou in 2016, both China and the US ratified the *Paris Agreement* to tackle the global climate change. It was not only a significant coordination effort in the establishment of a new-type major-country relationship between China and the US, but also an important measure for promoting global sustainable development. China should further study international legal systems and international trade, promote the improvement of international trade rules, and guide global trade towards fair competition and mutual benefits. China should also abide by international rules and better utilize the multilateral mechanisms of WTO to settle trade frictions. Efforts should be made to establish and improve warning mechanisms, including market warning, policy warning and government–enterprise communication. Efforts should be made to build the capacities of national industrial associations to organize, coordinate, negotiate, discipline and serve their respective industries. Both country should also enhance their roles in responding to accusations and guide and coordinate various parties involved in trade cases, as well as avoid unnecessary high-level frictions. Implementation of intellectual property strategy should be accelerated and the effort to build international brands intensified, to reduce technical trade frictions. China should also speed up its industrial restructuring and upgrading, promote its transformation from an extensive economy to an intensive economy, and correctly select target market segments and position its products to shape a new pattern of differentiated development and competition.

Tap into the role of development finance to promote new development of China–US economic and trade ties. Both China and the US need to tap into the unique role of development finance during economic restructuring and sustainable development. China needs to make full use of market tools to provide financial services for mid- and long-term infrastructure construction and social progress. The US needs institutions such as the newly founded Infrastructure Bank to fund both the maintenance and upgradation of the massive existing infrastructure and new infrastructure construction. Development financial institutions of the two countries can strengthen exchange and cooperate in research, planning, innovation and construction. Both may not only establish a development finance dialogue and cooperation mechanism on the national level, but on the basis of projects and between enterprises, as well as launch cooperation at the local level; and we may cooperate not only to support trade but also to make investment. It may not be reasonable to expect quick and huge gains, small yet continuous steps will carry bilateral cooperation deeper.

IV. Conclusion

Amid this era of globalization, economic and trade relations between China and the US are of paramount importance. As the international situation has changed, China–US economic and trade relations experienced quantitative changes, but qualitative change has not kicked in. China needs the US and *vice versa*. Neither can move on without the other's market, and as opportunities overweigh challenges in their relations, China–US economic and trade frictions tend to become regular. As the relations mature, both sides tend to hold higher expectations and become critical of the other.

China needs a peaceful and stable international environment to facilitate its development. It has not, does not currently, and in future will not pose any threat to the US. The United States, a great country, has been good at taking advantage of opportunities and innovations; establishing a unique democracy and balancing mechanism; forming a set of effective economic systems, and thereby making important contributions to the modern civilization of the world. The first group of Chinese overseas students was educated in the US and learned a great deal about modern science and technology. During World War II, China and the US formed a solid alliance that played a major role in winning the war. Over the past few decades, relations between the two countries have been steadily progressing. The largest developed country and the largest developing country in the world, the US and China have the resources and vision to contribute to the welfare of over 6 billion world population. A strategic partnership of reciprocity and mutual benefit between the two countries serve the fundamental interests of the Chinese and Americans as well as people of the whole world.

Though the two cooperate and compete with each other, cooperation leads to win–win results, while fighting holds out failure for both. Generally, relations between the two progress amid ups and downs. The Trans-Pacific Partnership (TPP) that the US promotes cannot replace the role of China–US economic and trade ties, and China's participation in the cooperation between BRICS countries is by no means a challenge for the status of the US in the international community. Over-politicization of such issues will only make them even more complicated. In market competition, each side needs to learn about the other. China and the US should both stand more in the other's shoes, seek common ground while reserving differences, increase consensus, base their plans for the future on reality, enhance cooperation and jointly promote economic globalization. This serves the long-term interest of China and the US, as well as the entire world.

Case Study: How Chinese and American Cultures Affect Decision Making

China–US relations call for consideration of cultural factors. Understanding of the cultural differences between the two countries can promote their cooperation.

1. *Characteristics of American culture*

The United States does not have a ministry of culture, but its strategic decision making has a strong cultural foundation. Its culture has a strong influence over the world, reflecting its unique cultural hegemony. This is not a coincidence, nor the results of a conspiracy; it is not only the embodiment of economic strength, but a unique cultural system. This system has four important features:

First, a special sense of competition was infused into it. American culture can be traced back to the European Renaissance Period. The Renaissance and a series of cultural and ideological revolutions caused by it in Europe infused adventurous spirit, democratic thought, legal ideas and the sense of competition into the blood of the colonists in North America. It can be said that American culture, from the very beginning, not only absorbed the value orientation of relatively advanced human civilization, but also inherited the hegemonic genes of capitalism and imperialism.[4]

Second, the government's role in culture administration is unclear. During the process of American cultural development, the government's role was invisible, but it existed everywhere. Even though the US federal government does not have a ministry of culture, most of its states and cities have cultural affairs offices. The US federal government and local governments mainly use taxation and other policies to support and control the cultural

---

[4] Yao Youzhi and Yan Qiying. *Heroic Spirit of Great Powers: Strategic Cultures of the World Power* (1st edition). Beijing: People's Liberation Army Publishing House, 2011, p. 35.

market. American culture is in fact supported directly and indirectly by the federal government and the cultural affairs offices in its states and cities, and it mainly gets indirect support. Section 501(c)(3) of the United States Internal Revenue Code clearly stipulates that various non-profit organizations, from labor unions to churches, from mutual insurance companies to cooperative retirement funds, from veterans organizations to community organizations, are exempt from federal income taxes and state corporate income taxes for the purpose of public interests. They can receive tax-reduction donations, and do not need to pay customs duties when they import artworks. These non-profit organizations are semi-public and semi-private, with some of the features of social enterprises or the social economy. They are public because they have some public purposes, and their profits are not shared by their shareholders and owners, but used for serving the society; they are private because their capital and operating ways are private. The capital source and management of these organizations combines some commercial elements; their guidelines are generally determined by the Board of Directors, but the Board of Directors is not specifically involved in planning; it is neither at the core of management nor at the periphery of decision making, but always able to create internal pressure and external tension.

Third, special emphasis on civic culture. US society emphasizes civic culture, judging culture, competitive culture and balanced culture; every citizen has a strong civic awareness. The US citizens' criticisms are very common and rooted in US society, and it is especially targeted at government policies and policy figures; the US citizens' criticisms and counter-criticisms are normal social behaviors, forming a society full of critics. In such a society, it is not easy for any person, policy or issue to get all the praise. However, the US is quite tolerant of minors, and especially good at using encouraging words and ways to guide children and adolescents, providing them with a sweet, happy living environment, where they strive based on self-reliance.

The US pays more attention to right balance and wealth balance, and encourage mutual conditioning of various forces. US government decision making is influenced and limited by US culture. Since the founding of the US, there has been a conflict between centralization and decentralization.

The result of the conflict is that diplomatic and military powers are highly concentrated in the federal government, while the economic and cultural activities are decentralized. For example, the US federal government does not make cultural budgets, but the states and cities all have local-level cultural development budgets. Cultural pluralism has always been the mainstream ideology of the US, strongly opposing centralization.

In addition to tax paying, the characteristics of the US civil society are also reflected in the following three aspects. First, donating activities. Not only wealthy and middle-class families but also low-income families make donations. For example, between 1969 and 2013, the value of personal donations increased from USD 72 billion to USD 180 billion, and more than 70% of the US families made donations. The average value of family donations accounts for 2% of the family income.[5] Second, volunteer activities. In the United States, there are about 93 million voluntary workers, who contribute to 20 billion working hours each year. The volunteers play an important role in social development. Third, charity. In the United States, charity is regarded as the mother of civil society. Just as Andrew Carnegie says, between the country and the market, and between the tax-financed government and the profit-funded enterprises, charity is the only antidote to unfair distribution of wealth; it can render reconciliation between the rich and the poor, and has positive significance for properly handling the relationships between people, between man and nature, and between human beings and themselves.

In US cultural theory and practice, the remarkable level of university cultural construction and common concern of communities are uniquely intertwined, forming a cultural ecology where high-brow literature and popular literature complement each other, and promoting the organic combination between culture and economy, and between culture and politics. For example, Disneyland is a very successful example of combining culture and business. It has meticulously created a dream world, combining art and business organically. It has been trying to do something impossible as a pleasure and fully realize the commercial value of art. It tells people that life comes from free Imagination, adventurous stimulation, warm interaction and lively images, exerting an important influence on Americans' lives as well as the lives of people around the world.

---

[5] *Giving USA 2004*, and *American for the Arts 2005*.

Fourth, special emphasis on cultural research and development. The United States has established an effective working mechanism of researching culture, innovating in ideas, displaying art, protecting genuine products and disseminating value. For example, the US has been trying to build universities into an intellectual platform that allows students to improve culture and create culture so as to guarantee human sustainable development. As early as the 1930s, the United States began to encourage and support the professionalization of university art, and suggested that the US universities should train both culture researchers and artists. US universities have the master's degree in art in the field of cultural research and development, and integrate literary creation, architectural design and dancing art into this degree, enabling the Master of Arts to become an internationally influential talent standard like the Master of Business Administration, Doctor of Laws and Doctor of Science and Industry.[6] US universities also have a rule of "a percentage point for art". When a university builds a new building, 1% of the budget should be spent on art (design) so as to give the building more cultural significance. The book publishing amount and turnover of University Press of America are relatively small, and respectively accounted for 1.25% of the total US book publishing amount and 2% of the total turnover of all US publishers, but University Press of America has greatly increased the diversity of books and had extensive academic influences.[7]

Reuters, Bloomberg, Bridge News and Dow Jones are using the Internet and other advanced technology to fully promote culture through innovation in financial information modes. In the future, economy, art, culture and society will be inseparable from the Internet, and people will further utilize the Internet to form new livelihood events. Just as the media says, the Internet will promote the real popularization of cultural practice, increase the diversity of service products and promote the research and development of culture, products, ideologies and systems.

In terms of sources of funding for cultural development, in addition to the government's incentive tax policies, enterprises' cultural sponsorship, non-profit organizations' fundraising and labor unions' active support, the

---

[6] Brenda Jubin. *Program in the Arts, 1911–1967*. New York: Carnegie Corporation, 1968.
[7] USA Statistical Service Center. *Book Industry Trends*, 2001, pp. 90–93.

United States also has hundreds of unique foundations, who have been playing a core role in unremittingly supporting cultural development.

In conclusion, the US cultural system involves indirect governmental support, strong financial support, non-profit organizations that are not subject to market discipline, a large number of rich donators, communities for expanding cultural diversity, enormous influences of elite culture and popular culture, and channels for cultural creation and dissemination. This system is neither fully independent of the government nor entirely dominated by the market, but right in stable balanced development.

2. *Characteristics of Chinese Culture*

Generally speaking, the traditional Chinese culture has the following five characteristics:

The Concept of Greatness. Traditional Chinese appreciate Greatness and look at greatness as the definition for the way of life and philosophic thought. Based on Greatness, there are a series of philosophic perspective, for example, Great one, Great unity, Great sound, Great beauty, Great truth, Great mind, Great learning, etc. From these philosophical treatises, we can see that Chinese traditional values attached great importance to the commonwealth, or collective interest. In light of the slack, small-scale peasant society, and in order to organize thousands of loosely connected peasant families to sustain in the social order and operate as complements of the autocratically centralized system, the conception of a great national unity emerged: all land under the sky belongs to the king; all the people within the country are king's subjects. The notion "just as there is only one sun in the sky, there is only one supreme ruler" may be said to be rooted in the people's mind. But in modern China, the scenario seems to take a different route: while the international society is paying more attention to globalization, China is directed toward seeking multifarious dimensions.

The Rational Attitude to Reality and the Culture of Human Relations. The ancient Chinese society was patriarchal, where the fundamental role was to give utmost priority to man. Yet, this attention on man was not attributed to man's personality and freedom, but to the human relations which referred to the basic interpersonal relations and the behavioral norms for people to observe. The culture of human relations attached importance to blood relationships. The society valued human relations and defined various social relations by morality, where morality functioned as a substitute for religion. This dominant position of morality in the traditional Chinese culture placed religion in a subordinate position.

The Chinese nation was possessed with the humanistic tradition; hence the rational disposition towards reality. Chinese were skeptical about next life as some religions advocate. For instance, Taoists advocated, "way of Heaven is nature", completely rejecting a god's intervention into human society. Under the influence of such culture, the spiritual life for the Chinese people was "sincerity, honesty and cultivation of their moral character" in order to fulfill a moral pursuit.

The notion of human relations finds its expression mainly in "the system of rites and music", for example, the three cardinal guides: ruler governs subject; father guides son; husband guides wife; the five constant virtues: benevolence, righteousness, propriety, wisdom and fidelity. These conceptions of traditional human relations serve not only a moral function, but also a political function. Ruling strictures in China's ancient society.

The culture of human relations has both merits and demerits. On the one hand, it helped strengthen the internal solidarity of the nation and society with blood ties and loyalties, thus mitigating social conflicts and stabilizing the society. On the other hand, it was a hindrance to individuality. As a result, the traditional culture valued morality, the rule of law and natural science.

One phenomenon that can be easily noticed in everyday life is the Chinese cultural influence on behavior, especially on interpersonal relationship. Chinese culture with its lack of business consciousness and emphasis on harmony and unification between people and de-emphasis of the individual, discourages people to act based on their own judgment or to express criticism. People get used to seeking agreement with others by compromise. Otherwise, the critic is very likely to be put in an isolated place and rendered unable to do anything. No matter who the individual might be, director or supervisor or manager, usually worries about attracting argument or isolation and avoids providing distinct proposals to others. To some extent this disposition may encourage people to succeed easily if the goal is same for everyone. The problem is that in situations of disagreement, it would present difficulty; because, although people do not express their different opinions to avoid harming the sense of harmony, they act out their disagreements by non-cooperative behavior.

But in web generation of China, things seem to change: they are paying more attention to regulatory behavior and seek to appeal individual characteristics; they do not try to duplicate something one has already done before. They do it their own way, and see where it's gone. They may not make as much money as they want to, but there is value in remaining true to themselves.

Continuity and Inclusiveness: China is among the four countries with an ancient civilization of 5,000 years. Yet the road to China's cultural development was far from being smooth. During its thousand years of development, China has undergone various ordeals. On the one hand, she withstood a number of foreign intrusions, for instance, the Hun's harassment in the Qin and the Han dynasties, the Mongolians and the Manchu's rule over the Central Plain. On the other hand, she also suffered from domestic suppression by the autocratic rule and civil strife, such as the First Emperor of Qin Ying Zheng's burning of the books and the massacre of Confucian scholars. Having gone through so much misery and so many ordeals, Chinese traditional culture has not only survived, but has also enriched and consolidated with time. The major reason behind the vitality and continuity of Chinese traditional culture is its magnanimity and inclusiveness. Just as all rivers meet in the sea, Chinese traditional culture has assimilated and incorporated various foreign cultures into its own. Exhaling the old and inhaling the new, it has been modified to be extensive and inclusive, eventually making itself more vigorous.

But things seem to change in the modern society of transformation: there are the major initiatives, modified and sometimes replaced, that powered the recovery and the changes in direction when the original initiatives failed to meet the stated objectives.

The Dialectical Pattern of Thought. The Chinese traditional dialectical pattern of thought is of ancient origin. Confucianism held that "endless multiplication constitutes changes", emphasizing the continuity and rationality of development and change. Taoism comprises Yin and Yang, and the unity of the two elements and their interaction as the two basic elements that propelled the ever-changing world. Lao Zi maintained that the fundamental form of motion lay in the transformation of contradictions, and it was a practical application of the dialectics to conquer the rigid with the pliable. Buddhism believed in the unity of the opposites, such as the whole and the part, the active and the passive and the generality and the individuality. In the 20th century, Mao Zedong further developed the traditional dialectical conception and put it into practice in the Chinese domestic revolution. Mao held that "in order to understand the development of a thing, we should study it internally and in its relations with other things; in other words, the development of things should be seen as an internal and necessary movement, while each thing in its movement is interrelated with and interacts with the things around it. The fundamental cause of the development of a thing is not external but internal".[8]

But traditional Chinese culture worshiped the sages and the men of virtue and kept a blind faith in the interpretation of classics. This way of thinking was stereotyped by taking as maxims the words of sages and men of virtue in the process of expounding their ideas, which seriously bound people's creative aspirations and thus became a kind of spiritual fodder.

Chinese intangible cultural heritage is battling to survive in modern society, and creating new industry. For example, *guqin*, the seven-stringed instrument, which represents one type of the art of thinking and dialogue and was played by ancient Chinese scholars for 1,000 of years, is regarded as the most elegant industry of the Eastern-culture. You may not see many people interested in this ancient art, which was once on the verge of extinction, and is now listed as an oral and intangible heritage of humanity by UNESO. According to a report from the Central Conservatory of Music, there are more than 20,000 people in China are playing or learning *guqin*, maintaining and exploring new possibilities for this age-old art.[9]

---

[8] Mao Zedong, *Selected Work of Mao Zedong*. Beijing: Foreign Language Press, 1967, p. 313.

[9] Life Culture, *China Dairy*, May 23, 2007.

The traditional Chinese culture will play a positive role for world peace. It is known that peace and development are still a predominant theme in today's world and increasingly challenged by the worsening environmental problems, widening wealth gaps, wars and crimes. The idea of "harmony" is not only deeply rooted in the Chinese nation, but is also respected and acclaimed by many nations in the world and held in high esteem by renowned thinkers around the global. In particular, Buddhist teaching, including the Five Commandments of "no killing, no stealing, no sexual misconduct, no lying and no intoxication" could heal a world harmed by terrorism, ecological and moral degradation.

3. *Summary*

We learn about the cultural characteristics of China and the US in order to promote cultural exchanges between the two countries. Each country or region has its own cultural characteristics. Social diversity and differences in lifestyle, culture and system cannot be a reason for hostility. On the basis of mutual respect, we should learn from each other's strong points to overcome our own weak points and enhance strategic mutual trust. We need to use the US cultural philosophy to communicate with the Americans in order to promote mutual understanding, trust and cooperation.

Human society is a colorful society. Oriental and Western cultures have different characteristics, but neither of them is superior to the other; the rise and fall of civilizations is related to historical opportunities and innovation ability. During the in-depth development of globalization, each country or region, as a member of the international community, has and continues to develop some common beliefs and views with others. In the process of modernization, the rule of law, innovation, credibility and democracy are universal values for economic society. Whether a country is powerful or not does not depend on its form of government, but depends on the ability of its government. The development of human society needs competition and cooperation among the countries in the world. Compared with the traditional society, modern society is a civil society based on the law, Internet and digitization, where each citizen represents the master of the country and the force of society, and "all the citizens are completely equal despite their differences in skill"; the government and its administrative staff, both of whom are entrusted by the citizens to implement the administration function, must be accountable to the citizens. In the market-oriented economy, fiduciary responsibility is the soul of the modern market; the citizens and the market participants composed of citizens are the basic force for the development of the market economy; all the citizens and market participants, taking law as the criterion and credibility as the bond, build modern social relations based on cultural differences. In modern international social governance, seeking common ground while reserving differences and learning from each other to achieve balanced development based on mutual tolerance and trust is an important basis for global good governance and sustainable development.

Epilogue

I have been preparing for this book for more than 10 years.

In the early spring of 2003, I resigned from my job in the US and came back to China to become a member of China Development Bank (CDB). I had been thinking about conducting specific operations instead of still focusing on research work but it turned out that my job at CDB was to do researches. I had to accept it and adapt myself to the new environment, striving to produce some fine writings. Over the years, I have led the preparation of more than 20 quite influential strategic plans and organized a number of researches, covering reform of CDB, national strategies and international cooperation. Among them, the "Post-Wenchuan Earthquake Reconstruction Plan and Development Strategy Research" received significant comments from then General Secretary, Premier, Vice Premier and State Councilor in 2008 and was taken into account in the formulation of related laws and policies; the paper "Improving Systems and Mechanisms to Promote the Healthy Development of Policy-based Financial Institution" received CDB's Special Contribution to Innovation Award in 2009; the Outline of DR Congo's Economic and Social Development Plan (2011–2050, draft) was spoken highly of by DR Congo's prime minister in 2011; the Southern Africa Cooperative Electric Power Development Plan received comments from President Xi Jinping in 2013; and the Strategic Plan for China–South America Agricultural Cooperation won second prize in the selection of outstanding research findings organized by the National Development and Reform Commission (NDRC) of China in 2015.

Besides office work, I also developed a research writing plan for my weekend and holidays — studying balanced thinking and apply it in analyzing China's sustainable development so as to provide academic support for the rejuvenation of the Chinese nation. To that end, I made some time-bound arrangements. Firstly, I wrote the book *Financial Empire: American Development and Enlightenment* (published by Hunan University Press) to introduce international experience in pursuing development that will help China develop. Secondly, I wrote the book *China's Rise: Development-oriented Finance and Sustainable Development* (published by Hunan University Press) based on my research into development-oriented finance as well as practices and theories with Chinese characteristics in this regard. Thirdly, after comparing China and developed countries in terms of rural development and to help address China's issues with agriculture, farmers and rural areas, I wrote the book *China Base: County-level Economy and Society* (published by China Financial Publishing House). With the great support and encouragement of my leaders, colleagues, friends and family, I have finished the book *China's Logic: The Balance Development*. I am really grateful to them for that.

How time flies! I have worked at CDB for 14 years. Before that, I worked with the World Bank, Research Office of the State Council and China Light Industry Press and I also worked hard in rural areas during the Cultural Revolution. All those add up to about 40 years. I am very close to retirement now. It is fair to say that of all the research outcomes I have produced, *China's Logic: The Balance Development* is the most important. In conducting the research, I tried to do the following:

First, be open-minded and seek truth from facts.

Second, be problem-oriented and try to find the way of thinking and measures to solve problems. I always think that China's development plans are right but it should focus more on the way of thinking, strategies and skills, or detail. Success is in detail. Over the years, the strategic plan objective of this country has been crystal clear but it has taken a lot to figure out how to achieve the objective. Mistakes are unavoidable for the lack of experience. It is acceptable that China makes mistakes but making mistakes repeatedly is not. The root cause of repeated mistakes is the wrong way of thinking. This is why I have been actively studying balanced thinking and exploring the important relationships that need to be dealt with properly for China's development in the medium and long term.

Third, focus on weaknesses that hinder economic development such as the desire for immediate benefits, underdeveloped market, lack of innovation, issues concerning agriculture, farmers and rural areas, and inadequate international cooperation. This is why I tried to use balanced thinking to find the imbalances, the weaknesses and the remedies.

My research is mainly about macroscopic analysis, with some column perspectives and case studies. The focus is on analyzing the causes of problems and how to solve them. My views are based on systematically studying and drawing upon the theories and practical experience of many other economists. In line with the law of balance, with the help of Internet technology and big data, and based on the realities of China as a developing country, I try to find new research perspectives, combining the vision for development with the means to achieve development and superstructure with economic base, and think about how to achieve sustainable economic growth in the long run.

I believe that it will help if we can tell a good story of sustainable development on the basis of studying and applying balanced thinking, combining theory with practice, and paying attention to innovative thinking. Some of my views may help China improve its economic policies now; some may be useful only in decades; some may still need to be discussed. If any errors are found in this book, I will take responsibility for that and corrections will be made.

In doing my research work over the years, I have referred to the research findings of many Chinese and foreign experts and received great support and encouragement from leaders, colleagues and friends at CDB, Research Office of the State Council, China Center for Modernization Research, and China Society of Economic and Social Systems as well as the careful guidance of editors at People's Press. In particular, I would like to thank Mr. Jiang Zhenghua, Vice Chairman of the Standing Committee of the 9th and 10th National People's Congress (NPC) and also my teacher, for writing the preface of this book, and my wife, children and relatives for supporting me all the time.

Life is short. A book may be nothing but it shows to some extent that someone's life is not wasted. It is fair to say that this research took up all my spare time and I have harvested a lot from it. I enjoy getting to understand and applying the law of equilibrium. Everything in this world is about balance. It is the key to development.

This book is dedicated to my beloved ones and my mother country.

Index

A

60 Articles, 201
abstract internal relationship, xlviii
abstract relationship, xlix
abundant resources, 343
accountability system, 194
acid rain, 93
action–response, xlvi
active credit activities, 117
active function, 179
active government, 178
active intervention, lviii
administrative efficiency, 13
advanced cooperatives, 198
advanced countries, 315
advanced factors, 241
advanced manufacturing industry, 236
advanced modern rural land market, 232
advanced plans, 298
advanced technologies, 28
advanced theories, 288
advocate of multilateralism, 304
Aegean Sea, 15
affordable housing, 188
Africa, 303
Africa and Latin America, 261, 268
African Continent, 342

African Development Bank, 336, 369
after-sales service, 296
after-sell services, 159
age of agricultural economy, 220
age of industrial civilization, x
age-old art, 392
aggregate demand, 100
aggregate supply, 100
agrarian age, x
agricultural age, 11, 26
agricultural civilization, 38, 55
agricultural industrialization, 173
agricultural labor, 201
agricultural land, 198
agricultural production, xviii, 47
agricultural productivity, 47
agricultural products, 23
agro-industrial system, 204
air pollution, 93
Akerlof, George A., 157
alienated technology, xviii, 236
all-around arrangement, 333
allocating resources globally, 250
all-round economic and trade
    exchanges, 290
all-weather pre-sale, 298
alms-giving, 17

American political culture, 303
ancestor stage, 5
ancestral land, 195
ancient African man, 10
ancient Asian man, 10
ancient Chinese scholars, 392
ancient Chinese society, 389
ancient civilizations, x
ancient currency system, 14
ancient Egypt, 12–13
ancient European man, 10
ancient human behavior, 11
ancient Indian culture, 13
ancient origin, 391
ancient Roman Empire, 13
ancient Silk Road, 33
ancient stage, 5
ancient times, 38
ancient world, 12
Anhui Province, 203
animal spirits, 151, 157
anthropoarchaeologists, 11
anthropoid stage, 5
anti-colonial movements, 40
anti-corruption, 33, 58, 60
anti-dumping measures, 320
anti-Fascist war, 75
anti-feudal, 40
anti-globalization, 29, 150
anti-imperialist, 40
anti-inflation, 58
anti-Japanese aggression war, 42
anti-merger terms, 319
anti-monopoly, 319
anti-speculative buying and selling
    activities, 58
anti-subsidy, 319
APEC free trade zone, 319
ape-man stage, 5

applied research, 238
appropriate operation, 131
arable land, 195
artificial intelligence, xliv
Aryans, 10
ASEAN-led mechanism, 319
Asian-African Conference, 274
Asian Development Bank (ADB), 334
Asian Infrastructure Currency Unit
    (AICU), 368
Asian Infrastructure Investment Bank
    (AIIB), 136, 188, 278, 332
Asia-Pacific allies, 305
Asia-Pacific economic cooperation
    and development, 305
Asia-Pacific market, 305
Asia-Pacific region, 305, 331, 367
Asia-Pacific Studies, 340
Asiatic ancient land ownership
    system, 223
Asiatic mode, 223, 229
Asiatic style, 223
asset-liability ratio, 153
asset management companies, 49
atomic level, xliv
automation, 24
Automation Age, 26
Autumn Harvest Uprising, 42
available labor force, 128
available resources, 295

B
Babylonians, 16
balanced approach, 4
balanced culture, 386
balanced development, lxii, 3, 62,
    102, 141
balanced diplomacy, xv, 262, 301,
    321

balanced economic and trade
  relations, 262, 286
balanced international trade, 262
balanced line, xii
balanced manner, 5
balanced proportion, 118
balanced state, xlii
balanced thinking, xvii, lviii
balanced thinking network, xiv
balanced way, 321
balanced way of thinking (BWT),
  xxxiv, liii, 301
balance force, xxxviii
balance law (BL), xxxiv, xiii
balance mechanism of society, 4
balance of credit income and credit
  risk, 168
balance of motion, xxxix
balance of payments (BOP), 118
balance-oriented way of thinking, lxi
balance-related knowledge, lv
balancing body, lx
balancing force, xlvii
bank governance, 373
bank-held securities, 144
banking regulatory mechanism, 307
banking union, 307
bank loans, 353
Basel Accords, 170
Basel model, 169
base money, 129, 142
base of development, xiii
basic conditions, 268, 305
basic economic system, 48
basic medical services, 65
basic pathway, lx
basic public service system, 66
basic research, 237
basic rules, 306

basic science, 237
basic social activities, 120
basic social organizations, 229
basic social security system, 34, 66
basis of connection, xiii
be digital, 247
behavioral rhythm, lii
behavioral strategy, 267
be home-based, 247
be individualized, 247
Belt and Road Initiative, 33, 271, 278,
  289–290, 292, 324
Belt and Road strategy, 295
be networked, 248
Bering Strait, 10, 291
Bernard Shaw, George, 267
be sharing, 247
be socialized, 248
Bible, 16
big balance, xix
big data, xvi, xxxiv, 30, 98, 155,
  161
biggest outbound investor, 283
big-pot system, 48
big powers, 261, 331
bilateral relations, 301, 312
bilateral relationships, 261
bilateral trade, 375
bill market, 142
biological chips, 28
biological phenomena, xliv
biological system, xliii
biological world, xlv
bionic engineering, 26
bit and atom, 248
blockchain settlement, xvi
Bloomberg, 388
Bodhisattva, 18
bodhi tree, 18

bond-based projects, 193
bond market, 93
bond portfolio, 143
borrowed characters, 14
Botswana, 341
bottom-up reform, 181
Boule, Marcellin, 7
brainstorming, xii
brand quality, xvi, 112, 145
Brazil, 303
Brexit, 307
BRIC mechanism, 311
BRICS Bank, 366
BRICS countries, 370, 384
BRICS currency, 372
BRICS monetary mechanism, 372
Bridge News, 388
bringing in, 292
British classical economics, 73
broad information base, 54
broken-down country, 44
brokerage system, 67
Bronze Age, 6, 21, 26
bronze coins, 14
bronze culture, 14
bronze ding, 14
bronze tools, 13
Buddhism, 15, 18–19, 312
Buddhist cultural heritage, 312
budgetary deficit, 190
building credibility, 287
bull-bear, 146
Bundi Dam, 348
bureaucratic capital, 39
business credit, 117
business layout, 295
business mode, 340
business network construction, 295
business risk control, 295

butterfly effect, 152
buyer's market, 96, 98

C
15th CPC National Congress, 278
21st century, 271, 308
21st-Century Marine Silk Road,
    289–290
calm mood, lxi
Canada, 307
capacity relocation, 102
Cape Horn, 291
capital-based risk management
    framework, 169
capital circulation, 167
capital economic circle, 332
capital flight, 370
capital flows, 318
capital investment, 287
capital inward and outward flows, 144
capitalist class, 72
capitalist market economy, 74
capitalist mode, 72
capitalist revolution, 72
capital market, 67
capital securitization, 121
Caribbean countries, 362
Caribbean region, 361
Carnegie, Andrew, 387
catching up, 178
causal chain, xliii
cause of existence, xiii
cave of Zhoukoudian, 10
cell phone disease, 28
centered around school districts, xviii
Central Asia, 289–290, 296
Central Conservatory, 392
central fiscal revenue, 60
centralized and planned ways, 178

Central Kingdom, 378
central limit theorem, xlvii
Central Plain, 391
central rule, xli
Chairman Mao, 234, 261
Changchun–Jilin–Tumen region, 333
Charlie Hebdo of France, 150
Chen Yun, 44, 59, 104, 197
Chen Yun thought, 104
Chen Zhi Wu, 134
Chiang Kai-shek, 42
chief architect, 277
China, 51, 286
China–Africa strategic cooperation, 361
China–Africa strategic partnership, 278
China Development Bank (CDB), 186, 341
China–EU relations, 308
China Investment Corporation, 49
China land reform, 197
China model, 323
China–Mongolia–Russia high-speed corridor, 332
China–Myanmar–India Railway, 291
China–Pakistan Economic Corridor (CPEC), 278
China–Pakistan oil project, 186
China–Pakistan Railway, 291
China rural collective construction land system, 176
China–Russia and China–Myanmar pipelines, 291
China–Russia–Canada–America line, 291
China–Russia oil project, 186
China–Russia–South Korea–Japan free trade zone, 333

China's modern history, 3
China's multinational companies, 278
China's new democratic revolution, xxxvii
China Society of Social and Economic Analysis, 340
China–South America Agricultural Cooperation, 395
China's political party system, 90
China Three Gorges Corporation, 341, 360
China–Turkmenistan natural gas project, 186
China–US economic and trade frictions, 384
China–US economic and trade relations, 384
China–US relations, 305, 385
China–US trade frictions, 381
Chinese ancient society, 221
Chinese characteristics, 56, 76
Chinese civilization, xxxvii, 15, 32
Chinese commercial banks, 170, 353
Chinese Culture, 389
Chinese dream, 76
Chinese e-commerce platforms, 296
Chinese enterprises, 293
Chinese–foreign joint ventures, 48
Chinese hydropower projects, 352
Chinese ideological system, xxxviii
Chinese industrial products, 293
Chinese intangible cultural heritage, 392
Chinese modernization process, xxxvii
Chinese monetary conditions, 127
Chinese nation, 14
Chinese revolution, 42

Chinese rural economy, 215
Chinese rural land system, 202
Chinese traditional culture, 61, 91
Chinese traditional dialectical pattern of thought, 391
Chinese traditional industrial product, 296
Chinese traditional values, 389
Chinese united front, 42
Chinese Workers, 42
Chongqing Land Quota, 173
Chongqing municipal government, 173
Christian Bible, 16
Christian doctrines, 17
Christianity, 15, 17, 317
circular mode, 380
circular trade, 380
city commercial banks, 170
civic culture, 386
civilized human society, 119
civil regional aircraft, 313
civil works, 356
clan management, 219
clan tribal society, 221
classical economics, 73
classical land systems, 219
classical religions, 15
clean energy, 97
closed rural communities, 40
co-building, 297
co-existence, 334
Cold War, 273, 275, 302, 316
Cold War thinking, 286
collective construction land, 209
collective farms, xviii
collectively-owned, 208
collectively-owned agricultural production bases, 195

collectively-owned rural land, 213
collective ownership, 213
collective self, 200
collectivization movement, 196
colonialism, 38
colonial trade, 23
colonist countries, 274
colorful society, 393
Columbus, 11
Columbus, Christopher, 72
co-management, 297
comfortable and functional consumption environment, 151
commercial laws, 94
Committee on Payment and Settlement Systems (CPSS), 124
commodity exchange, 119
commodity markets, 48
commodity-possessors, 128
common destiny, 267
common development, 286, 288, 290, 340
common grounds, 286, 305, 375, 381
common interests, 281
Communist Party of China (CPC), xi, 41, 90
communist wind, 46, 201
community of common destiny, 285
comparative advantages, 98, 278
competition of values and strategies, 280
competitive advantages, 246
competitive culture, 386
competitive markets, 94
competitive player, 271
complete cycle, 172
complete picture, 364
complex and ever-changing circumstance, 133

complex behaviors, xlvi
complex global changes, 376
complex global situations, 95
complex situation, 269, 308
complex structures, xli
complex world, xlvii
complicated political situation, 321
complicated related guarantee risk,
    162
complication, xlix
compound indicatives, 14
comprehensive balance, xx, 105, 107,
    359
comprehensive competitiveness, 133
comprehensive consideration, lxi
comprehensive development, v
comprehensive development level, xi
comprehensive earnings, 24
comprehensive equilibrium, 103
comprehensive function, 179
comprehensive interactions, 29
comprehensively deepening reform,
    108
comprehensive overview, lii
comprehensive reform, 49
comprehensive relationship, 288
comprehensive support, 288
comprehensive systemic risk, 153
computer science, xliv
concept of balance, liv
Concept of Greatness, 389
condition of activity, xiii
Confidence-building Measures in
    Asia (CICA), 334
Confucianism, 18–19, 316, 391
Congo River, 341–342
consolidated public safe payment
    system, 133
Constantinople, 17

constitutional democracy, 41
construction course, 356
constructive capacity, 4, 35
consultancy and planning, 371
consumer economy, 101
consumption culture, 151
consumption demands, 91
contemporary philosophy, lv
continental Europe, 306
contingency reserve fund, 369
continuous adjustments, lx
continuous local evolution, 11
contract-based land use, 205
contracted land, 196
contraction period, 98
contract protection system, 134
contract right, 207–208
contract terms, 206
contractual rights, 206
contradictory motion, xiii
control channel, 143
conventional money, 135
conversion process, 248
cooperation, 335
cooperation approaches, 289
cooperation ideas, 289
cooperation in planning, 340
cooperation mechanisms, 289
cooperation models, 289
cooperation platform, 289
cooperation space, 289
cooperative interaction, xiv
cooperative service centers, 216
coordinated development, xi
coordination mechanism, 162
core change, xlvii
core Hegelian philosophy, lvii
core manufacturing, 97
core strategic interests, 304

correspondence, xiii
corresponding nature, lii
corruption problems, 92
corrupt officials, 60
cost-effective means, 364
cost-effectiveness, 102
Cote d'Ivoire, 294
counterbalance, xiii
countermeasures, 146
country-focused and people-centered
    projects, 188
country-level governance reform, xv
county-level companies, 189
county-level departments, 173
CPC Central Committee Leading
    Group, 207
Created by China, 328
creative ideas, 247
credit accumulation, 116
credit activities, 115, 120
credit balance, xii
credit-based exchanges, 119
credit capital, 111
credit-centered, 137
credit condition, 127
credit currency, 120
credit economy, 117, 119
credit-evaluation mechanism, 162
credit exchange, 118
credit imbalance, 181
credit infrastructure network, 113
credit infrastructures, 137
credit integrity, 119
credit record, 113
credit services, 117
credit subjectivity, 119
criminal law, 94
crisis-driven Basel III, 169
critical thinking, 153

critical turning, 145
cross-border circulation, 293
cross-border e-commence mode,
    292
cross-border e-commerce platform, 295
cross-border e-commerce service, 298
cross-sectional studies, 238
cross-sector, 24
crowd funding, 253
cultural administrative system, 49
cultural affairs offices, 385
cultural and ideological progress, 64
cultural and social impact, 356
cultural backgrounds, 377
cultural characteristics, 393
cultural consumption, 101
cultural cooperation ties, 312
cultural differences, 150, 287, 385,
    394
cultural diversification, 29
cultural diversity, 12, 150, 259, 389
cultural exchanges, 284, 286, 325,
    393
cultural foundation, 385
cultural gap, 46
cultural heritage, 28
cultural interactions, 22
cultural market, 386
cultural pluralism, 387
cultural psychology, 158
cultural research and development, 388
Cultural Revolution, xxxvii, 46, 57,
    68, 195, 197
cultural rhythm, lii
cultural similarities, 316
culture of balance, liv
culture researchers, 388
currency, 120
currency devaluation, 370

current global geopolitical situation, 302

current international order, 335

current world landscape and order, 302

customer asset management system, 141

customer-centered integrated marketing, 140

customer satisfaction, 140

cut-throat price, 159

cyber physical system (CPS), 240

cycle, 127

D

3D printers, 247

3D scanners, 249

3D structure, 249

Dai Bingguo, 339

daily operations, 373

dam construction, 356

darkest period, 314

Das Kapital, 73

data protection, 240

debt crisis, 306

de-capacity, 102

decision-making mechanism, 191

deepening reform, 207

deep knowledge, xvii

deeply rooted questions, 210

deep-rooted tribal and religious, 355

deep-seated problems, 108, 153, 261

deep technology, xvii

deep-water zone, 81

deep-water zone of reform, 153

deindustrialization, 244

delivered value, 140

delivery to the so-called manufacturing bases. This causes the, 380

Delphi method, 346

democracy innovation, x

democracy movement, 303

democratic decision-making mechanism, 54

Democratic Republic of the Congo (DRC), 341

democratic revolutions, 3

democratic system, 321

democratic thought, 379

democratization, 37

Deng Xiaoping, 52, 58–59, 106, 234, 275

Denisovans, 7

deposit reserve system, 142

derived characters, 14

designed in California and made in China, 380

desktop factories, 247–248, 252

Destroying the Old and Establishing the New, 56

devastating epidemics, 21

developed countries, x, 62, 316, 318, 326

developed countries-led TPP and TTIP, 318

developed economies, 303

developed market economy, 151

developing countries, 100, 279

developmental economy, vi

Development Bank, 333

Development Bank of Shanghai Cooperation Organization, 333

development-by-phase plan, 349

development finance, 383

development logic, xvi

development-oriented financial institution, 187

development-oriented poverty relief, 85

development path, 286
development require, 381
Development Research Center of the
    State Council, 336
development strategies, 89, 178
Development Strategy Research, 395
device disease, 28
dialectical pattern, 391
dialectical thinking, lviii, lix
different civilizations, 268
different civilizations, regions, 334
different cultural spirit, 12
differential pricing, 358
differentiated advantages, 141
different interests, 281
different land ownerships, 209
different languages, 282
different legal systems, 209
different political systems, 377
different species, 9
different value standards, 377
digital information, 249
digitalization, 30
digital revolution, 303
digital tools, 247
diplomatic consultations, 324
diplomatic environment, 275
diplomatic relations, 263
diplomatic thought, 290
diplomatic work, 284
direct financing, 167
Disneyland, 387
disputes of economic interests, 280
distinctive characteristics, 11
distinctive financial ecosystem, 112
distribution costs, 311
distribution networks, 25
distribution system, 48
diverse financial ecology, 138

diverse thoughts, 29
diversified development, 308
diversified world, 76
Doctor of Laws, 388
Doctrine of the Mean, 377
Doha Round negotiation process,
    317
domestic development, 288, 321
domestic financial institutions, 282
domestic market, 288
domestic ruler, 39
domestic social conflict, 98
dominant force, 380
dominant form, 260
double-digit rate, 287
double-track price system, 47
Dow Jones, 388
DPRK, 310
Dravidian cities, 13
DR Congo, 395
Dresden University of Technology,
    337
driving forces, 35, 318–319
drought-resistant plants, 220
dual economies, 77
dual system, xxi
Duck-Woo, Nam, 336
dynamic balance, xlvi, lxii
dynamic balancing activities, liii
dynamic equilibrium, 3
dynamic Renaissance movement, 72
dynamic stability, xxxiii, xxxix,
dynamic status, lx

E
early and mid-term stage, 345
early warning mechanism, 155
earth system, xlv
East Africa, 21

East Asia, 289, 305
Eastern civilization, 32
Eastern Europe, 273
Eastern Han Dynasty, 14
Eastern Roman Empire, 17
East India, 23
e-cash, 133
ecological balance, xxxvi, 33
ecological civilization, 62
ecological relationships, 262
e-commerce websites, 159
economic aggregate, xvi, 378
Economic and Social Development
    Plan, 395
economic and trade relations, 285, 290
economic bifurcation, 111
economic bubble, 120, 122
Economic Community, 306
Economic Co-operation and
    development (OECD), 96
economic crisis, 178
economic efficiency, 98
economic field, lxii, 37
economic games, xx
economic geographical spaces, 96
economic globalization, 293, 375
economic growth rate, 90
economic integration, 29
economic miracle, 309
economic nationalism, 281
economic order, 57, 179
economic quality, 98
economic risks, 146
economic roadmap, 104
economic rules, 191
economic security, 371
economic strength, 385
economic structures, 332
economic system, 107

economic win–win cooperation, 263,
    286
ecosystem, 344
Edictum Mediolanense, 17
Edison, Thomas, 24
educational cooperation, 284
effective channel, 184
effective government–market
    interaction mechanism, 182
effective governments, 178
effective working mechanism, 388
Eight-Power Allied Forces, 38
Einstein, Albert, 234
elastic coefficient method, 346
Electric Age, 26
electric equipment, 24
electric power, 351
electronic money, xvi
electronic technologies, 24
elementary education, 99
emergency bases, 154
emergency education, 154
emergency insurance fund system,
    155
emergency measures, 307
emergency organization system, 154
emergency reserve, 155
emergency response mechanism, 154,
    327
emergent public events, 154
emerging countries, 296
emerging industries, 332
emerging markets, 302, 318
emerging powers, 259
e-money, 129
Emperor Wu, 72
Emperor Xiaowen, 223
employment protection, 281
encashment ability, 130

ending point, 380
endogenous force, xvi, xlv
endogenous part, xlv
endogenous self-forming force, xlv
end power price, 354
end-to-end projects, 298
enemy-occupied areas, 42
energy security, 100
English Bourgeois Revolution, 55
enhance consensus, 375
enlightenment, 55
enormous market, 288
enterprise contract and responsibility
    system, 47
enterprise credit, 117
enterprise joint stock system, 47
entire Inga Hydropower Project, 350
entire value chain, 364–365
entry-level, 235
environmental assessment, 356
environmental governance, 51
environmental protection, 62, 97, 281,
    371
environment protection, 363
epistemology, lv
equality and mutual benefit, 266
equal land allocation, 208
equal land distribution, 223
equipment maintenance, 251
e-science, 233
ethnic conflicts, 314
EU, 303, 306
EU fiscal policies, 306
EU-led Basel II, 169
Euphrates River, 12
European brotherhood, 306
European Central Bank, 306
European Coal and Steel Community,
    306

European Community, 306
European Continent, 306–307
European debt problems, 307
European Deposit Insurance Scheme
    (EDIS), 307
European financial market, 306
European Renaissance Period, 385
European Union, 25
EU–US relations, 308
ever-changing market system, 111
everyone's mind, 20
excessive liquidity, 131
excess production capacity, 50, 94
exchange control risk, 357
exchange rate, 376
exchange rate risk, 357
exchanging resources, 269
existing deep-rooted conflicts, 368
expanding domestic demand and
    supply, 289
expert management, 340
explosive force, li
export-oriented economy, 334
extensive development, 103
extensive growth, 107
extensive growth model, 89, 92
extensive manufacturing, 246
extensive pattern, 99
extensive production, 91
external connection, xiv
external constraints, 21
external economy, 83
external environment, lii
external factors, 90
external interaction mechanisms, 217
external interactions, xxxviii
external resources, 344
external structures, lxii
external world, xlvi, lii

extra insurance cost, 251
extreme ideas and means, 69

F
13th Five Year Plan, 87
factor-driven, 98
Factory to Business (F2B), 295
fair social environment, xxi
faith-based salvation, 17
faith in afterlife, 17
faith in Allah, 17
faith in angels, 17
faith in classics, 17
faith in predestination, 17
fake and shoddy goods, 60
false borrower, 165
false guarantee, 165
false project, 165
false transaction, 165
family-based emergency response
   training, 154
family planning policy, 315
family workshops, 252
Far East, 14
far-off countries, 351
far-reaching impact, 117, 308
fast-developing enterprise, 294
fast-growing debt, 190
favorable conditions, 295, 382
favorable natural environment, 313
favorable policy, 143
favorable weather conditions, 313
Fei Xiaotong, 214
fertility rate, 315
feudal empire, 38
feudal forces, 39
feudalism, 38
feudal society, 221
feudal system, 39, 220–221

financial analysis, 353
financial assets, 101, 130
financial capital, xvi, 130, 136
financial claim, 130
financial companies, 160
financial conditions, 127
financial credit evaluation, 112
financial crisis, 156
financial deepening, 210
financial derivatives, 131, 171
financial diplomacy, 287
financial ecology, 160
financial futures, 121
financial information modes, 388
financial infrastructures, 111
financial innovations, 132
financial institutions, 138
financial intermediaries, 132
financial market instability, 371
financial option, 121
financial regulation ideology, 113
financial regulation system, 168
financial resources
   adjustment system, 49
financial security, 112, 371
financial service strategy, 141
financial structure reform, 137
financial swap, 121
financial system, 49
financial system reform, 113
financial variables, 141
financing plans, 371
fine example, 309
first come first go, 161
First Emperor, 391
First Five Year Plan, 45
first flushing toilet, 15
first group of Chinese overseas
   students, 384

first-hand information, 156
first KMT–CPC civil war, 42
first National Congress of KMT, 42
first place, 379
first supranational central banking
   system, 306
First World, 275
five constant virtues, 390
Five in One, 63
Five Principles, 289
Five Principles of Peaceful
   Coexistence, 271
flat organization, x
flattened and sun-dried papyrus, 13
flexible relationship, 288
flexible specialization, 298
flexibly specialized, 151
fluctuation factors, 100
food safety, 65
food security, 371
food supply system, 144
Ford automobile production line, 24
Ford management system, 24
foreign aid, 283
foreign capital, 40
foreign concessions, 39
foreign direct investment (FDI), 279
foreign floating capital, 144
foreign multinational companies, 282
foreign trade and economic thoughts,
   285
foreign trade enterprises, 299
for-itself, xiii
formation of a market rules system,
   323
Fortaleza Declaration and Action
   Plan, 366
foundation of the universe, xiii

Four-Pronged Comprehensive
   Strategy, 50
four-trillion-yuan investment, 49
four-trillion-yuan stimulus package,
   50
fragmented information, 158
fragment psychology, 158
fraudulent guarantee, 164
free competition, 74
free laborers, 128
freelance work, x
Free Trade Zone Development
   Strategy, 92
Free Trade Zones (FTZs), xi, 47, 51,
   325
French Revolution, 55
frequent related guarantee, 164
fresh air, 93
frigid climate, 310
frontline battlefields, 42
Fukuyama's, Francis, (USA), 182
full economic cycle risk evaluation,
   172
functional component, xli
functional planning, 212
functional transformation, 179
fundamental conflict, 100
fundamental goal, 305
fundamental ideological revolution,
   55
fundamental intention, 317
fundamentality, xiii
fundamental motivation, 33
fundamental nature, lii
fundamental of the universe, xxxix
fundamental role, 19
fundamental system, 228
fund centralization, 142

funding capacity, 23
fundraising mechanism, 373

G
G20 Summit, 382
gambling acts, 151, 157
game theory, xl
Gansu Province, 14
Gardner, Howard, lvi
gene decoding, xvii
General Agreement on Tariffs and
    Trade, 279
general dialectical method, lvi
general dialectics, lix, lxi
general knowledge, xlii
general laws, 94
General Layout, 63
General Secretary, 395
General Secretary Xi Jinping, 50
general strategic interests, 304
genetic code, xli
geo-economy, 344, 357
geographical basis, 91
geographical conditions, xxxv
geographical locations, 299
geographic locations, 11
geological disaster, 356
geopolitical boundaries, 306
geopolitical landscape, 259
geopolitical position, 311
geopolitical situation, 331
geopolitical strengths, 302
geopolitics section, 355
German classical philosophy, 73
German government, 237
German Nation, 316
Germany's Industry 4.0, 236
Gini coefficient, 93
global balance, 329

global climate change, 382
global competition, 246
global division, 324
global economic imbalances, 95
global economic integration, 107,
    259, 324
global economic order, 95
global economy, 317
global environment network, 262
global financial crisis, 122
global gene pool, 262
global geopolitical uncertainties, 95
global governance mechanism, 312
global governance reform, xix, 309,
    367
global governance structure, 95
global health network, 262
global infrastructure, 95
global integration, 31
global interoperability, 293
globalization, 3
global landscape, 316
global M&A, 25
global market, 248, 293
global modernization, 4
global network, 30
global peace and development, 260
global perspective, 168
global relief network, 262
global resource allocation, 30
global safety and development, 286
global superpower, 302
global supply chain, 318
global good governance, 394
going global, 292
golden mean, xxxv, lviii,
gold standard, 120
good balance force, lxi
Good Hope, 291

governance capability, 51, 100
governance diversification, 31
governance pluralism, 259
governance structure, 335
governance system, 51
governance transformation, 102
government administrative function, 179
governmental income, 116
governmental organization, 86
government background, 281
government body culture, 185
government credit, 117
government economy, 178
government-guided phase, 262
government intervention, 3, 71
government-led, 292
government-led and market-based means, 188
government-led foreign aids, 262, 272
government-led market, 74
government-led special debt, 167
government's perspective, 74
grand drama, 22
gravitational force, xlv
gravitational waves, xvii, li
Great beauty, 389
Great Cultural Revolution, lviii, 104
Great Depression, 66
Greater Inga Hydropower Project, 352–353
Greater Inga Hydropower Station, 348
great investment potential, 362
Great Leap Forward, 4, 45, 57, 104, 200
Great mind, 389
greatness in simplicity, xli
Great one, 389

great rejuvenation, 62
Great Rift Valley, 6
Great sound, 389
Great tourist market, 361
Great truth, 389
Great Tumen Initiative, 334
Great unity, 389
Greece, 306
Greek, 308
Greek debt crisis, 307
Greek islands, 15
green economy, 62
grid power price, 354
gross domestic product (GDP), 90, 283
group polarization, 152
growth transformation, vi
guarantee of life, xiii, xxxix
guerrilla battlefields, 42
guiding principle, 321

H
handcraft trade, 23
handicrafts making experience, 101
Han dynasties, 72, 391
Hansen, Alvin H., 78
hard landing, 192
harmonious world, 18
harmonized social progress, 119
harmony, 393
health and old-age care, 97
health insurance reform, 304
healthy consumption culture, 101
healthy development of society, 266
healthy international environment, 268
heavy chemical industry products, 24
heavy industry, 310
Hegelian idealism, lviii
Heidelberg man, 10

Heisenberg, Werner, 154
hematopoietic function, 85
Herd behavior, 156
herd effect, 151, 156
heterogeneity, xlvii
hetero-organization, xliv
hierarchical causal chain, xliv
hieroglyph of Egypt, 12
hieroglyph of India, 13
hieroglyphs, 13
high-end personnel and workforce,
    303
high-end valve products, 298
high-end valves, 296
higher-level governments, 217
higher-level medium, 112
higher-level social exchange medium,
    145
high-level measures, 169
high-level meetings, 324
high-level visits, 313, 324
highly-politicized planned economy,
    58
high-performance pipeline, 293
high-quality beaches, 361
high-quality fund supply, 132
high-risk mode, 287
high-speed growth, 287, 315
high-speed railway network, 50
high standard, 318
high-tech exports, 378
high value-added, 243
Hindu, 10
hinterland areas, 365
hiring long-term laborers, 199
historical awareness of autonomy, 306
historical events, 312
historical experiences, 37
historical factors, 5

historical issue, 312
historical materialism, 73, 105
historical sites, 356
historical task, 276
Holy Roman Empire, 316
homestead system reform, 208
homo sapiens, 7
homo sapiens stage, 5
horizontal economic relations, 267
horizontal trade, 380
hostage incidents, 281
household-based economy, 218
household contract responsibility
    system, 47, 68, 197, 202
household contract right, 205
household farming firm, xviii
household industries, 80
household managed land, 195, 213,
    215
household management, 219
household registration system, 51,
    230
Hugo, Victor, 306
Hui Neng, 18
Hu Jintao, 59, 278
human, 266
human activities, 111
human balance, xxxvi
human behavior, lii, lv
human brain, 158
human capital, 129, 303
human civilization, lvi, 12, 38,
    266–267, 310
human culture, 32
human development, 21
human evolution, 6
human instinct, 21
human intelligence, 240
humanistic philosophy, 19

human life, 28
human resource development, 371
human rights, 363
human society, xlv, 32, 268, 366
human sustainable development, 388
Hundred Days' Reform, 55
Hundred Days' Reform movement, 39
hybrid creation, 249
hydrogen bombs, 57, 234
hyperinflation, 45

I
Ice Age, 8
idealism, lvii
idealistic utopian socialism, lviii
idea of balance, lv
ideographic and phonographic
    elements, 14
ideological and behavioral conflicts,
    150
ideological and institutional
    foundation, 212
ideological and policy system, xv
ideological basis, 260
ideological building, xxxiv
ideological country, 379
ideological determination, 145
ideological emancipation, 55–56
ideological field, 37
ideological perspective, 127
ideological revolution, lvii
ideological system, 271, 285
imbalance, xiii
immature legal system, 284
imperialism, 38
implementation system, 174
importance of alliances, 304
important historical stage, 46
important strategic interests, 304

inclusive finance, 188
incomplete credit system, 161
incomplete interpretations, xlviii
increase–decrease linking land quota,
    175
increment optimization, 99
independent development path, 43
independent force, 308
independent scientific innovation, 234
independent variables, xlii, xliv, xlvi
in-depth development, 293
in-depth development of
    globalization, 393
in-depth discussion, 295
in-depth financial cooperation, 334
in-depth globalization, 34
in-depth ideological revolution, 55
in-depth integration, 236
in-depth R&D, 29
in-depth research, vi
in-depth study, 108
India, 303
Indian Buddhism, 19
Indian history, 13
Indian Ocean, 71
Indian subcontinent, 14
Indirect and serial guarantee, 163
indirect governmental support, 389
individual behaviors, xlvii
individual credit, 117
individualized consumption demands,
    30
individualized demand, 250
indomitable will of Russia, 268
Indus river basin, 13
Indus script, 13
Industrial Age, 22, 26
industrial and agricultural
    development, 91

industrial chain, 253
industrial civilization, 38
industrial cultural heritage, 246
industrialization, 28, 37
industrial park, 248
Industrial Revolution, 37–38, 55
industrial society, x
industrial structure, 105
industrial take-off, 310
industrial technologies, 23
industrious spirit, 268
industry associations, 185
Indus Valley civilization, 14
infinity of hierarchy, xliv
inflation risk, 357
Information Age, 26, 30
information economy, 129
information integration, 29, 155
information metric, 248
information network technology, 149, 259
information risk, 112
information society, 158
information technology-based service industry, 312
informatization, 28
Infrastructure Bank, 383
infrastructure construction, 97, 289, 332
infrastructure development, 367, 371
infrastructure projects, 193
Inga–Bas Congo Province, 350
Inga–Calabar power transmission line, 352
Inga Hydropower Project, 341, 351, 357
Inga Hydropower Station Project, 348
Inga–Katanga, 350
Inga–Kinshasa, 350

Inga Phase II Hydropower Stations, 348
initial public offering (IPO), 144
initial stage, 100
initial tax, 222
in-itself, xiii
innate social qualities, 269
inner peace, 33
innovation activities, vi
Innovation Award, 395
innovation awareness, 268
innovation balance, xv
innovation capabilities, 233
innovation capacity, xii, xvi
innovation-driven, 98
innovation-driven development, 62, 64
innovation supporting system building, xvii
innovative development, 103
innovative mechanism, 295
innovative operational mechanisms, 373
innovative tools, 32
inorganic life, 26
in-stage improvement and development, 237
instinctive need, 266
instinct of balance, lv
institutional change, 316
institutionalization, 205
institutional transformation, 102
insufficient mutual understanding, 378
insufficient preparation, 378
intangible assets, 134
intangible values, 112, 145
integrated European economic and monetary union, 307

integrated regional organization, 306
integrated rental and tax, 223
integrate ideology, 316
integration of the reindustrialization,
    242
integration process, 306
integrity mechanism, 124
intellectual property, 376
intellectual property protection, 281,
    327
intelligent labor, 28
intelligent machines, xvii
intelligent robots, 28
intensive development, 103
intensive economies, 367
intensive growth, 107
intensive management model, 103
intensive pattern, 99
interaction mechanism, xiv, 80
interactive behavior, 268
interactivity, xlvii
Inter-American Development Bank
    (IDB), 369
inter-bank borrowing, 167
intercontinental transportation, 23
interest rate, 141
intergovernmental cooperation, 287
intergovernmental interest-free loans,
    280
intermediary objectives, 141
internal and external balance, xxxvi
internal and external powers, 182
internal cognitive ability, liv
internal-combustion engines, 24
internal components, xliv
internal connection, xiv
internal demand and supply, 145
internal driver, 115
internal economy, 83

internal elements, xlvi
internal nourishment of the spirit, 15
internal refraining, 21
internal relations, xxxviii
internal special contradiction, xlviii
internal structure, xl
internal value, 119
internal wars, 17
international anti-fascist alliance, 42
international balance, xv
international business, 24
international collaboration, vi
international community, 262, 286, 384
international competition, 37
international cooperation, xiii, 238, 269
international cooperation mechanism,
    284
international distributors, 298
international division of labor, 329
international economic and trade
    strategy, 262
international environment, 33
international exploration, 321
international financial crisis, 281, 347
international financial reform, 333
international financial system, 333
international geopolitical landscape,
    302
international geopolitical risk, 98
international governance, 293
international governance regulations,
    318
international governance structure, 312
internationalized industrial standards,
    238
international layout, 302
international market, 288
International Monetary Fund (IMF),
    136, 150, 368

international multi-polarization
    process, 310
international network, 25
international operations, 282
International Organization of Securities
    Commissions (IOSCO), 124
international political order, 326
international politics, 259
international regulations, 329
international service standards, 30
international situation, 322
international standards, 93
international strategic layout, 261
international trade, 318
international trade front, 299
international travel, 101
international vision, xii
Internet-based cross-border logistics,
    293
Internet business, 254
Internet companies, 160
Internet disease, 28
Internet finance, 112, 135
Internet Finance Risk, 160
Internet information-based economy,
    111
Internet Plus, 236–237, 240
Internet technology, 160, 295
Internet technology + traditional
    manufacturing, 297
Internet thinking, 98
interpersonal interactions, 21, 26, 32
inter-regional infrastructure
    construction, 374
intuitive reflection, xliii
investment activities, 318
investment-driven, 98
investment model, 284
investment risk, 284, 357

invisible barrier, 321
invisible hand, xx, 72, 76
Iron Age, 26
Islam, 15, 17
IT-based economic development, 24

J
Japan, 303
Japanese invasion, 40
Japanese people, 309
Jerusalem, 16
Jiang Zemin, 59, 278
Jiang Zhenghua, 339, 397
job opportunity, 141
Johannesburg Stock Exchange, 314
joint action, 304
joint construction, 288
joint development, 286, 288
joint efforts, 332
joint-equity cooperation, 206
joint-equity cooperative companies,
    195
joint force, 289
joint infrastructure projects, 291
joint liability guarantee, 163
joint planning, 288
joint-venture agricultural companies,
    219
joint-venture companies, 215
joint ventures, 23
Judaism, 15
judging culture, 386

K
156 key construction projects, 45
Kang Youwei, 39
Kapilavastu, 18
karma, 18
Katz, Stanley, 336

keeping policies unchanged, 205
Kenya, 294
key industries, 246
Keynes, John Maynard, 74
Keynesianism, 152
Keynesian macroeconomics, 74
key point, lxi
key strategic resources, 96
KMT–CPC civil war, 40
KMT–CPC cooperation, 42
KMT–CPC struggle, 42
knowledge-based economies, xi
knowledge economy, 129
knowledge sharing, 247
Korea, 45
Korean War, 273
Kunming–Bangkok Highway, 291
Kuomintang (KMT), 41
Kush Mountains, 11

L
labor-intensive industries, 243
labor movements, 73
labor-power, 128
labor standards, 327
Lake Taihu, 93
land assets, 67
land circulation market, 216
land contract, 205
land contractual relationships, 205
land cultivation administration, 226
land equalization system, 224
land finance, 194, 213
land occupation, 356
land of state-owned farms, 213
land ownership, 198, 231
land private ownership, 222
Land Quota, 174
land quota holder, 175

land quota revenue, 175
land quota system, 176
land reclamation, 175
land reform, 195
land-related incidents, 196
land-related income, 190
land right confirmation, 206
land right diversification, 230
land rights, 205, 231
landscapes, 356
land system, 219
land to the tillers, 195, 216
land transaction, 176
land transfer, 217
land use, 198
land use right, 210
Lao Zi, 392
large-scale, 296
large-scale construction, 276
large-scale manufacturing tools,
    252
large-scale mechanization, 25
largest country, 310
largest export destination, 375
largest foreign holder, 375
largest hydropower infrastructure
    project, 343
largest outbound investor, 279
late-mover advantage, 185
latent energy, 1
late-starter's strengths, 111
latest Ice Age, 8
Latin America, 280, 324
Latin American countries, 377
Latin letters, 13
law of balance, 56, 105
law of contradictory movement of
    things, xxxix
law of diminishing marginal return, 97

law of economic growth rate regression, 97
law of equilibrium, xlii, 3
laws of economics, 105
leading group, 51
leading position, 376
leaning to one side, 273
least developed countries, 279
leaving Europe and turning to Asia, 310
left-behind children, 85
leftist, 57
left-leaning errors, 46
leftover problems, 314
left wing, 68
legal awareness, 94
legal concept, 379
legal environment, 240
legalization, 37
legal support, 373
legal systems, 282
legitimate successors, 18
lending money for interest, 199
Lenin, 74
less-developed data base, 161
letting a hundred flowers blossom, 64
letting a hundred schools of thought contend, 64
leverage ratio, 144
Liang Qichao, 39
Liangzhu Culture, 14
life games, xx
Lighting Africa, 343
Li Keqiang, 60, 339
limited arable land, 213
limited centrist economy, 78
Lin Luan, 289
Lin Yutang, 158
listed companies, 164
livelihood projects, 246

livelihood-related business, 188
living environment, 32
living environment of human beings, v
living matter, xvii
living things, xliii
loan-cheat risk, 164
loan limit, 142
loans and investment, 371
local fiscal revenue, 60
local government financing vehicle (LGFV), 152, 189
local government investment, 194
local government ownership, 211
localization of Buddhism, 18
localization of Marxism, 43
localized management mechanism, 67
localized operation, 295
local Mayans, 15
logical concepts, xlix
logic of thinking, lvi
Logistics Economy, 92
long economic cycle, 96
long-effective risk management mechanism, 140
long Ice Age, 8
long-inhibited productivity, 205
longitudinal issues, 301
long run, 372
long-term and arduous task, 85
long-term arduous task, xxxiv
long-term benefits, 246
long-term care, 85
long-term common interests, 312
long-term development strategy, 245
long-term goals, 105
long-term interests, 280
long-term investment, 371
long-term plan, 87

long-term policies, 145
long-term process, 259
long-term relationship, 260, 269, 288
long-term risk management
    mechanism, 172
long-term stability, 206
long-term steady, xx
long-term thing, 266
long-term use right of land, 211
low-carbon development, 235
low-carbon urbanization, 235
lowest-energy orbit, xl
lowest power price, 359
low-income families, 387
low-income people, 85
low-risk mode, 287
low yield per unit area, 310

M
Maastricht, 306
Maastricht Treaty, 306
Machine-made products, 23
machinery industry, 39
Mackinder, Halford, 290
macro-control, 142
macro-control policies, 143
macroeconomic control capacity, 182
macro-economic risks, 190
macro level, 100
made in China, 244, 328
Made in China 2025 strategy, 236
Magna Carta, 306
Mahabharata, 13
Ma Hong, 336
main balancing force, 301
mainstream ideology, 387
maintaining stability, 59
main themes, x
Majiayao culture Linjia, 14

major contradictions, 39
Malawi, 294, 341
Malaysia Airlines, 150
management right, 207–208
managerial rules, 138
man-made life, 28
man-made satellite, 57, 234
manual work, 28
Mao Zedong, lviii, 45, 57–58, 106,
    272
Mao Zedong Thought, 3, 43, 52
marginal analysis, xiv, 73
marginal cost, 160
marginal income ratio, 161
marginal returns, 97
marginal revolution theory, 73
marginal utility, 97
marginal utility theory, 73
Marine powers, 23
marine resources and channels, 305
Maritime Silk Road, 21
market abnormality, 181
market-based campaign, 4
market-based means, 54
market-based way, 186
market behavior, 157
market builder, 184
market building, 210
market competition, 3, 71, 375
market construction, 380
market credit, xii
market credit activity, xvi
market credit climate, 146
market credit regulations, 147
market credit regulation system, 111
market credit system, xv, 100, 111,
    115, 137
market credit system building, xvi
market demands, 95

market demand–supply mechanism, 168
market economic order, 146
market economy, 267, 394
market economy and politics, 269
market expectation, 143
market foundation, 285
market influence, 94
market integration, 29
market intermediaries, 185
marketization, 37
market-led stage, 262
market liberalization, 319
market objectives, 141
market operation, 344
market-oriented, 102, 292
market-oriented characteristics, 94
market-oriented development, 102, 136, 151
market-oriented economy, 46, 54
market-oriented incentive mechanism, 174
market-oriented operation, 67
market participants, xvii, 394
market players, 156
market preconditions, 112
market promotion, 177
market rules, 145
market segmentation, 138, 141
market segments, 141
market shares, 140
market stability, 146
market standards, 112
market survey, 254
Marshall, Alfred, 73
Martin Luther, 17
Marx, Karl, lvii, 113, 126, 222
Marxism, 42, 73
Marxism–Leninism, 52

mass–energy unity, xli
mass events, 152
massive industrialization, 28
mass movement, 58
Master of Arts, 388
Matamba, 349
material and cultural demand, 100
material basis, 96
material enjoyment, lx
materialism, lvii
materialist, lviii
materialist dialectics, xxxv
materialized consciousness, 111
material progress, 64
material society, 158
material world, liii
mathematical model, xlii, xliii, lix
mathematical reflection, xliii
mature economy, 99
May Fourth Movement, 41–42, 55
Mayan calendar, 15
medical services, 320
Mediterranean, 71
Mediterranean basin, 14
medium and long term, 287
medium and long-term investment, 186, 344
medium- and low-level, 243
medium-end product market, 244
medium function, 179
medium-term and long-term plans, 85
Meiji Restoration, 39
mental activities, liv
mental world, liii
mercantilism, 72
merchant caravans, 289
mergers and acquisitions (M&A), 24, 283
meritorious statesman, 227

Mesopotamia, 12
Mesopotamian civilization, 12
metallurgy techniques, 21
meta-synthesis, xlii
method of innovation, lvi
mianzi, 158
micro and small enterprises (MSEs),
   167
micro basis, 142
micro level, 100
micro-manufacturing, 247
mid-19th century, 38
mid- and long-term strategic projects,
   188
Middle Ages, 17
middle ancient state, 5
middle-class families, 387
Middle East, 15, 296
middle ground, xiv
middle-income developing country,
   313
middle path, xxxviii
Middle Pleistocene, 10
middle section, lix
middle zone, lviii
migrant workers, 214
migration problem, 356
military and paramilitary alliance,
   321
military products, 24
Milky Way, xiii
Ming dynasty, 71, 289
mining industry, 204
Ministry of Land and Resources, 213
Minoan Civilization, 15
Minoans living, 15
mixed economies, xiv, 3, 71, 78, 286
mixed ownership, 71, 83
mixed-ownership economy, 82

mobile services, 98
mode of market operation, 250
moderately developed country, 313
moderately prosperous society, xi
moderately-sized farmer economy,
   215
moderately small-scale peasant
   economy, xviii
moderate-scale land management,
   218
moderate scale management of land,
   232
modern biology, 19
modern civilization, 39, 55
modern civilization centers, 55
modern collective agricultural
   companies, 215
modern commercial companies, 23
modern countryside, 101
modern economic corridor, 290
modern education system, 40
modern globalization, 23
modern governance level, xxii
modern human age, 6
modern human behaviors, 11
modern industrial civilization, 38, 308
modern industrial system, 244
modern industry, x
modern international social
   governance, 394
modernization process, 361
modern land system, 209
modern logistics, 97–98
modern party, xxi
modern philosophy, lv
modern public service system, 180
modern science, lvi
modern service industry, 332
modern social relations, 394

modern society, 99, 268–269, 391
modern time, 290
modern trade system, 23
Mohammed, 17
monetary and capital markets, 166
monetary market infrastructures, 142
monetary policy dissemination
    mechanism, 142
money base, 145
money-oriented activities, 59
Mongolians, 391
monotheism, 15
moral bankruptcy, 65
moral enlightenment, 18
moral guidance, 19
moral high ground, 310
morality, 117
moral quality, xlviii
moral restrictions, 124
moral risk, 112
mortgage, 230
mortgage loans, 132
most attractive beach destinations, 361
most economically developed
    country, 313
most influential countries, 314
most intelligent animal, 34
most valuable human qualities, 269
movement of chasing wealth, 60
Mozambique, 294, 341
multi-angled trading philosophy, 263
multi-business operation, 25
multiculture, 284
multi-dimensional financial
    structures, 139
multi-dimensional monitoring system,
    162
multifaceted diplomacy, 263
multifaceted nature, 304

multilateral and bilateral interaction,
    325
multilateral arena, 325
multilateral free trade circle, 332
multilateral mechanisms, 383
multilateral political and economic
    cooperation institutions, 285
multilateral relations, 301
multilateral relationships, 261
multi-layered, 333
multi-layered tax systems, 66
multi-level system, 246
multinational groups, 324
multi-party cooperation, 377
multi-party system, 41
multiple different entities, 232
multiple equilibriums, 78
multiple factors, 111
multiple lines, 11
multiple ownerships, 78
multiple societies, 312
multiple systems, 259, 322
multi-polarization, 302
municipal bond market, 194
municipal government of Tianjin, 338
must-read classics, 17
mutual aid groups, 198
mutual benefits, 286–287, 324, 361,
    379
mutual complementarity, 377
mutual dependence, xli
mutual help, 290
mutual learning, 286
mutually-complementary
    development, 255
mutual non-aggression, 274
mutual reliance, 266
mutual trust, xv, 269, 286
mutual trust and benefit, 288

mutual trust and cooperation, 260,
   265, 286, 289, 292, 334
mutual understanding, 266

N
Namibia, 341
Nanchang Uprising, 42
National balance sheet, 47
National Balance Sheet of China, 190
national capital, 39
national capitalism, 38, 40
national crisis, 39
national cultures, 32
national debt management system,
   194
National Development and Reform
   Commission (NDRC), 395
national development goals, 288
national economy, 287
National Equities Exchange and
   Quotations (NEEQ), 294
national glory, 312
national governance modernization, x
national identity, 377
national innovation fund system, 67
national interests, 95, 107, 304
national land foundation, 144
national land fund, 67
national liberation, 38
national master emergency plans, 154
National People's Congress (NPC),
   397
national reconciliation, 314
national rejuvenation, x, 86, 276
national religion, 379
national reunification, x
national security, 263, 281, 286, 381
national strategy, 286
national supreme law, 197

national unity, 91
national well-being, 263, 286
natural and geographic conditions, 12
natural cognitive aptitude, 1
natural environment, 28
natural force, xiii
natural geographic pattern, 248
natural phenomena, xlix
natural place, xl
natural positions, xxxix
natural purpose, xl
natural resources, 91, 100, 180, 303
Neanderthal River Valley, 7
Neanderthals, 7
Neanderthal species, 7
near future, 378
near-term construction, 344
near-term target year, 347
negative changes, 354
negative effects, 121
negative list, 47
negative-list management mode, 51
negative reaction, xlix
negative roles, 122
neighboring countries, 305
neoclassical economics, 73
neoclassicism, 152
Neolithic Age, 8, 14, 26
Nepal, 18
network configuration, 248
network society, 286
new, 311
new balance, lix
new biology era, 29
new biotechnologies, xxxiv
new center of the age, 305
new challenges, 370
New China, 37
new Cold War, 150

new communities, 86
new conflicts, 29
new content, 334
new continent, 5
new core competitiveness, 293
New Culture Movement, 55
new cycle of human development, 28
New Democratic Revolution, 3, 43, 51, 195
New Development Bank, 366
New Development Bank BRICS, 333
New Development Bank (NDB), 136, 278, 366
new diplomatic image, 304
new driving force, 235
new economic drivers, 111
new economic form, 82
New Eurasian Continental Bridge, 291
new Eurasia–North America passage, 333
new financial institution, 368
new forces, 368
new global geopolitical environment, 289
new global manufacturing industry, 322
new global trade and investment rules, 326
new growth points, 99, 107
new handcraft, 247
new holistic global entity, 32
new home-based manufacturing, 29
new household manufacturing, 92
new immigrants, 303
new industrialization, 173
new industrial revolution, 245, 277
new industry, 392
new international financial institution, 338

new international trade rules, 295
new long waves, 96
new market player, 82
new market system, 101
new medium, 325
new normal, 90, 107, 187
new normality, 151
new path of industrialization, 246
new platform, 369
new pragmatic framework, 289
new psychological attributes, 158
new round of cross-regional economic and trade integration, 319
new round of international competition, 282
new rules of origin, 320
new rural property right structure, 205
new security threats, 329
new strategic partnership, 312
new technical regulations and standards, 282
new technical revolution, 96
New Testament, 16
Newton, Isaac, 234
new-type consumer markets, 101–102
new-type family handcraft, 247
new-type home-based manufacturing industry, 247–249, 255
new-type home-based manufacturing workshops, 252
new-type industrial support, 255
new-type industry, 97
new-type major-country relationship, 382
new-type manufacturing, 101
new type of partnership, 263
new-type urbanization, 101, 255
new urbanization, 92
next-generation trade, 328

Nile river basin, 12
nine squares, 219
nine squares land system, 229
nine squares system, 221
nitrogen oxide pollution, 93
Nobel Prize, 157
non-agricultural industries, 80
non-agricultural purposes, 204
non-cooperative behavior, 391
non-correspondence, 154
non-financial outbound direct
    investment, 279
non-functional components, xli
non-governmental cooperation, 287
non-governmental organizations, 284,
    337
non-human society, 32
non-interference, 274
non-legal awareness, 94
non-liquid assets, 131
non-living matter, xvii
non-living things, li, lvii
non-performing asset (NPA), 102
non-profit organizations, 389
non-public sector, 79
non-tillers, 216
non-traditional political groups, 304
non-traditional security, 301, 329
normal credit activities, 116
normal international order, 326
normalization, 298
normal operation, 100
normal sense, 254
North Africa, 21
North America, 10, 291
North American countries, 316
North American Free Trade
    Agreement (NAFTA), 25

North American Free Trade Zone,
    316
Northeast Asia, 290, 331
Northeast Asia Economic Forum, 336
Northeast Asian community, 332
Northeast Asian countries, 332
Northeast Asian Development Bank
    (NEADB), 331
Northern Song Dynasty, 72
Northern Warlords government,
    41–42
Northern Wei Dynasty, 223
North Pacific economic circles, 333
North–South, 335
North–South gap, 169, 318
northwestern India, 11
Northwest Passage, 291
Norway, 307
novel drivers, 98
novel opportunity, 98
novel speed, 98
novel structure, 98
novel system, 98
nuclear waste, 28
nucleic acids, xli

O

objective system, 141
objective world, liii
obligation contribution, 12
observable and re-adjustable
    indicators, 141
October Revolution of Russia, 41
old-age care, 65
old-age pension, 65
old balance, lxi
old democratic revolution, 52
old ideas, 55

old northeast industrial bases, 333
old regime, 39
old systems, 55
Old Testament, 16
On Contradiction, xxxvii
one axis with two branches, 187
one central task, 62
one hegemon and multiple great
  powers, 301
one-stop purchasing platform, 296
one-stop services, 25
online shopping, 101
Online to Offline (O2O), 295
on self-organization, xliv
ontology, lv
open and dynamic environment,
  xxxviii
opening up, 33
opening up in economic and trade
  relations, 377
opening-up strategy, 278, 287
open society, 268
operating objectives, 141
operating risk, 112
operating system (OS), 158
operational mechanism, 154, 306, 373
operation mechanism, 216
Opium War, 3, 38, 51, 53
opposing centralization, 387
opposite direction, 268
opposite philosophy systems, lvii
optimized development, 212
optimum state, li
oracle bones, 14
orderphilicity by living creatures, xl
organic integration, 244
organic interaction between, xiv
organic political–economic union,
  306

organic unity, liii
organizational core, 140
organizational innovation, xvii
Oriental and Western cultures, 393
oriental culture, 312
oriental religions, 16, 20
original balance, lix
original causal relationship, xliii
original company, 81
original innovation, xvii
original natural, 356
origin of things, xiii
Orthodox Eastern Church, 17
outdated infrastructures, 341
Outline of Land Law of China, 196
outsourcing jobs, 312
outstanding problems, 37
overall and systematic consideration,
  364
overall balance, 4, 105
overall characteristics, xlvii
overall economy, 98
overall form, 14
overall morphology level, xliv
overall planning, xx, 246
overall situation, 321
overall strategy, 245
overall urban–rural planning, xii
overall well-off society, 60
overarching design, 289
over-centralization, 139
over-confident, 379
overcorrection, liv
over-correctness, liv
overdevelopment, xx
over-politicization, 384
oversea Chinese-invested enterprises,
  278
overseas layout, 295

overseas risks, 281
ownership structure, 211, 213

P
Pacific, 305
Paleolithic Age, 10, 26
Palestinian region, 16
paper money, 72
parent companies, 164
Paris Agreement, 382
Park Geun-hye, 337
partial equilibrium, xiv
partial equilibrium analysis, 73
partial oversupply, 100
participants, 151
particle motion, 154
Party ideologies, 285
Party self-discipline, 50
Passive credit activities, 117
passive follow-up out, 318
payment risk, 159
Payment Systems (PS), 124
peaceful coexistence, 285, 288, 301
peaceful competition, x
peaceful development, 287, 315
peaceful multi-polarization, 309
peasants, 42
peer-to-peer (P2P), 160
people communes, 196, 200
people-oriented, 171
people's commune movement, xxxvii
People's Republic of China, 43
per-capita agricultural land, 213
per capita arable land, 303
per capita figures, 378
perfect long-effective management
    mechanism, 140
performance evaluation, 140
periodic changes, 1

permanent development relationship,
    298
permanent member, 272, 311
permanent tenancy system, 230
perpetual-motion machine, 111
personal freedom, 377
personal information, 30
personal interests, 95
PetroChina, 297
petty gain, 159
Pharaohs, 13
Phase 1 Hydropower Station, 350
philosophical category, lv
philosophic thought, 389
phono-semantic compounds, 14
physical form, xlix
physical power, 8
physical production, 287
physical world, xlii, xlv, lvii
pictographic elements, 14
Pilot Work, 207
pioneer of regional political,
    economic and social integration,
    308
plain financial system, 138
planned economy, 54, 123
planning-led development, 188
planning principles, 344
planning tasks, 344
Plus Internet, 242
policies, 57
policy and technical tool kits, 143
policy communication, 184
policy development, 316
policy information, 143
policy orientation, 143
policy systematization, 171
political and cultural interactions,
    366

political and economic integration, 292
political behavior, 267
political climate, 355
political consultation, 377
political cooperation mechanism, 284
political diversification, 29
political entities, 22
political equality, 286
political expression, 267
political field, 37
political games, xx
political headaches, 331
political ideology, 19
political intention, 311
political means, 54
political mobilization, 191
political mobilization capabilities, 91
political movements, 4
political organizational capability, 90
political organizations, 171
political pluralism, x, 286
Political Risk, 355
political security, 355
political significance, 308
political system reform, xxi
political tasks, 56
political trust deficit, 312
political wisdom, 266
politicize economic and trade issues, 377
polling system, 34
Polo, Marco, 12
poor economic investment climate, 284
poor outcome, 97
poor population, 85
population density, 303
population growth, 303
position of balance, xxxix
positive and cooperative attitude, 329

positive factors, 382
positive reaction, xlix
positive role, 393
post-war recovery and development, 302
Post-Wenchuan Earthquake Reconstruction Plan, 395
potential customers, 253
potential economic growth rate, 90
potential market, 311
poverty alleviation, 85
poverty-alleviation campaign, 84
poverty-stricken people, 84, 86
poverty threshold, 85
Power Development Cooperation Plan, 341
power financial market, 358
powerful warships, 38
power industry, 341
power output, 354
power plant, 24
power transmission and transformation facilities, 353
power transmission and transformation project, 350
power transmission line, 351
pragmatic style, 268
predynastic period, 12
Premier Zhou Enlai, 261, 274
price scissors, 199
price stability, 141
pricing mechanism, 47
primary distribution, lii
primary factors, 377
primary industry, 99
primary productive force, 234
primary stage, 381
primary stage of socialism, 62
primitive accumulation, 129

primitive agricultural economy, 222
primitive commune system, 220
primitive globalization, 5, 9, 25
primitive handy, 7
primitive homo sapiens, 7
primitive land public, 220
primitive man, 7
principal contradiction, xlviii
principle of giving more and taking less, 269
principle of justice before interests, 269
Principles for Financial Market Infrastructures (PFMIS), 124
prioritized development, 212
prioritizing neighboring countries, 290
private economy, 72
private investment, 287
private land and public land systems, 221
private land ownership, 196
private ownership, 3, 71, 201, 220
private ownership system, 222
procedural laws, 94
process risk, 112
production, 151
production–exchange–consumption, 127
productive forces, 66
productivity, 4
professionalization of university art, 388
progressive cyclic motion, xxxiii
project construction risks, 357
project construction sequence, 351
project investment, 354
prolonged process, 9
promised land, 16
promotes sustainable, vi

proper economic rhythm, lii
proper interaction, xiv
proper pace, 308
property right, 120
property right reform, 51
prosperous society, 4
Protestantism, 17, 23
psychological imbalance, 92
psychological world, liii
public administration, 200
public awareness, 154
public confidence, 145
public interest, 281
public involvement, 117
public land ownership, 200
public opinion, 285
public ownership, 3, 71, 203
public power, 95
public–private cooperation, 367
public–private interaction, 301
public sector, 79
public services, 102
public service system, 181
Pure Land Sect, 18–19
pyramids, 13

Q
Qing dynasty, 39
Qing government, 39
Qin Ying Zheng, 391
qualitative change, xxxiv, 237
qualitative prescription, li
quality programs, 344
quantitative change, xxxiv, li, 237
Quran, 17

R
10 relationships, 75
Ramayana, 13

rapidly changing events, 152
rare resources, xvii
rash actions, xii
Rational Attitude, 389
rational consumption policies, 101
rationalization, 37
rational thinking, lv, 34
raw materials, 248
Raymond, 292
Raymond Co, 293
re-adjustment, lxii
real and effective competition, 266
real economy, xvi, 65, 120, 122, 132
real environment, lx
real estate bubbles, 191
real estate market, 152
real estate registration, 153
realistic existences, 120
reality capture, 248
real-time capital monitoring system,
    172
real world, 247
reasonable development and
    utilization, 343
reasonable land use, 218
reasonable pricing, 251
reason for existence, xxxix
rebalance, xiii
receive duty-free access, 320
reciprocal investment, 325
reclaimed farms, 213
Rectification Movement, 55
recycling of resources, 33
Red Army, 42
rediscount interest rate, 142
rediscount mechanism, 142
Red Sea, 16
Red Sea–Mediterranean high-speed
    railway project, 291

reductive release, xliii, xliv
refinement and integration, 298
reflective nature, lii
reflectivity, xiv
reform and opening up, xxxvii, 37,
    46, 91, 104, 262, 327
reform and opening-up policy, 89
reform of global governance, 326
regional and international affairs, 286
regional and sub-regional
    development banks, 335
regional economic development, 345
regional economic integration, 293
regional industrial division and
    cooperation, 324
regional integration, 308–309, 312
regional organizations, 25
regional structure, 105
regression analysis method, 347
re-guarantee company, 67
regular balance, xxxv
regular farmers, 227
regular social behavior, 159
regulation-based, xxi
regulatory coordination, 171
regulatory mechanism, 162, 278
regulatory philosophy, 171
regulatory system reform, 139
reindustrialization, 242–243
relationship, 335
relative, 218
relaxing environment, lxi
relevant policies, 293
reliability, 134
religion-like value system, 19
religious antagonisms, 150
Renaissance, 23, 55
renewable resources, xx
Renminbi exchange rate, 378

rental rise, 228
repair service, 251
Research Center, 336
resentment, 159
reserve system, 373
reserving differences, 381
resource consumption, 314
resource integration, 344
restricted development, 212
retail terminals, 299
return to Asia, 305
Reuters, 388
revolutionary means, 57
revolutionary pioneers, 54
revolutionary united front, 43
revolution of 1911, 39
reward and punishment system, xxxv
rewarding system, 327
rhythm change, li
rhythmical movement, l
rhythmic motion, xiv
rhythmic nature, lii
rhythmic pace, lxi
rich donators, 389
right balance, 386
rightist, 57
right wing, 68
rigorous attitude, 268
Rigveda, 13
risk-bearing capacity, 153
risk compensation, 140
risk management and control, 285
risk prevention measures, 307
risk sharing, 340
risk-warning mechanism, 162
river basin economy, 146, 261
RMB exchange rate formation
    mechanism, 47, 49
RMB internationalization, 112

Road and Belt Initiative, xix
roadmap, 80, 295
Roman Catholic Church, 17
Roman Empire, 17
royal families, 227
rule making, xvi
rule of law system, 102
rural areas, 64
rural centralized schools, 86
rural collective construction, 173
rural collective ownership, 211
rural construction land, 195
rural enterprises, 204
rural infrastructure, 216
rural land collective ownership, 205,
    207
rural land collectivization, 198
rural land cooperative service centers,
    218
rural land ownership, 208
rural land system, 195, 197
rural land system reform, 204
rural poor residents, xviii
rural population, 91
rural society, 40
Russia, 303

S

1989 student movement, 58
sacred temple, 17
safeguarding national interests, 287
safety net, 34
Sakyamuni, 18
sale- and after-sale responses and
    services, 298
sale restriction, 146
sales centers, 298
same fate, xli
same source, xli

Samuelson, Paul A., 78
Sang Hongyang, 72
scaled land management, 215
scarce resource, 266
scientific and technological
    development, 313
scientific and technological
    innovation oriented country, 234
scientific and technological
    revolution, 55
scientific socialism, 42
scientization, 37
search engines, 247
secondary distribution, lii, 66
secondary industry, 99
second KMT–CPC civil war, 42
second KMT–CPC cooperation, 42
second largest trade partner, 375
Second Opium War, 38
Second World, 276
securities market, 94
securitization, 120, 194
security and risk control, 371
security system, 92
seeking truth from facts, 43
self-adaptation, xliv
self-adjustment, xl, 240
self-balance, xl
self-balancing, xiii
self-centered focus, 304
self-confidence, 60
self-constraints, 100
self-control, 192
self-cultivation, xxxvi
self-cycling, 67
self-employed, 144
self-evolution, xliv
self-forming process, xlv
self-generating, xiii

self-growth, 67
self-improvement, 39, 266
self-inflicted setbacks, 35
self-learning, xliv
self-motivation, 100
self-organization phenomenon, xl
self-owned brands, 234
self-reflection, 240
self-regulation, 363
self-reliance, 64, 105
self-structure of intrinsic forces, xiii
self-sufficient economy, 22
seller's market, 98
selling indulgences, 17
semi-feudal and semi-colonial
    society, x
semi-private, 386
semi-public, 386
sense of balance, liv
sensitivity analysis, 354
September 11 attacks, 302
serving society, 35
SF Express, 101
shadow banks, 190
Shanghai Cooperation Organization
    (SCO), 186
shanty towns, liii
shared culture, 297
shared interests, 287
shared village land, 213
sharing and win–win, 297
sharing economy, x, 92, 297
Shia sects, 17
Shiller, Robert, 157
shortage economy, 96
short cycle, 96
short history, 282
short supply, 48
short-term and direct objectives, 141

short-term assets, 131
short-term borrowing and long-term
   investment, 191
short-term financial assets, 131
short-term interest rate, 142
short-term policies, 145
short-term saving and long-term loan,
   191
short waves, 96
Shuowen Jiezi, 14
Sikila Island, 348
Silk Road, 21
Silk Road Economic Belt, 289
Silk Road Fund, 278
Silk Road strategy, 332
simple basic elements, xli
simple indicatives, 14
simple kernel, xlvii
simple style, 268
simplification, xlix
Sinai Desert, 16
sincerity-based relationship, 269
Single Day sales, 159
Single Resolution Mechanism
   (SRM), 307
Single's Day, 101
Single Supervisory Mechanism
   (SSM), 307
Sino-African cooperation, 281
Sino-Caribbean cooperation, 361
Sino-French War, 38
Sino-Japanese relations, 310
Sino-Japanese War, 38
Sinopec, 297
Sino-US economic and trade
   relations, 375
Sino-US relationship, 261
six-pronged reform plan, 51
Sixth BRICS Summit, 366

Sixth Patriarch of Zen, 18
Sixth Scientific and Technological
   Revolution, 233
sliding world economy, 151
slow-growth, 287
small agricultural economy, 228
small- and medium-sized banks, 51
small- and medium-sized
   cooperatives, 68
small- and medium-sized enterprises
   (SMEs), 153
small function, 179
small-scale agricultural economy, 220
smart architecture, 240
smart city, 240
smart economy, 240
smart environmental protection, 240
smart home, 240
smart manufacturing, 236, 245
smart medical services, 240
smart society, 241
smart technology, 240
smart transportation, 240
smart world, 241
Smith, Adam, 72
social and economic cooperation, 145
social and environmental risks, 356
social capital, 136
social cost, 180
social credit, 117
social credit building, 117
social credit system, 146
social development, 117
social distribution, 179
social equity, 30
social governance, 229
social harmony and stability, 91
social honesty, 59
social instability, 92

socialism, 3, 38
socialism with Chinese
    characteristics, 4
socialist China, 321
socialist construction, 107
socialist democracy, 62
socialist economic system, 75
socialist economic theory, 73
Socialist Education Movement, 45
Socialist Harmonious Society, 59
socialist industrialization, 45
socialist market economy, xxxviii, 79,
    100, 123, 137, 381
socialist market system, 328
socialist planned economy, 75
socialist revolution theory, 73
socialist system, 76
sociality, 117
socialized economy, 74
social management, 179
social outcome, 1
social participation, 34, 86
social preconditions, 112
social problems, 93
social progresses, 118
social psychology, 158
social responsibility, 340
social security, 179
social values, 118
socioeconomic scenario, 123
soft constraint, 192
soft landing, 146
soft power, 303
software industries, 312
soil erosion, 356
soil fertility, 204
solar calendar, 12
solar system, xiii, xlv
solar year, 15

soul of American spirit, 379
sound social development, 133
South, 303
South Africa, 313, 341
South America, 291, 296
Southeast Asia, 290, 296
Southeast Asian Nations (ASEAN),
    25
Southern Africa, 341
Southern Africa Cooperative Electric
    Power Development Plan, 395
Southern African countries, 341
Southern African Development
    Community (SADC), 341
Southern African power network, 344
Southern African Power Pool (SAPP),
    341
southern tip of Africa, 291
southern tip of Chile, 291
South Korea, 337
South–South, 335
South–South cooperation, 275, 279,
    343, 361, 369
Soviet economic model, 44
Soviet empire's collapse, 75
Soviet model, 43
Soviet Russia, 41
Soviet Union, lviii, 42, 44, 74,
    273–274, 317
Soviet Union government, 46
Spaghetti bowl, 377
Spain, 308
special bronze culture, 14
Special Contribution, 395
special customer base, 253
Special Drawing Rights, 372
Special Economic Zones (SEZs), xi,
    47
specialized farming households, 218

specialized investigation, 34
specialized markets, 80
specialized vocational management, 24
special market, 250
special nature, xlviii, 273
special scarce resource, 265
specific world, xlviii
sphenogram of Mesopotamia, 12
spiral manner, 380
spirit of sharing, 297
spiritual nourishment, 34
spiritual nurturing, 18
spiritual power, liv
spiritual support, 20
spiritual world, lvii, 21
splendid civilization, 289
splendid isolation, 306
Spring and Autumn Annals, 222
Stability and Growth Pact (SGP), 306
stable and long-term rural land contractual relationships, 207
standard credit activities, 135
standard credit order, 117
standardization, 134, 298
standardized credit, 135
Standing Committee of the 9th, 397
stand on reality, 375
starting point, 380
starting-up stage, 251
start-up companies, 253
state-controlled companies, 164
State Council, 186
state credit, 119, 188
state monopoly capitalism, 40
state-owned enterprises (SOEs), 49, 82
state-owned farms, 195
state-owned giant banks, 139

state-owned land, 194, 223
state-owned sectors, xvi
state ownership, 211
state religion, 17
state-run enterprises, 287
state-run farms, xviii
stationing garrisons to grow food, 219
Stavrianos, Leften Stavros, 303
steam engine, 23
steam engine-powered industries, 25
step-by-step advance, 344
stock market, 93
stock market entry and exit mechanism, 94
Stone Age, 6
store-renting mode, 298
storing oil and developing new energy sources, 303
strategic cooperative partnership, 312
strategic credit, 358
strategic culture, 379
strategic emerging industry, 236
strategic goals, 103
strategic leeway, 307
strategic level, 295
strategic motive force, xv
strategic mutual trust, 373
strategic mutual trust, policy support, 340
strategic opportunities, 104
strategic options, 358
strategic partners, 379
strategic partnership, 286
strategic perspective, 287
strategic planning, 316
strategic plan objective, 396
strategic platform, 334
strategic support, 311
strategic training, 238

strategic vision, 289
Strategy of Eastern European
  Expansion, 308
strengthen economic cooperation, xxii
strict system, xiii
strong financial support, 389
structural adjustments, 76
structural changes, 237
structural nature, lii
student loans, 188
subconscious mind, 1
sub-contracting, 206
subjective initiative, xxxvii
subjective world, liii
subjectivity, 117
subordinated bond, 132
sub-regional cooperation, 334
sub-regional coordination mechanism,
  324
sub-regional development, 333
sub-regional international financial
  institution, 332
sub-Saharan Africa, 314
substantive laws, 94
suburban collective produce bases, 213
sudden non-programmed decisions,
  152
Sumer, 12
summarization, xlix
Sunni, 17
Sun Yat-sen, 39
super free trade area, 318
superimposed problems, 108
super power, 378
superstructure, xi, 108
supplementary support, 288
supplement to the World Bank, 368
supply chain, 296
supply–demand structure, xvii

supply-side reform, 102
supporting systems, 217
surface water, 93
sustainability, 4
sustainable development, v, 35, 103,
  297, 344, 394
sustainable economic development, x
swap, 146
swap mechanism, 372
swimming pools, 15
Switzerland, 307
symmetrical balance, xl
symmetrical nature, lii
symmetry, xiii
synchronous development, 62
synchronous tendency, 151
synergetic balance, xli
synthetic chemical products, 24
system, xiii
systematic change, xiii
systematic financial risk, 101
systematic nature, lii
systematic risks, 153
system elements, xlv
system of equal land distribution, 219
system of multi-party cooperation and
  political consultation, 90
system of permanent tenancy, 219,
  226
system of stationing garrisons to grow
  food, 225
system of tax distribution, 47
system's continuous development, li
systems engineering, xiv
system theory, xl

T
Taiping Rebellion, 39
Tang dynasty, 72, 289

tangible form, 134
Tan Sitong, 39
Tanzam Railway, 275
Taoism, 18, 316, 391
Tao-te Ching, 19
target year, 345
tariff barriers, 327
taxation risk, 357
taxation system, 101, 222
tax distribution, 49
tax partnership, 13
tax policy, 102
tax-reduction donations, 386
Taylorism, 24
technical applicability, 171
technical conditions, 100
technically calculated risks, 169
technical operation process, 248
technology fetishism, 236
technology gap, 46
tenancy, 230
tenancy system, 227
Ten Commandments, 16
Ten Major Relationships, 57
territorial integrity, 274
tertiary distribution, lii
tertiary industry, 99
the anti-rightist struggle, 45
theoretical innovation, 56
theory of Three Worlds, 272, 288
think broader, 293
thinking method of balance, 54, 62
thinking methodology, xii
Third Industrial Revolution, 233
third- or fourth-tier cities, 153
third-party collection/payment
    platforms, 162
third-party guarantor, 161
third-party payment, 160

Third Plenary Session, 50, 61, 205
Third World, 273, 276
Third World countries, 276, 283, 288
three big mountains, 3, 44
three campaigns, 45, 104
three cardinal guides, 390
Three Gorges Project, 352
Three Rural Issues, xii, 84
three-step development strategy, 52,
    62
three subcontinents, 290
three-tier ownership system, 201
tiandi, 227, 230
Tianjin, 336
Tianjin-based Nankai University, 336
tianmian, 227, 230
Tibetan region, 242
tightened economic operation, 100
tightened economy, 100
Tigris River, 12
time-efficiency, 142
time–space unity, xli
top 50 universities, 303
top-down reform, 181
top-level conference, 205
Toshiki Kaifu, 337
total fertility rate, 303
total social financing, 141
tough and time-consuming campaign,
    85
tough negotiations, 307
tourist industry, 361
tourist infrastructure, 361
tourist resources, 362
tourist spending, 365
trade and logistics costs, 367
trade conditions, 380
trade friction, 281
trade frictions, 381, 383

trade imbalance, 376
trade limitations, 240
trade-off process, 335
trade-offs, lviii
trade protectionism, 370
trading activities, 277
trading platform, 218
traditional Chinese culture, 389
traditional Christianity, 20
traditional consumption, 101
traditional culture, 285
traditional feudal economy, 40
traditional foreign trade, 299
traditional growth points, 99, 107
traditional handcrafts, 23
traditional industrial products +
    Internet, 295
traditional industry, 241
traditional large-scale industrial
    chain, 248
traditional local features, 30
traditional manual work, 28
traditional monetary administration
    tools, 129
traditional philosophy, lv
traditional powers, 259, 302
traditional social relations, 269
traditional values, 94
traditional way, 312
training programs, 371
Transatlantic Trade and Investment
    Partnership (TTIP), 302
transferring management right, 207
transition of economic model, 99
transnational banking groups, 283
transnational e-commerce service, 298
Trans-Pacific Partnership, 384
Trans-Pacific Partnership Agreement
    (TPP), 302

transportation infrastructure, 91
Truman, Harry S., 379
trust mechanism, 59
trustworthy partnership, 287
TV disease, 28
two basic points, 62
two circles and one corridor, 332
two economic systems, 54
two elements, 392
two intermediate zones, 274
two parallel markets, 273
two-party representative democracy,
    377
two skin phenomenon, 234
two superpowers, 274
typical herd mentality, 156

U
Uganda, 294
Ukraine, 307
Ukrainian crisis, 150
ultimate objectives, 141
ultimate sameness of things, xli
ultimate truth, liii
ultra-high indexes, 294
uncertain social nature, 152
unconscious credit activities, 117
undergoing changes, 32
underground railway, 28
underlying causes, 95
underlying reasons, 376
unified management, 298
unified purchase, 204
unipolar and multipolar trends, 302
unique computer world, xlviii
unique cultural hegemony, 385
unique cultural system, 385
unique linearograph, 15
unique socioeconomic system, 14

unique writing system, 14
united front, 272
united in diversity, 306
United Nations, 16
United State, 308
Universal Gravitation, 234
universal value, 107
unlawful credit activities, 117
unprecedented conflicts, v
UN Security Council, 272, 311
unstable political situation, 284
upper reaches, 348
urban and rural structure, 105
urban areas, 64
urban collective ownership, 211
urban development, 92
urban groundwater, 93
urban rail transit, 97
urban–rural bifurcation, 209
urban–rural dual economic structure,
    92
urban–rural integration development,
    85
urban state-owned land, 209
US cultural system, 389
US federal government, 385
US-led Basel I, 169
US strategic objectives, 304
utopian socialism, 73

V
valuable diplomatic resources, 283
value-added, 380
value alliance, 321
value-centered integrated supply
    chain, 140
value communication, 140
value-creating effect, 130
value of land, xviii

value orientation, 305
various factors act, lxii
vector representation, xlii
vegetation damage, 356
vertical and cross-coverage of
    financial services, 142
vertical trade, 380
Vietnam, 294
village collectives, 175
virtual capital, 122
virtual economy, xvi, 65, 120, 122, 132
virtual form, 134
virtual reality, x
virtual world, 247
visible hand, xx
vocational training, 239

W
Wang Dayuan, 289
Wang Qishan, 339
Wang Shuzu, 338
warehousing centers, 298
warning and prevention mechanism,
    371
Washington Consensus, 377
water conservancy, 97, 223
water conservation facilities, 200
way of expression, lvi
weak productivity, 100
weak serves as a prey to the strong,
    379
wealth balance, 386
Wealth of Nations, 72
well-being of rural areas, xviii
well-built market credit system, xii
well-chosen cross-border product
    library, 296
well-developed education, 303
well-developed finance, 313

well-developed petrochemical, 313
well-established property protection
    system, 134
well-regularized manner, 207
well-trained outstanding talents, xxi
Wen Jiabao, 339
Wenzhou models, 80
West Asia, 14
Western civilization, 16
Western countries, 323
Westernization Movement, 55
Western market economy system, 123
Western powers, 38
Western religions, 15–16, 20
West India, 23
wholly foreign ventures, 48
win–win cooperation, 260
win–win outcomes, 286
win–win results, 340, 361, 381, 384
win–win spirit, 332
working-age population, 97
world, 303, 310, 343
world assembly factory, 381
World Bank, 136, 150, 326, 368
World Development Report, 182
world empire, 316
world factory, 311, 319
world landscape, 259
world power, 39
world stability, 316

world trade, 318
World Trade Organization (WTO),
    47, 277
World War I, 177
World War II, 276, 306, 310, 368,
    379, 384

X
Xiaogang Village, 203
Xibaipo Village, 197
Xi Jinping, 60, 279

Y
Yangtze River Economic Belt, xi
Yangtze River Economic Zone, 92
Yin and Yang, 391
young adults and middle-aged people,
    85
Yuan Dynasty, 289
Yuan Shikai, 41

Z
Zambia, 294
Zbigniew Brzezinski, 290, 309
zero tariff, 320
Zhang Gaoli, 338
Zheng He, 71, 289
Zhou Enlai, 58
Zuo Zhuan, 222